Asymmetric Synthesis
of Natural Products

Asymmetric Synthesis of Natural Products

Ari Koskinen
Department of Chemistry
University of Oulu
Oulu, Finland

JOHN WILEY & SONS
Chichester · New York · Brisbane · Toronto · Singapore

Other Wiley Editorial Offices

John Wiley & Sons, Inc., 605 Third Avenue,
New York, NY 10158-0012, USA

Jacaranda Wiley Ltd, G.P.O. Box 859, Brisbane,
Queensland 4001, Australia

John Wiley & Sons (Canada) Ltd, 22 Worcester Road,
Rexdale, Ontario M9W 1L1, Canada

John Wiley & Sons (SEA) Pte Ltd, 37 Jalan Pemimpin #05-04,
Block B, Union Industrial Building, Singapore 2057

Library of Congress Cataloging-in-Publication Data

Koskinen, Ari M. P.
     Asymmetric synthesis of natural products   /   Ari M. P. Koskinen.
        p.   cm.
     Includes bibliographical references and index.
     ISBN 0 471 93966 8 (cloth); ISBN 0 471 93848 3 (pbk)
        1. Natural products—Synthesis.   2. Asymmetry (Chemistry)
     I. Title.
     QD415.K66   1993
     547.7′0459—dc20   92–43969
                                                                   CIP

British Library Cataloguing in Publication Data

A catalogue record for this book is available from the British Library

ISBN 0 471 93966 8 (cloth)
ISBN 0 471 93848 3 (pbk)

Typeset in 10/12pt Times from author's disks by Text Processing Department,
John Wiley & Sons Ltd, Chichester
Printed and bound in Great Britain by Bookcraft (Bath) Ltd, Midsomer-Norton

Contents

Preface

This book is based on a one-semester, 24 hour lecture course given over the past six years at the University of Helsinki, Finland, University of Surrey, England, and University of Oulu, Finland. The course is intended for senior undergraduate and beginning graduate students. It is also hoped that the book will be useful for practising research workers who want to refresh their knowledge on the field.

The basic idea of a course combining asymmetric synthesis and natural product chemistry came from Professor Tapio A. Hase early in 1987 when discussing how best to cover both the fundamentals and latest developments in asymmetric synthesis in a stimulating way. As natural product synthesis is the logical field of application for asymmetric transformations, I decided to try out the concept. Over the years it has worked well, and the course has developed into an enjoyable one, both for the students and the teacher. For the evolution, I must thank the many students at the three universities for their helpful comments and suggestions.

The book begins with a brief introduction to the general field and its allied applications. Chapter 2 covers the basic thermodynamics and terminology as well as processes for asymmetric synthesis. Chapter 3 forms the main body of the individual asymmetric reactions which are covered both in terms of theory and applications. The rest of the book, Chapters 4 to 10, covers the individual natural product classes. I have tried to give a brief overview of the structural varieties and biosynthetic pathways leading to these compounds, as well as the practical (mainly pharmacological) importance of a number of representative compounds. To keep the reading lighter, I have also included some rather amusing anecdotes from the past. The syntheses of the individual natural product types are covered with examples, giving some general methods for the particular natural products. I have deliberately not included repetitions of long sequences of reactions which are not pertinent to the subject—these can be found in the references and in recent literature. The references are not exhaustive; quite the contrary, I have tried to keep the number as low as possible without sacrificing the context. For all omissions of important work, or references, I express my apologies. I will also warmly welcome all comments for possible future editions.

I wish to thank Professor Tapio A. Hase for the impetus for coming up with the course, and my mentors Professor Mauri Lounasmaa (Helsinki University of Technology, Finland) and Professor Henry Rapoport (University of California, Berkeley, USA) for leading me to the wonderful world of natural product chemistry and asymmetric synthesis.

Finally, my wife Päivi, and three daughters, Tiina, Joanna and Heidi, have taken a lot of grief during the writing process. Without their everlasting understanding and love, the whole project could not have been accomplished.

A.M.P.K.

List of Common Abbreviations

Ac	Acetyl
AIBN	Azoisobutyronitrile
anh.	Anhydrous
aq.	Aqueous
atm	Atmosphere
ATP	Adenosine triphosphate
9-BBN	9-Borabicyclo[3.3.1]nonane
BINAP	2,2′-Bisdiphenylphosphino)-1,1′-binaphthyl
BMDA	Bromomagnesium diisopropylamide
BMS	Borane–dimethyl sulfide
Bn	Benzyl
BOC	*tert*-Butoxycarbonyl
n-Bu	*n*-Butyl
s-Bu	*sec*-Butyl
t-Bu	*tert*-Butyl
Bz	Benzoyl
18C6	18-Crown-6
CAN	Ceric ammonium nitrate
cat	Catalytic amount
CB	Catecholborane
CBS	Corey–Bakshi–Shibata oxazaborolidine
Cbz	Benzyloxycarbonyl (carbobenzyloxy)
CDI	1,1′-Carbonyldiimidazole
Cp	Cyclopentadienyl
CSA	Camphorsulfonic acid
DABCO	1,4-Diazabicyclo[2.2.2]octane
DAIB	3-*exo*-(dimethylamino)isoborneol
DAST	Diethylamino sulfur trifluoride
DBU	1,8-Diazabicyclo[5.4.0]undec-7-ene
DCC	Dicyclohexylcarbodiimide

DDQ	2,3-Dichloro-5,6-dicyano-1,4-benzoquinone
DEAD	Diethyl azodicarboxylate
DET	Diethyl tartrate
DIBAL	Diisobutylaluminium hydride
DIPEA	Diisopropylethylamine (Hunig's base)
DIPT	Diisopropyl tartrate
DMAD	Dimethyl azodicarboxylate
DMAP	4-Dimethylaminopyridine
DME	1,4-Dimethoxyethane
DMF	N,N-Dimethylformamide
DMPU	N,N'-Dimethylpropyleneurea
DMSO	Dimethyl sulfoxide
DPPA	Diphenylphosphoryl azide
Et	Ethyl
HMDS	Hexamethyldisilazane
HMPA	Hexamethylphosphoric triamide $(Me_2N)_3P=O$
HMPT	Hexamethylphosphorous triamide $(Me_2N)_3P$
HOBt	Hydroxybenzotriazole
Im	Imidazole
$(Ipc)_2BH$	Diisopinocampheylborane
KAPA	Potassium 3-aminopropyl amide
L-Selectride	Lithium tri-sec-butylborohydride
LA	Lewis acid
LAH	Lithium aluminium hydride
LDA	Lithium diisopropylamide
LiTMP	Lithium 2,2,6,6-tetramethylpiperidide
mCPBA	$meta$-Chloroperoxybenzoic acid
Me	Methyl
MEM	Methoxyethoxymethyl
MOM	Methoxymethyl
MoOPH	Oxodiperoxymolybdenum(pyridine)hexamethylphosphoramide
Ms	Mesyl (methanesulfonyl)
MTPA	α-Methoxy-α-trifluoromethylphenylacetic acid
NADH	Nicotinamide dinucleotide hydride
NADPH	Nicotinamide-adenine dinucleotide phosphate
NBS	N-Bromosuccinimide
NCS	N-Chlorosuccinimide
NMMO, NMO	N-Methylmorpholine N-oxide
Nu	Nucleophile
OP, OPP	Pyrophosphate (in biosynthetic schemes)
PCC	Pyridinium chlorochromate
PDC	Pyridinium dichromate
Ph	Phenyl
PLE	Porcine liver esterase
PMB	p-Methoxybenzyl
PPTS	Pyridinium p-toluenesulfonate
Pr	Propyl

PTSA	Pyridinium *p*-toluenesulfonate
py	Pyridine
Red-Al	Sodium dihydrobis(2-methoxyethoxy) aluminate
SEM	Trimethylsilylethoxymethyl
TADDOL	$\alpha,\alpha,\alpha',\alpha'$-Tetraaryl-2,2-dimethyl-1,3-dioxolane 4,5-dimethanol
TASF	Tris(dimethylamino)sulfur (trimethylsilyl)difluoride
TBAI	Tetrabutylammonium iodide
TBS	*tert*-Butyldimethylsilyl
TBDPS	*tert*-Butyldiphenylsilyl
TEA	Triethylamine
TES	Triethylsilyl
Tf	Trifluoromethanesulfonyl
TFA	Trifluoroacetic acid
TFAA	Trifluoroacetic anhydride
THF	Tetrahydrofuran
THP	Tetrahydropyranyl
TIPS	Triisopropylsilyl
TMANO	Trimethylamine *N*-oxide
TMEDA	N,N,N',N'-Tetramethylethylenediamine
TMS	Trimethylsilyl
TPAP	Tetra-*n*-propylammonium perruthenate
TPP	Triphenylphosphine
Tr	Trityl
Trisyl	Trimethylphenylsulfonyl
Ts	Tosyl (*p*-toluenesulfonyl)
xyl	Xylene
% ee	Enantiomeric excess
% de	Diastereomeric excess
% o.a.	Overall yield
X_c, X_n	Chiral auxiliary

1 Introduction

Natural product chemistry covers the chemistry of naturally occurring organic compounds: their biosynthesis, function in their own environment, metabolism, and more conventional branches of chemistry such as structure elucidation and synthesis. The purpose of this text is to familiarize the reader with the most common classes of natural products, with particular emphasis on the methods currently available for their asymmetric synthesis.

We shall begin by defining a few common concepts often encountered in connection with natural product chemistry: endogenous and exogenous substances, primary and secondary metabolism, and detoxification.

Endogenous substances are compounds produced as a result of the normal functioning of an organism. Amino acids, many carbohydrates, peptide and steroid hormones, neurotransmitters etc. produced by the body are typical endogenous substances. *Exogenous compounds* are compounds coming from the outside of the organism, such as drugs and many environmental pollutants. Exogenous compounds are also known as *xenobiotics*.

Primary metabolism is the system of biochemical reactions whose products are vital for the living organism. Photosynthetic plants convert carbon dioxide to primary metabolites, carbohydrates, amino acids and other compounds ubiquitous to all forms of life. Primary metabolic pathways often function in cycles, such as the carbon fixation cycle (Figure 1.1), which converts carbon dioxide into glyceraldehyde 3-phosphate, the building block for carbohydrates, fatty acids and amino acids.

Secondary metabolism refers to the functions of an organism yielding products that are not necessary for the essential biochemical events. Secondary metabolites are thus compounds which are often species dependent. The actual role of secondary metabolites is still largely unclear. The photosynthetic plants convert carbon dioxide and water into

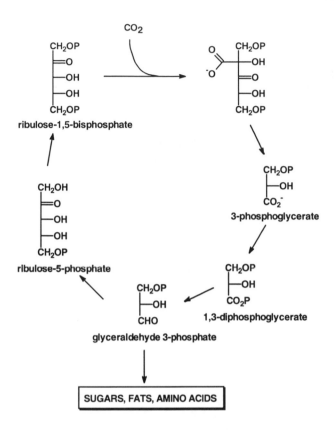

Figure 1.1

simple carbohydrates (monosaccharides), which can be combined to make more complex polysaccharides and glycosides. Further breakdown of the simple carbohydrates leads to pyruvic acid, which itself functions as the precursor to shikimic acid and thereby the aromatic compounds present in nature. Decarboxylation of pyruvic acid gives acetic acid, which functions as the biogenetic precursor to practically all the remaining types of natural products. Condensation reactions lead to polyketides, mevalonic acid acts as a link between acetate and isoprenoids (terpenes), and also amino acids and thereby peptides and alkaloids are formed from acetic acid (Scheme 1.1).

Prostaglandins are an example of mammalian secondary metabolites. They are biosynthesized from an unsaturated fatty acid, arachidonic acid, via enzymatic oxygenation producing an intermediate prostaglandin endoperoxide, prostaglandin H_2 (Scheme 1.2). An alternative dioxygenation leads to 5-hydroperoxyeicosatetraenoic acid, 5-HPETE, the precursor of leukotrienes. Prostaglandins and leukotrienes participate in several physiological events in cells and organs.

The distinction between primary and secondary metabolites is often difficult: pyrrolidine-2-carboxylic acid (proline) is a primary metabolite, whereas piperidine-2-carboxylic acid (pipecolic acid) belongs to the secondary metabolites (Figure 1.2).

The reaction path leading to a particular natural product is called the biosynthetic pathway, and the corresponding event is known as the biogenesis. Different plant and animal species can employ dramatically different biosynthetic pathways to produce the

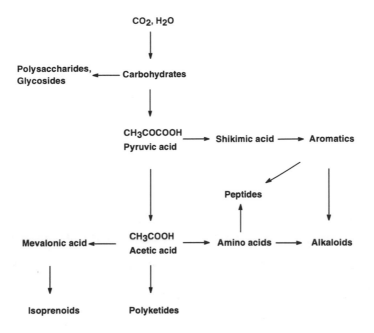

Scheme 1.1

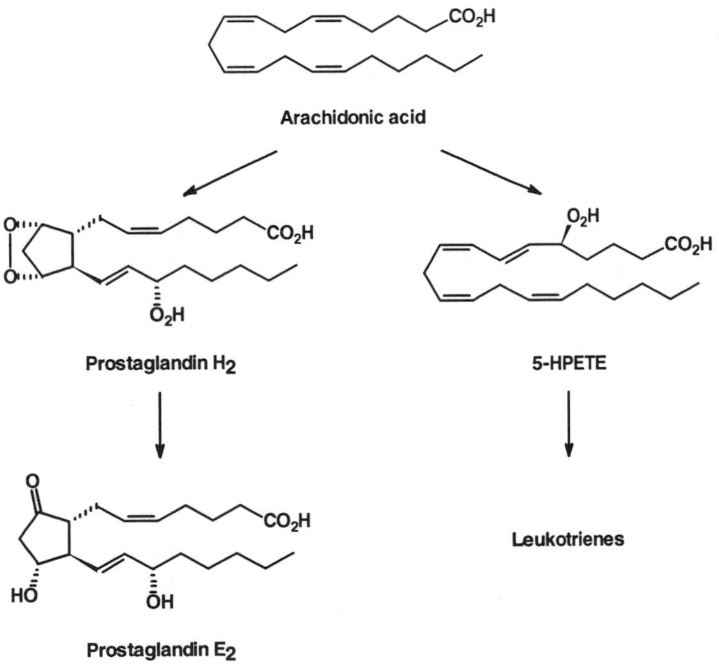

Scheme 1.2

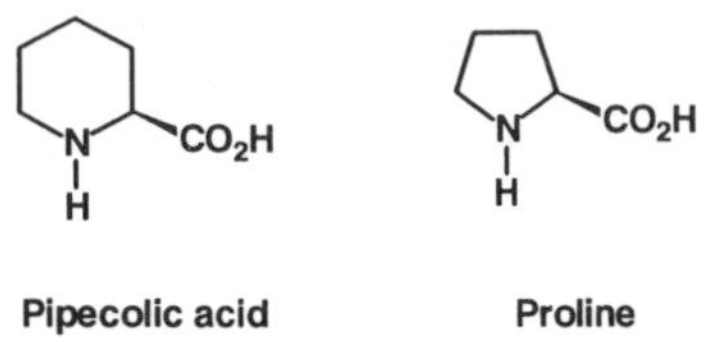

Pipecolic acid   **Proline**

Figure 1.2

same metabolite. This feature can be employed in the classification of plants in terms of their chemotaxonomy.

Elimination of foreign, often toxic, compounds employs metabolic reactions. In such cases one commonly speaks of *detoxification processes*. A major part of detoxification requires oxidative transformations, and in this respect the liver plays a central role. Liver contains a multitude of oxidases capable of converting many different types of compounds to more highly oxidized products, which are broken down and eventually secreted.

Biochemical reactions and their control with all the associated intricacies do not belong to the subject matter of this text. To give the reader a sufficient general understanding, enzymatic reactions of special value to natural product chemistry will be discussed in connection with the individual natural product types.

1.1 Some Properties of Natural Products

Throughout the known history mankind has been interested in naturally occurring compounds. Simple aqueous extractions of flowers, plants and even insects have been used to isolate compounds whose taste, colour and odour could be used for various purposes. Also, healing creams and liniments were produced from plant extracts in practically all ancient cultures. South American Indian hunters used, and still use, plant extracts as arrow poisons. With this plethora of applications, it is no surprise that with the development of chemical methods the natural products have gained increasing importance in various aspects of human endeavour.

The majority of natural products have been isolated from plant origins, mainly due to the ease of the isolation process. The most common procedure still in use is, in outline, as follows. The plant material is divided according to the plant parts (the leaves are separated from the roots and stem etc.), and the material is dried and ground to a suitable particle size. This dry material is then extracted with a suitable solvent (e.g. methanol or chloroform), and the organic extract is then concentrated. The crude extract may contain hundreds of compounds, and earlier their separation was based on crystallization or distillation techniques. The development of modern chromatographic methods has facilitated the separation processes, and in practice nearly all the components can be isolated in pure form.

Natural products are usually given a trivial name derived from the plant origin (e.g. muscarine from the mushroom *Amanita muscaria*, fly agaric). In some cases the name describes the physiological action, as in the cases of vomitine, emetine and putrescine.

Taste, odour and colour of organic natural products are usually the properties most easily detected. The relationships between aroma compounds can often be quite surprising, and chemically closely related structures can have quite different properties.

Strawberries and pineapple contain large quantities of furaneol (Figure 1.3), whose close relative can be found in cloudberries and (canned) mango. Sulfur containing compounds (Figure 1.4) are abundant in onions. The strong lachrymatory action of onions is caused by decomposition products, and the aroma compounds of fresh and boiled onions are chemically different. The aroma compounds of garlic are closely related to those of the onion. Sulfur compounds are found also in passion fruits and asparagus. Perhaps the organic compound richest in sulfur occurs in red algae.

Furaneol
strawberry, pineapple

cloudberry, canned mango

Figure 1.3

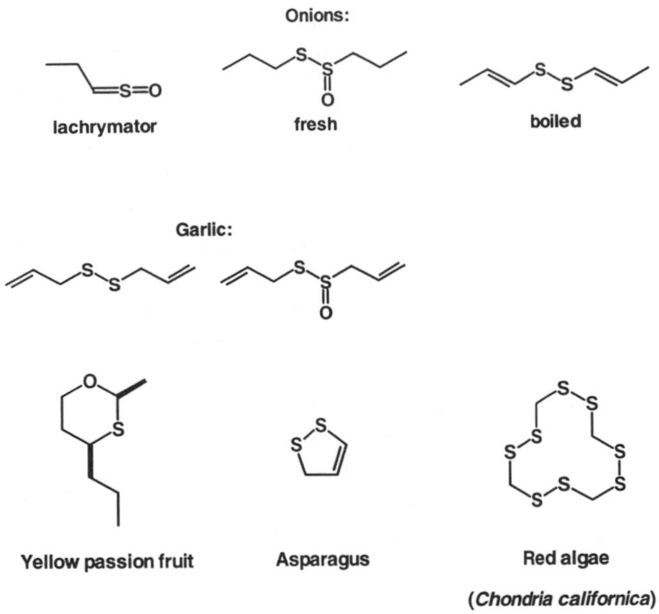

Onions:

lachrymator   fresh   boiled

Garlic:

Yellow passion fruit   Asparagus   Red algae

(*Chondria californica*)

Figure 1.4

Other examples of related structures with different aromas are the aroma compounds of raw potato, green pepper and popcorn (Figure 1.5). Also some very unpleasant smell and taste effects can be caused by organic natural products. The 'aroma' compound of sunburnt beer is structurally related to the odour substances of the North American skunk and cat's urine (Figure 1.5).

The taste effects of many organic compounds can also depend on the solvent. For instance, 2-acetyl-3-methylthiophene dissolved in syrup induces a honey-like taste, whereas in coffee it gives a 'nutty' flavour.

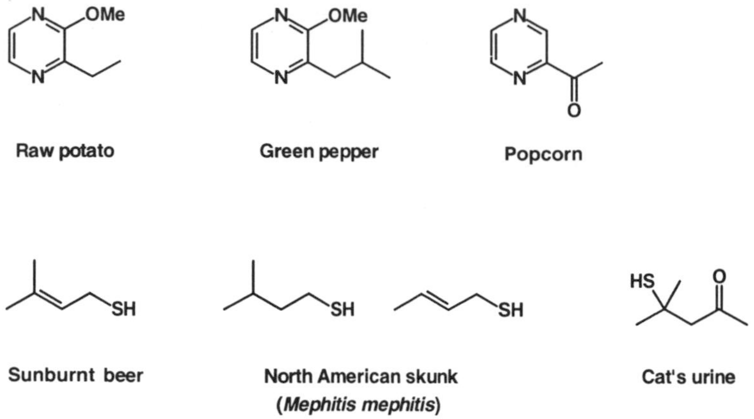

Raw potato   Green pepper   Popcorn

Sunburnt beer   North American skunk   Cat's urine
                              (*Mephitis mephitis*)

Figure 1.5

Why plants produce secondary metabolites is still largely unknown and subject to speculation. In many cases the importance of a particular substance to the plant is not known. It has often been suggested that the plant simply excretes part of its waste products in the form of natural products. This is not an appealing suggestion since the natural products often exhibit very complicated structures. Recent developments in biology have given us some hints in understanding the importance of these compounds. Many natural products have a regulatory role (e.g. growth hormones). Some function as chemical defence agents against pests; the strongest ones may be lethal. The role of certain compounds is to act as chemical messenger molecules between species of the same genus.

Many green plants produce gibberellins (Figure 1.6), which function as growth hormones. As soon as the structures of the compounds became known, the synthetic efforts led to practical preparations of many of their congeners. Currently, several gibberellins are produced in bulk for agriculture and plant development. Certain gibberellins are also widely used in brewing to shorten the time needed for malting.

Some natural products act as chemical policemen. Aphidicolin (Figure 1.6), a complex terpenoid, reduces the appetite of aphids. This has been put to use in crop protection. The most effective compounds of this class are so potent that 10 g of material are sufficient to treat one hectare for the whole of the growth period.

All animals can communicate with members of the same species. Voice, touch and sight are not the only possible methods for transferring messages. Perhaps the most sensitive sense is based on chemical recognition: the senses of smell and taste are fairly sensitive among men, but dogs, for example, have considerably better abilities in handling information transmitted in these ways. Even small insects transmit messages dressed in chemical form. These often rather simple organic molecules are called pheromones (Figure 1.7). Many such compounds affecting the sexual behaviour of insects are widely used in plant protection.

Many plants and animals produce chemicals with a toxic or even lethal action on other species. Snake venoms are usually rather small peptides, but other small organic molecules are known with such characteristics. The brightly coloured, small South American frogs of the *Phyllobates* and *Dendrobates* species have been used by hunters to

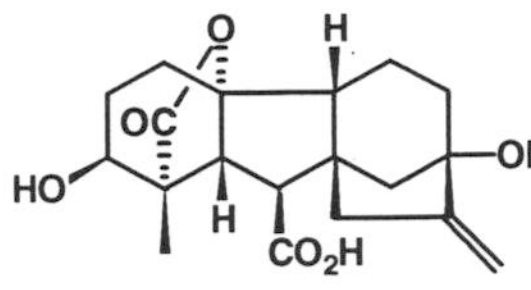

Gibberellic acid A₃   Aphidicolin

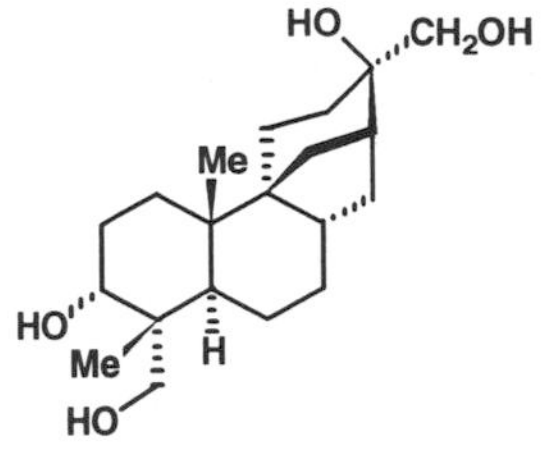

Figure 1.6

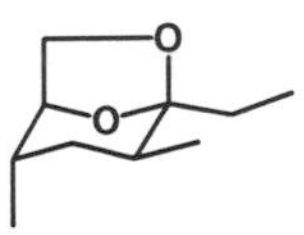

Multistriatin 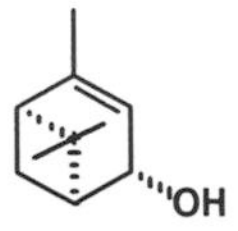   *cis*-Verbenol   Bombycol
                         (*Ips typographus*)   (*Bombyx mori*)

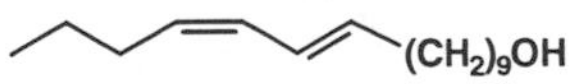

Figure 1.7

envenomate their poison darts. Batrachotoxins, pumiliotoxins, histrionicotoxins and gephyrotoxins (Figure 1.8) are the major frog poisons whose use persists even today. Sea cucumber produces holothurines, which can cause blindness in man. Alas, the most potent of these toxins, ricin (a peptide) and saxitoxin (an alkaloid), have found their use as chemical warfare agents (compounds W and TZ, respectively).

1.2 Natural Products as Drugs

Natural products have played a key role in the development of medicinal chemistry. Even today, a large number of new chemical entities are arrived at through the help of natural products. In the early days, medicines were isolated from plant material. Later, man learned to utilize fermentation to produce the medicinal agents, and within the last decade or so the methods of molecular biology have enabled the programming of cells to produce several variants of compounds known earlier. These different approaches are not alternatives, rather they complement each other to give the natural product or medicinal chemist a wider spectrum of tools with which to work.

In the following, we shall take a brief look at the history of drugs as far as natural products are concerned. The oldest information on drugs goes back to China in the Bronze Age. The first medical herbal, the *Pen-Tsao*, was compiled by Shen Nung (*ca* 2800 BC) and describes 365 drugs used in those days. The Chinese are renowned for being the people who first familiarized themselves with the noble teachings of alchemy. The honour of being the first alchemist is often attributed to Li Shao-chun, and the first textbook of

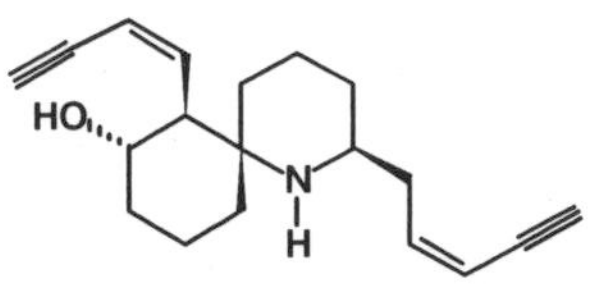

Histrionicotoxin

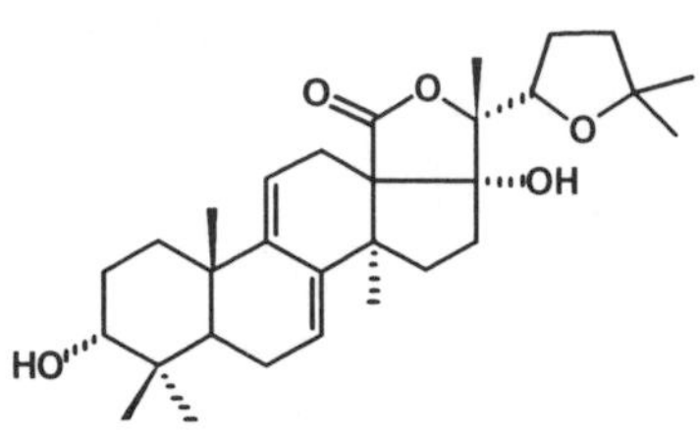

22,25-oxidoholothurinogen

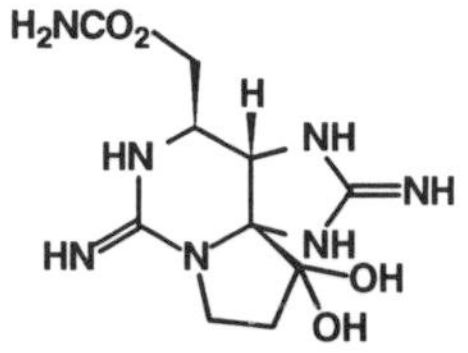

Saxitoxin hydrate

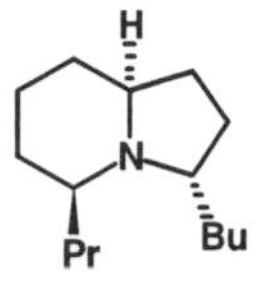

Gephyrotoxin 223AB

Figure 1.8

alchemy is the *Ts'an t'ung Ch'i* from *ca* 120 BC. Thus it is no surprise that the first known drug ephedra (the 'horsetail' plant), came from China, isolated from a plant known as *ma huang* and described by Shen Nung. Ephedra has been used for thousands of years as a stimulant, a remedy for respiratory diseases, and to induce fever and perspiration and to depress cough. Ephedra was also included in the Greek pharmacopoeia. In Western medicine, the active principle (ephedrine) was isolated in chemically pure form in the latter part of the 19th century. Ginseng (man shaped root) was also known in those days. Among other effects, it was supposed to delay ageing and restore sexual powers. Ginseng also alleviates diabetes and stabilizes blood pressure.

The Sumerians and Babylonians also had a highly developed understanding of the natural sciences. The Mesopotamian women, who isolated ethereal oils from plants, can be considered to be the first known chemists. The oils were apparently used as perfumes, and the art of distillation was probably learned as early as 3600 BC. The Greek Theophrastos (*ca* 370–285 BC) is renowned as giving the first explicit details on the distillation of odours.

The next important step comes from the Egyptians: the so called Papyrus of Ebers (from *ca* 1500 BC, named after the German Egyptologist Georg Ebers) was transcribed in 1937 by the Norwegian doctor Ebbel. The Papyrus of Ebers describes several preparations in use even today, such as opium, castor oil and liver (vitamin A). The Egyptians also used 'rotten bread' to treat infections—this clearly has a striking connection to our current understanding and use of compounds produced by moulds and fungi (such as penicillin) as antibacterial agents.

On the north shores of the Mediterranean, we meet the Greek father of medicine, Hippocrates (*ca* 460–370 BC), and the Roman physicians Dioscorides and Galen. Pedanius Dioscorides was an army physician who studied the medical uses of hundreds of plants and wrote probably the first systematic *materia medica* during the first century.

Already, Diocles of Carystus (in the fourth century BC) had collected similar information on medicinal plants, but the works of Dioscorides are attributed as the first thorough and systematic studies of their kind. Together with Pliny (Caius Plinius Secundus, the 'Elder Pliny', 23–79), Dioscorides also described the medicinal properties of wines. Among the effects noted in those days were quickening of the heartbeat and the injurious effects of its continued use. Galen (*ca* 129–200) is considered to be the forefather of experimental physiology. He was convinced that all illnesses could be cured with mixtures of concoctions, as long as one could find out the necessary proportions. A typical concoction from the antique was teriak (Greek for counterpoison), which often contained more than 60 components, such as opium, dried snake meat, cinnamon, pepper, onions, fennel and cardamom. Teriak reached other parts of the world very slowly; for instance, Scandinavia got its first teriak in the 18th century, and as late as the 20th century one could still find such concoctions from some small chemists' shops in Central Europe.

The colourful Middle Ages brought with them the signature theory, according to which Nature has itself marked the suitable medicines for all illnesses; liver shaped leaves sign; for example, that they can be used to treat illnesses of the liver. Philippus Aureolus Theophrastus Bombastus von Hohenheim (born in 1493), better known as Paracelsus, was the town doctor of Basel from 1526, and also a lecturer of medicine at the University of Basel. The much disputed writings of Paracelsus were the first of their kind in the field of *iatrochemistry*, which aimed solely at curing illnesses. Some earlier indications towards similar goals can be found in the writings of Raymond Lull (Raymundus Lullus) and the Arabian al-Razi. Paracelsus, however, so furiously resisted traditional practices of medicine that he was forced to leave Basel. After living a life of a vagabond, he died in Salzburg in 1541.

During the 17th century, the Jesuits brought with them from South America the bark of the china tree (1632, for the treatment of malaria) and some plant concoctions developed by the Incan Indians. As a reflection of these events, one can today find a large proportion of medicinal agents being derived from alkaloids. During the same century, chemistry also started to gain respect as a natural science. The first university chemistry laboratory was opened in 1609 in the University of Marburg, and where Johannes Hartmann with his students produced pharmaceutical products.

In the 19th century, the development of organic chemistry took on rapidly. The isolation and identification of natural products started to be more systematic. In 1820, Pelletier and Caventou isolated from the china tree the active compound against malaria—quinine. This sparked a rapidly growing interest in isolating the chemical constituents of the medicinal plants.

The art of organic synthesis was transmitted from the apothecaries to the expert chemists, and at the same time the quality of the products improved. Pure chemical entities started to replace old, dried isolates and decocts (extracts). The first such compounds were naturally occurring nitrogenous compounds, alkaloids, which were easy to isolate by repetitive extractions and could be purified in their salt form by crystallization.

Gerardus Johannes Mulder had in the 1830s first observed the existence of proteins. Hoppe-Seyler was able to obtain crystals of a protein, haemoglobin, in 1864. Five years later, in 1869, Friedrich Miescher found the chemical carrier of heredity, nuclein, whose deoxyribonucleic acid structure remained obscure until the works of Albrecht Kossel in the 1890s.

Towards the end of the 19th century, microbiology developed into a separate scientific discipline. Robert Koch showed that living organisms can cause an infection (1876). Pasteur and Joubert showed in 1877 that bacteria can antagonize each other's growth. Vuillemin defined the concept of antibiosis (Greek *anti* = against, *bios* = life) in 1889.

The widest known story of antibiotics produced by microbes must be that of penicillin, which was first isolated by the Scotsman Alexander Fleming in 1929. The development of penicillin into a drug was slow, and only during World War II through the efforts of the Americans Florey and Chain was the drug made available in large quantities to the Allied soldiers. However, this was not the first commercial antibacterial drug to be produced by microbes. Gosio had already shown in 1898 that mycophenolic acid (Figure 1.9), produced by the mould *Penicillium brevicompacticum*, inhibited the growth of the bacterium *Bacillus anthracis*.

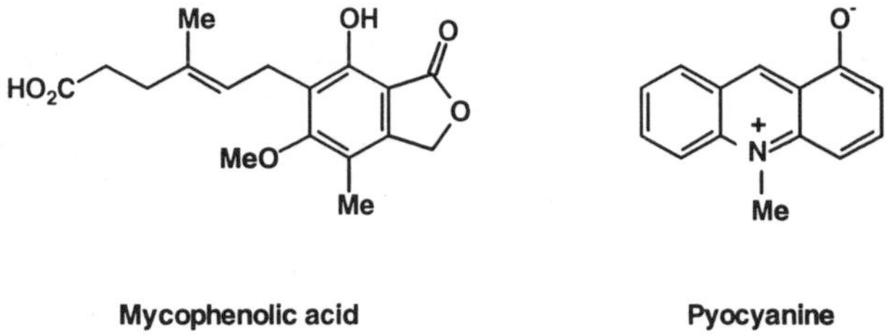

Mycophenolic acid   Pyocyanine

Figure 1.9

Actinomycin D   Streptomycin

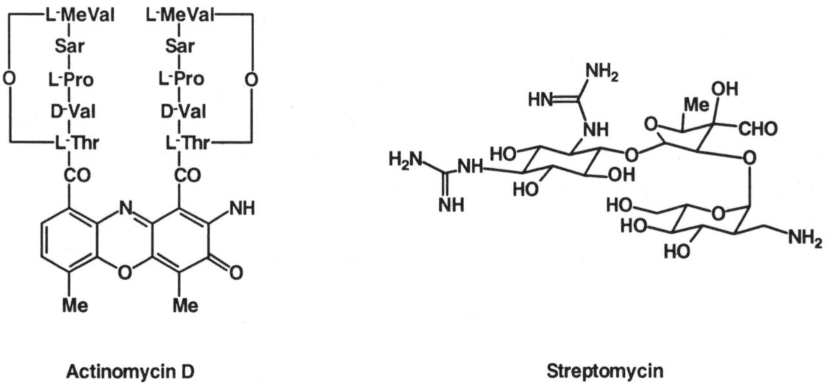

Figure 1.10

The first industrially produced and therapeutically used antibiotic was a crude extract from *Pseudomonas aeruginosa*, which contained the enzyme pyocyanase and an antibiotic, pyocyanine (Figure 1.9), whose structure was not elucidated until 1929. Waksman and Woodruff isolated actinomycin (Figure 1.10) in 1940; 12 years later this compound became the first cytostatically active compound for humans (cytostats inhibit the growth of tumours). Waksman also isolated streptomycin (Figure 1.10) in 1944 (particularly useful for the treatment of tuberculosis).

As a source for drugs, Nature provides a wide spectrum of compounds which in themselves can be used to treat diseases. These compounds can be classified as fine chemicals, with typically a high degree of refining and thus a high price. The *Catharanthus* alkaloids, important for the treatment of several forms of cancer, can cost hundreds of thousands of dollars per kilogram. However, one should not forget the importance of simpler compounds produced by Nature. These can be valuable starting materials or intermediates for the synthesis of other products. Many amino acids, lipids and carbohydrates are being produced in ton quantities (often for the price of only a few dollars per kilogram, or even less). Table 1.1 lists some commercially important drugs whose production is based on natural products isolated from plants.

Table 1.1 Some drugs of plant origin.

Compound	Origin	Medicinal use
Steroids:		
Hormones (95% of diosgenin)	*Dioscorea* (Mexican yams)	Contraceptives, anabolic steroids, corticosteroids
Digitalis glycosides (digitoxin, digoxin)	*Digitalis purpurea,* foxglove	Cardenolides
Alkaloids:		
Opium alkaloids (morphine, codeine)	opium poppy	Pain relief
Catharanthus alkaloids	Madagascan periwinkle	Cancer
Pilocarpine	*Pilocarpus* species	Glaucoma
Colchicine	Autumn crocus	Gout
Cocaine	Coca leaves	Local anaesthesia

1.3 Structures of Natural Products

The elucidation of the biosynthetic pathways, the varying properties and the medicinal uses of natural products would be in themselves good enough reasons to study the synthesis of natural products. However, added momentum is gained by the enormous variations in the structures, and especially by the occurrence of structures whose complexities surpassed the wildest imaginations of the chemists in the times when these compounds were isolated.

Structural complexity has been the driving force for the development of spectroscopic and spectrometric means of structure elucidation. Detailed information on UV chromophores was obtained and efficient correlations (e.g. the Scott and Woodward rules) were formulated in connection with the work on steroid structures. Infrared spectroscopy gained a similar impetus from the study of ketones and lactones of varying size. The fragmentation patterns in mass spectrometry were uncovered mainly through natural product work, and ^{1}H and ^{13}C NMR spectrometric studies on ever more complex structures have aided the development of good correlations between structure on the one hand and chemical shifts and coupling constants on the other.

Understanding the biosynthetic reaction mechanisms has had an important role in the development of the general theory of physical and mechanistic organic chemistry. For instance, the biosynthetic pathways leading to terpenes often involve rather deep seated cationic rearrangements. In other cases, photochemically induced rearrangement reactions give structures with unforeseen carbon skeleta. Nucleophilic and electrophilic alkylation reactions abound in Nature. Some remarkable general patterns can be seen in the stereochemical outcomes of such reactions, as evidenced by the nearly exclusive existence of a single, diastereomeric series of aldol products in the biogenesis of macrolide antibiotics. Acylation reactions are also common, and perhaps millions of years ago Nature found ways of achieving such reactions using a nucleophilic acylating reagent (umpolung), a concept only relatively recently introduced to synthetic organic chemistry.[1-3] The H-2 of the thiazolium ring in thiamine pyrophosphate (Scheme 1.3) is acidic enough to undergo deprotonation,[4] and reaction with pyruvate gives hydroxyethylthiamine pyrophosphate, the biological 'active aldehyde'. As an enamine, this can react with aldehydes to give the acetoin product and thus liberate the thiamine pyrophosphate.

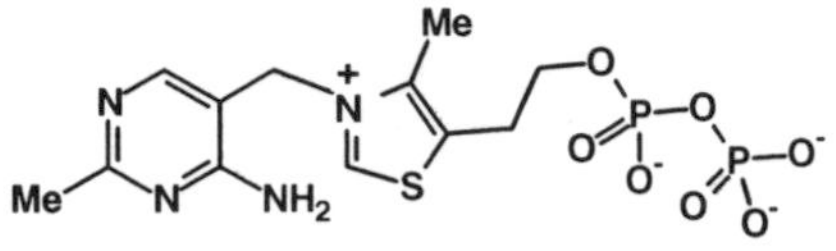

Thiamine pyrophosphate (vitamin B$_1$)

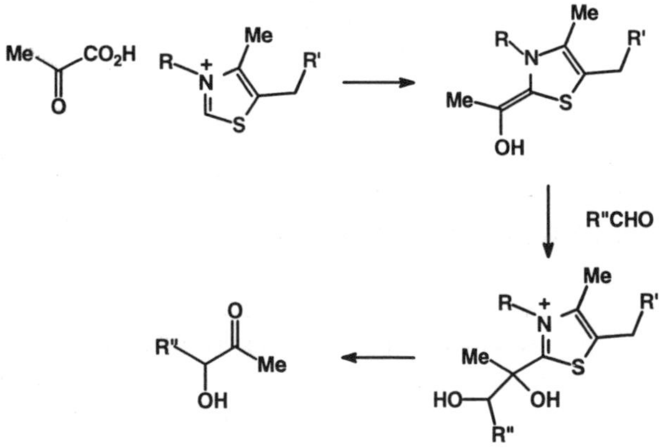

Scheme 1.3

In some cases the mechanisms of action of natural compounds have been worked on before the actual natural products have been isolated. One striking example of such a case is the development of the Bergman cyclization (Scheme 1.4) of enediynes to aromatic systems.[5] This reaction found its natural counterpart nearly two decades after it had been developed in the laboratory. Nature produces a number of what are now known as enediyne antibiotics which cleave DNA with high efficiency.[6] As a result the DNA cannot be replicated, and therefore some of these compounds are under scrutiny as anticancer candidates.

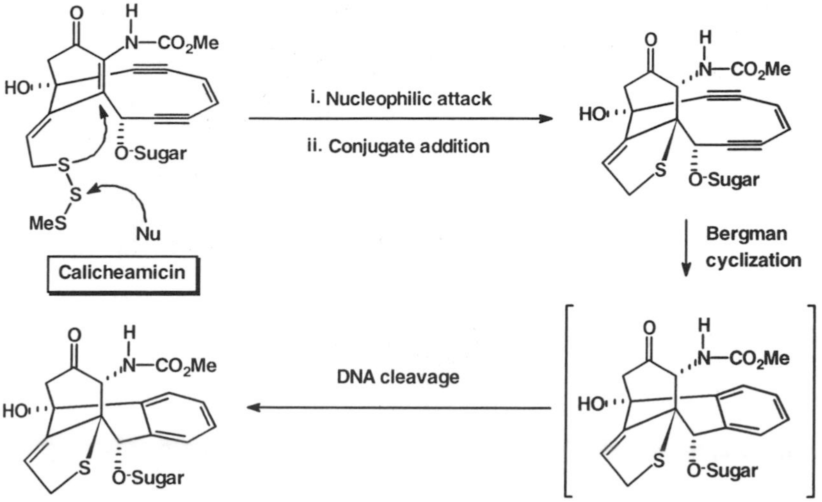

Scheme 1.4

From the point of view of stability, natural products often contain remarkably labile structural units. This is of course a complicating matter as far as isolation and structural work are concerned, but at the same time it provides challenges to the experimental techniques without which the development of synthetic chemistry would have missed major contributions. Only a few decades ago, few chemists would have dared even to think of synthesizing a conjugated polyene with, say, a dozen double bonds. Nature does it, and these compounds are relatively stable in their respective environments, as evidenced by such compounds as the polyenes and enynes. Aurantosides (Figure 1.11), which contain a dichlorinated polyene and a tricarbonyl unit, were the first tetramic acid glycosides to be isolated from marine organisms.[7] These orange pigments were isolated from the marine sponge *Theonella*, and they possess cytotoxic activity. Compounds charged with hydroxyl groups are also found among the natural products. Ryanodol (Figure 1.11) has a remarkable structure consisting of 20 carbon atoms and incorporating altogether seven hydroxyl groups, one of which is involved in a hemiketal.

Increasing ring strain brings with it increasing reactivity and thus also lability. Three, four and five membered rings are, however, quite common in natural products (Figure 1.12). The β-lactam story would have suffered a lot without the bold minds of the chemists who dared to attack the four membered lactam ring structure. Strain is also displayed in other kinds of structures. Silphinene (Figure 1.12) is a terpene which contains a broken window version of the fenestranes.

Peptide structures are today often considered to be rather mundane, being made of only 20 amino acid building blocks by rather simple condensation reactions—not quite so! Nature often does the unexpected and provides surprises. Nature abounds with peptides which contain amino acids other than the 'natural' 20. The immunosuppressant cyclosporin A (Figure 1.13) is a case in point. The rare amino acid MeBmt has been the target of several synthetic approaches, and its structural analogues have given more insight into understanding the mechanism of action of this important peptide. Nature takes this game even further. Echinocandins (Figure 1.13) are antifungal, cyclic lipopeptides produced by the *Aspergillus* species.[8,9]

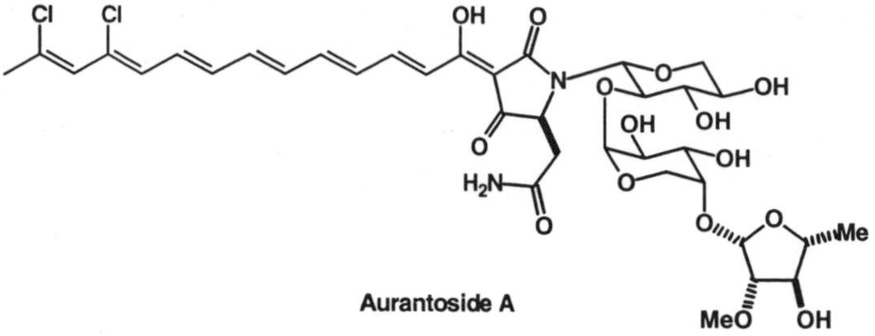

Aurantoside A

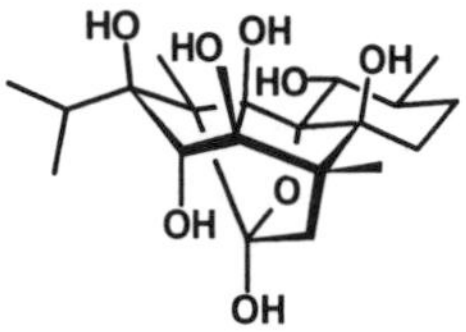

Ryanodol

Figure 1.11

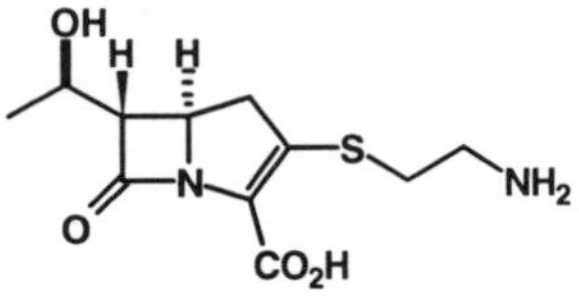

Thienamycin

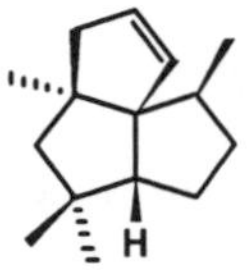

Silphinene

Figure 1.12

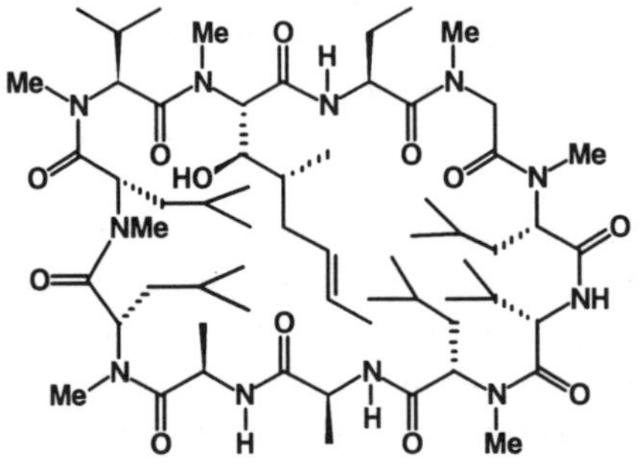

Cyclosporin A

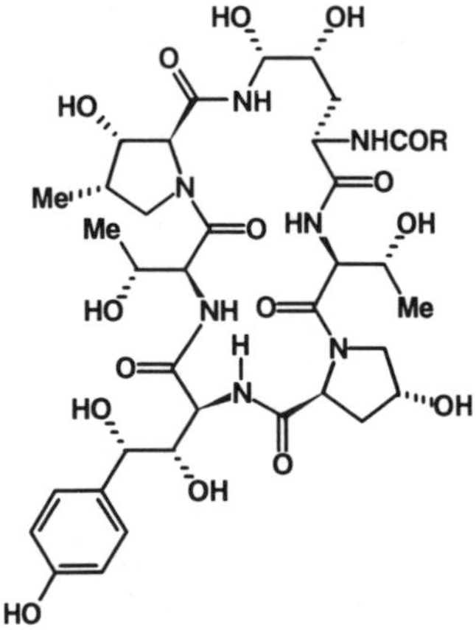

Echinocandin B

Figure 1.13

The increasing challenge of molecular complexity is also manifested in the choice of target structures for synthesis. Woodward and Eschenmoser took on an enormous task to achieve the synthesis of vitamin B_{12} (Figure 1.14), a notoriously complex and sensitive molecule of the utmost importance in the living system. They achieved the synthesis in what turned out to be perhaps the most spectacular synthetic endeavour of this century.[10,11]

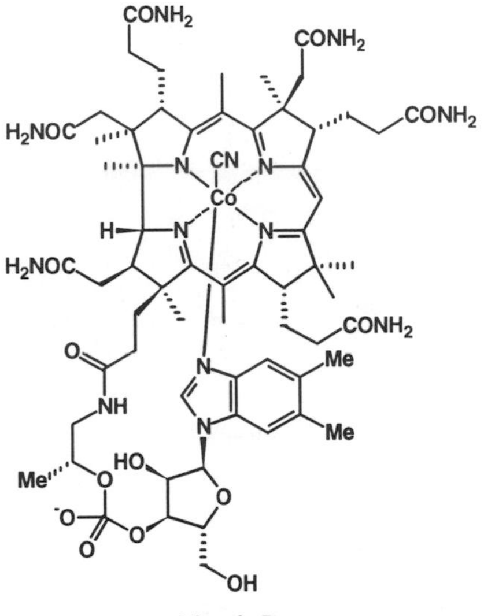

Vitamin B$_{12}$

Figure 1.14

Much more elaborate structures are also found within the polyketides. Oxygenation, etherification and several other structural modifications are often encountered. As a consequence of the biosynthetic route employed for the polyketides, the oxygenation often leads to 1,3- and also 1,2-dioxygenated units. This is displayed by the tremendous variation and structural complexity of the polyether, macrolide and spiroketal antibiotics. The most remarkable structure discovered so far is displayed by palytoxin (Figure 1.15), one of the most poisonous substances known today.[12] The toxin was isolated from the Hawaiian soft coral *Palythoa toxica*, which was found in a single tidal pool of only six feet by two feet, and just 20 inches deep! Palytoxin's story began in 1961 and it took nearly three decades before the complete structure, including stereostructure, was finally established through a combination of spectroscopic and synthetic methods.[13,14] Palytoxin contains no less than 64 chiral centres and seven double bonds capable of *E/Z* isomerism, giving rise to the possibility of $2^{71} = 2.4 \times 10^{21}$ isomers! This number is close to Avogadro's number!

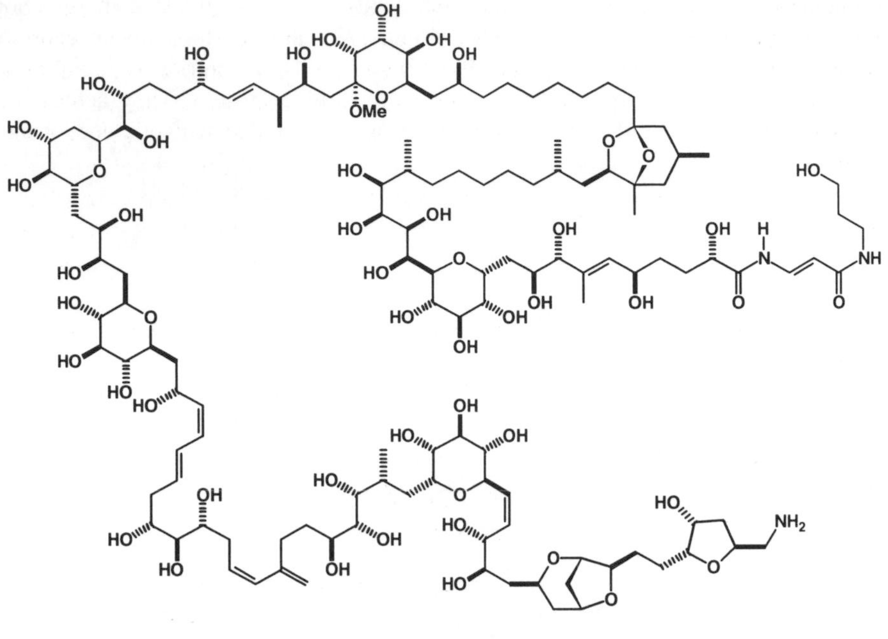

Palytoxin

Figure 1.15

1.4 Synthesis of Natural Products

Synthesis has played a major role in natural product chemistry. Until the evolution of spectroscopic methods, independent synthesis of a compound was thought to give the ultimate proof for the proposed structure. The proponents of X-ray crystallography and spectroscopy have often claimed that with the advent of these powerful techniques the need for synthesis is abolished. Even a quick look at the recent literature shows, however, that no matter how sophisticated are the instrumental methods used, there is still the ambiguity of experimental error, and even today structures elucidated through instrumental means are revised as a result of synthesis. This is increasingly true as the level of complexity of compounds increases.

It is certainly true that in the majority of cases the instrumental methods provide a rapid and reliable means of establishing the structure of a new compound. In most cases it is also possible to ascertain the absolute stereochemistry of the compound. If this is the case, why should synthesis continue to play a key role in natural product chemistry? The answer lies in the everlasting need for new and efficient methods for the synthesis of more and more complicated structural arrays. As we have already seen, rather minor modifications of the molecular structure can induce quite dramatic changes in the characteristics of the molecules. This allows the chemist to design ways of fine-tuning the properties of the molecules in order to obtain highly defined and sophisticated materials for various purposes. For instance, in medicinal chemistry one of the main objectives of drug development is to find a suitable drug candidate which possesses the desired pharmacological activity (preferably towards a single receptor) and is completely devoid of any side effects. For practical purposes, it may be necessary to be able to vary the solubility and penetration of the compound in order to obtain a formulation that is more convenient to use. If we were to rely solely on natural products, we would probably be expecting too much from Nature; the massive numbers of compounds of a related structural series that need to be tested usually far outnumber those we could in all fairness expect to find from natural sources. Therefore, we need to have synthetic access to the analogues of the natural products, and this work usually also requires synthetic access to the natural product itself.

We have already seen some quite demanding structures whose synthesis would not have been feasible just a few decades ago. The efforts to attack such complex functional arrays have resulted in the development of plenty of efficient methods gaining access to compounds containing such structural units. The field of synthetic method development is highly advanced at present, but at the same time still highly lacking in many respects. The full importance of chirality has only relatively recently been appreciated, and the single most important effort in synthetic method development at present is devoted to asymmetric synthesis. The main purpose of this text is to highlight this area of chemistry, as exemplified by the applications to natural product chemistry or arising from the needs within it.

References

1. Seebach, D. and Enders, D. *Angew. Chem.* **87**, 1–18 (1975).
2. Lever, O.W. *Tetrahedron* **32**, 1943–1971 (1976).
3. Hase, T.A. (Ed.) *Umpoled Synthons: A Survey of Sources and Uses in Synthesis* John Wiley & Sons: New York, 1987.
4. Breslow, R. *J. Am. Chem. Soc.* **80**, 3719–3726 (1958).
5. Bergman, R.G. *Acc. Chem. Res.* **6**, 25-31 (1973).
6. Nicolaou, K.C. and Dai, W.-M. *Angew. Chem., Int. Ed. Engl.* **31**, 1387–1414 (1991).
7. Matsunaga, S., Fusetani, N., and Kato, Y. *J. Am. Chem. Soc.* **113**, 9690–9692 (1991).
8. Benz, F., Knusel, F., Nuesch, J., Treichler, H., Voser, W., Nyfeler, R., and Keller-Schierlein, W. *Helv. Chim. Acta* **57**, 2459–2477 (1974).
9. Keller-Juslen, C., Kuhn, M., Loosli, H.R., Petcher, T.J., Weber, H.P., and von Wartung, A. *Tetrahedron Lett.* 4147–4150 (1976).
10. Woodward, R.B. *Pure Appl. Chem.* **17**, 519–547 (1968).
11. Woodward, R.B. *Pure Appl. Chem.* **33**, 147–177 (1973).
12. Moore, R.E. *Prog. Chem. Org. Nat. Prod.* **48**, 81 (1985).
13. Anon. *Chem. Eng. News* Sep. 18, 23–24 (1989).
14. Armstrong, R.W., Beau, J.-M., Cheon, S.H., Christ, W.J., Fujioka, H., Ham, W.-H., Hawkins, L.D., Jin, H., Kang, S.H., Kishi, Y., Martinelli, M.J., McWorther, Jr. W.W., Mizuno, M., Nakata, M., Stutz, A.E., Talamas, F.X., Taniguchi, M., Tino, J.A., Ueda, K., Uenishi, J.-I., White, J.B., and Yonaga, M. *J. Am. Chem. Soc.* **111**, 7525–7530 (1989).

2 Chirality, Topology and Asymmetric Synthesis

Carbon atoms carrying four different substituents possess a unique property. The substituents can be arranged in two alternative ways to bring about two forms of the molecule with the same constitution. In Figure 2.1, two molecules of the same constitution (CHXYZ) are depicted so that in each case the smallest substituent, hydrogen, lies behind the plane of the paper. The substituent X is drawn in each case in the plane, pointing up, and the other two substituents occupy positions either on the left or right of the central carbon atom. Looking along the bond from the central carbon atom towards the hydrogen at the back, one finds that the two molecules differ in the way the remaining three substituents are arranged in space: in **A** the substituents X,Y and Z follow a clockwise rotation, whereas in **B** the rotation is counter-clockwise.

The two forms of the molecule are related as hands to each other, being non-superimposable mirror images of each other. They are called *chiral* (from Greek *cheir* =

A   B

Figure 2.1

hand), and the central carbon atom is known as the *chiral* or *stereogenic centre*. In this case, the whole molecule does not possess any element of symmetry (except identity), and the molecule is also *asymmetric*. However, asymmetry is not a necessary requirement for chirality. Dissymmetric molecules which lack one or more elements of symmetry can also be chiral, and the requirement for chirality can be defined as follows: molecules which do not possess rotation–reflection axes (S axes) are chiral or dissymmetric. Based on point groups, those molecules which belong to the C_n or D_n point groups are chiral. For instance, the compound *trans*-2,5-dimethylpiperidine (Figure 2.2) contains a two-fold rotation axis, belongs to the point group C_2, and is chiral.

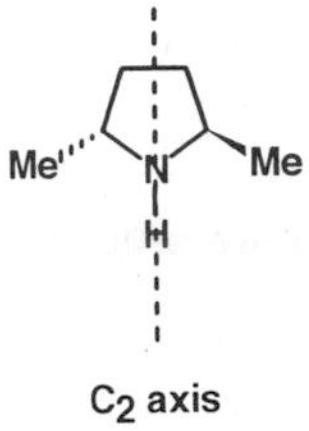

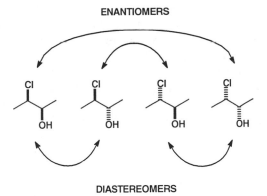

C$_2$ axis

Figure 2.2

If the molecule contains more than one chiral centre, there emerges the possibility of another form of stereoisomerism. Stereoisomeric molecules which cannot be superimposed by any symmetry operations are called diastereomers. Thus, for 2-chloro-3-hydroxybutane (Figure 2.3), one can draw four different structures: two pairs of enantiomeric compounds and two pairs of diastereomeric compounds.

ENANTIOMERS

DIASTEREOMERS

Figure 2.3

Based on this definition, the *cis* and *trans* forms of 1,2-, 1,3- and 1,4-disubstituted cyclohexanes are also diastereomers, although in the last case the possibility of optical isomerism cannot exist. It should also be pointed out that diastereomeric relationships can be found by inspecting a single molecule, whereas the property of enantiomerism always necessitates comparison with another molecule. Furthermore, in both cases the implication is that at least one other form (for diastereomers often several) does exist.

In reactions involving one or more chiral centres, we are interested in bringing about transformations which produce one stereoisomeric form in abundance over the other possible ones. One speaks of *stereoselective reactions* if the outcome of the reaction is non-statistical, and *stereospecific reactions* if the product is produced in one enantiomeric form only. All stereospecific reactions are necessarily stereoselective, but selectivity alone is not a sufficient criterion for specificity.

Dissolving metal reduction of 4-*tert*-butylcyclohexanone (Scheme 2.1) is a classic example of a stereoselective reaction; the more stable equatorial alcohol isomer predominates in the reaction products (98 % of the alcohol product).[1] Since the *tert*-butyl group effectively locks the cyclohexane into one chair form with the bulky substituent equatorial, the reduction occurs through delivery of the hydride from the upper face of the cyclohexanone. We shall return to more thorough rationalization of the reduction of cyclohexanones in Chapter 3. Although this reduction exemplifies the case of a substrate controlled process (the stereochemical bias resides in the starting material), much effort is currently being devoted to developing efficient methodologies for reagent controlled stereoselective reactions.

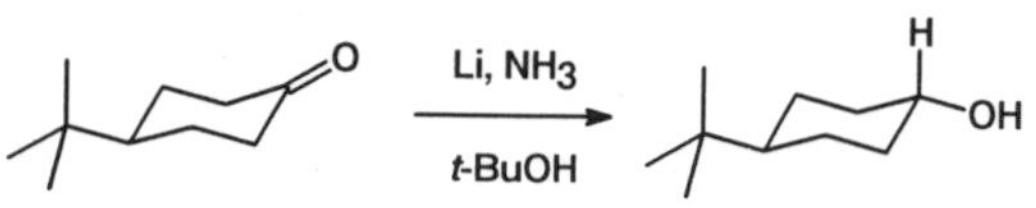

Scheme 2.1

The stereochemistry of the hydroboration reaction is determined by the geometry of the starting alkene. The *E* and *Z* isomers of the alkene give different products, as shown in Scheme 2.2. Both reactions are stereospecific because the mechanism of hydroboration necessarily places the hydride and boron on the same face of the double bond (the attack on the upper face only is depicted, in both reactions both enantiomers are formed). Specificity is thus a property embedded in the starting material and is typically mediated by the mechanism of the reaction.

In Scheme 2.2 the two faces of the double bond react with equal facility, and thus a racemic mixture is produced. This is obvious, since we are using an achiral reagent, and thus there is no source of asymmetric information in the reaction. But let us consider

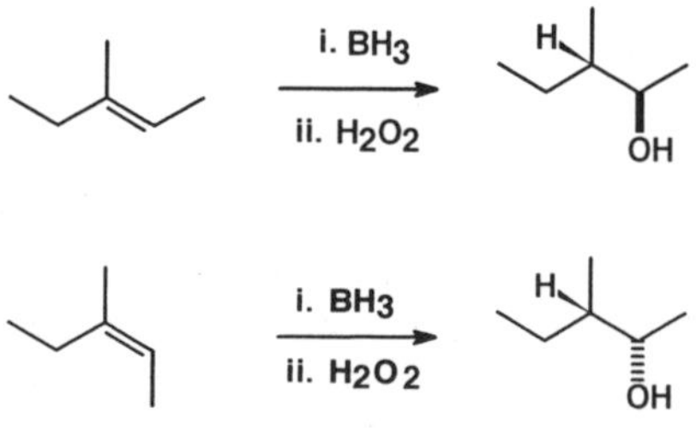

Scheme 2.2

whether the reaction could, at least in principle, be persuaded to give an enhancement in optical activity.

We will need to consider the topology of the molecules in order to be able to deduce what kinds of stereocontrolled processes a certain molecule can undergo.[2] In general, two groups are said to be topologically equivalent if they can be interchanged by rotation about any *n*-fold axis of rotation C_n to give a structure indistinguishable from the original. Thus, the two methyl groups of *trans*-2,5-dimethylpiperidine are equivalent.

A similar definition applies for equivalent molecular faces: two faces of a molecule are equivalent if the plane defining the faces contains a coplanar axis of symmetry. Such faces are not restricted to achiral molecules, but the attack by a reagent, whether chiral or not, at equivalent faces leads to equivalent transition states, and thereby also identical results. Thus, addition of a nucleophile Nu⁻ to the equivalent faces of the chiral, C_2 symmetric *trans*-2,6-dimethylcyclohexanone creates a new centre which is not chiral (Scheme 2.3).

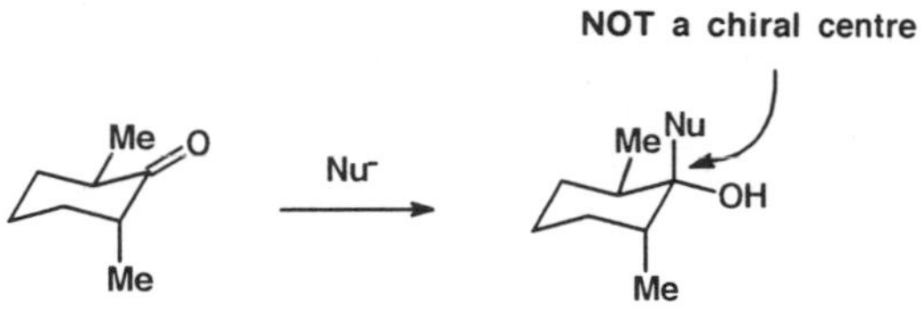

Scheme 2.3

If two groups can be interchanged by the operation of a rotation–reflection axis, the environments of the groups are enantiomeric, and the groups are said to be enantiotopic. Similarly, diastereotopic groups are defined as groups whose environments are diastereomeric. Atoms containing enantiotopic groups are also called prochiral, as the replacement of one of the enantiotopic groups will give a chiral compound.[3,4] In a similar fashion, the replacement of a diastereotopic atom will give rise to diastereomers. As a corollary to the definition, enantiotopic groups can only occur in achiral molecules.

Let us inspect a molecule CXYHH, whose geminally bound hydrogen atoms are enantiotopic. The prochiral substituents are distinguished in nomenclature through a modification of the sequence rules. Substitution of one of the hydrogen atoms with a dummy atom in such a fashion that it gains a higher sequence rule order (H'>H) than the other, followed by application of the normal procedure for assigning the *R/S* descriptors, gives the labels pro-*R* and pro-*S* to the geminal hydrogen atoms (Figure 2.4).

Figure 2.5 shows some enantiotopic arrangements. The cyclobutanone is an interesting example, showing both enantiotopic and equivalent groups. If we consider the planar conformation of cyclobutanone, the hydrogen atoms at C-3 are equivalent. Hydrogen atoms H-1 and H-4 can be interconverted by rotation of the molecule by a two-fold rotation axis, and are thus equivalent. Similarly, the hydrogens H-2 and H-3 are equivalent. The relationship between these pairs of equivalent atoms is enantiotopic. We can say that H-1/H-4 and H-2/H-3 form two enantiotopic sets of equivalent atom, or, vice versa, that H-1/H-2 and H-3/H-4 form two equivalent sets of enantiotopic atoms.

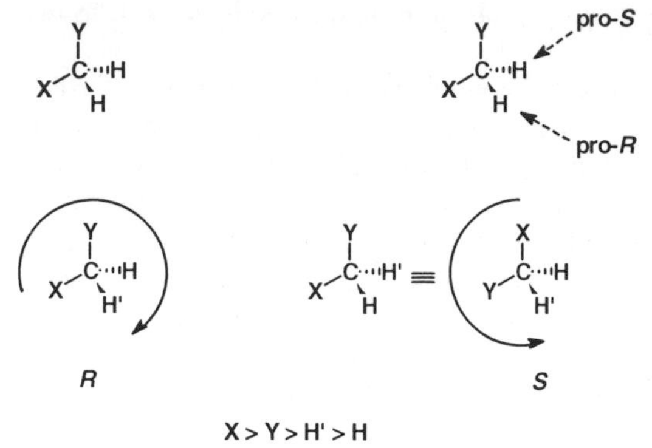

$$X > Y > H' > H$$

Figure 2.4

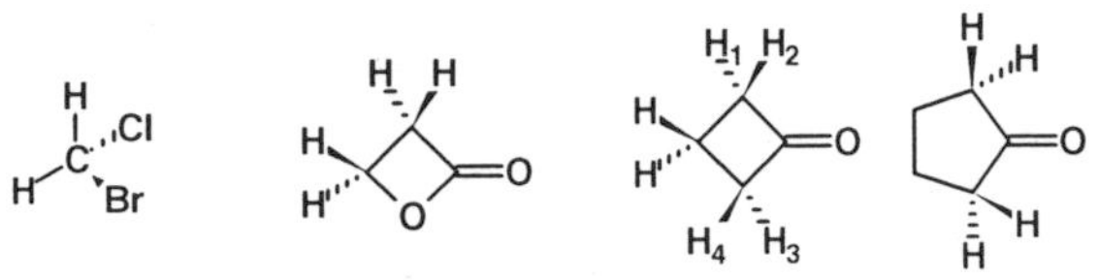

Figure 2.5

It is important to emphasize at this point the fact that the prochirality descriptor does not have any correlation with the absolute configuration of a product formed from the substitution of a prochiral atom. The replacement of the pro-R methoxy group of acetophenone dimethyl acetal (Scheme 2.4) gives either the R or S absolute stereochemical descriptor for the product, depending on the relative sequence rule orders of the replacing atom or group and the existing groups.

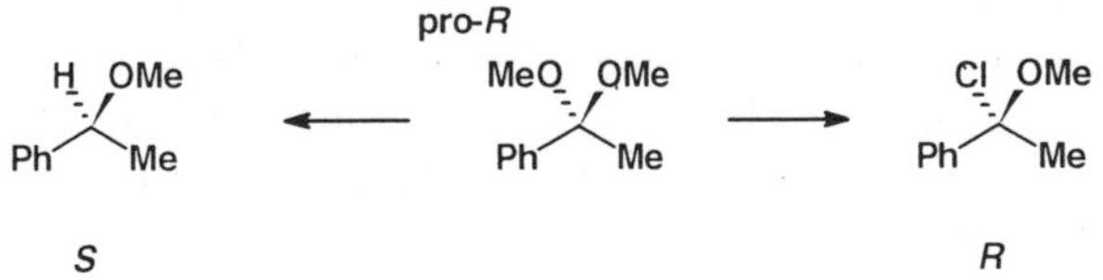

Scheme 2.4

We can define diastereotopicity as follows. Groups or atoms are said to be diastereotopic if they reside in diastereomeric environments and cannot be interchanged by any symmetry operation. In other words, replacement of one of two diastereotopic atoms leads to the formation of diastereomers. Some typical examples are shown in Figure 2.6. Molecular dissymmetry is not a criterion for the presence of diastereotopic groups. In the chair conformation of cyclohexane (D_{3d}), the six pairs of diastereotopic hydrogens (axial and equatorial) are interchangeable by either C_2 or C_3 and the pairs are therefore equivalent. The two examples of N-benzyl-2,6-dimethylpiperidine show quite

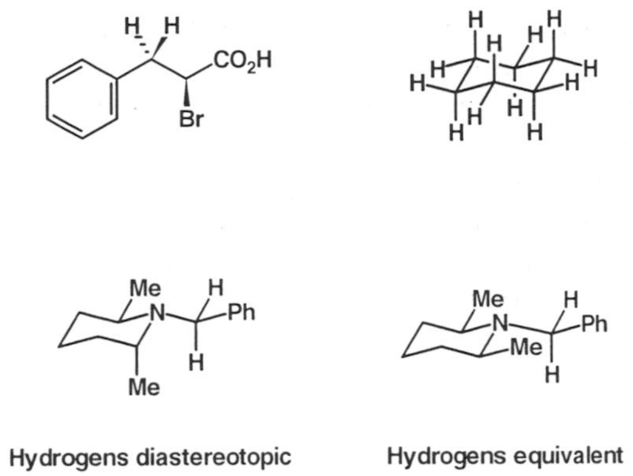

Hydrogens diastereotopic   Hydrogens equivalent

Figure 2.6

drastic differences in their NMR spectra.[5] The benzylic protons in the *trans*-dimethyl compound show a clear AB system, whereas the *cis*-dimethyl compound shows a singlet for the two protons. It is a common feature of diastereotopic atoms that they can be distinguished by the appearance of their NMR signals quite easily.

Reactions involving enantiotopic groups can exhibit enantiotopic selectivity. Such reactions are brought about by the use of a chiral reagent, and in some cases can lead to quite high levels of enantioselectivity. The advantage of enantiotopic selectivity is that the symmetry of the starting material is converted into asymmetry in the chirality generating step. The symmetry in the starting material can be taken into account in planning the synthesis of the chiral product.

The use of chiral bases provides an interesting means of achieving enantiotopic selection. An early example was provided by the conversion of cyclohexene epoxide to the corresponding allylic alcohol by rearrangement brought about by the C_2 symmetric chiral base derived from phenethylamine (Scheme 2.5).[6] The reaction proceeds through abstraction of the proton on the same face of the cyclohexane ring as the epoxide oxygen. As the two protons on the methylene groups on either side of the epoxide ring are enantiotopic, a chiral base (or presumably an aggregated form of it) will be able to distinguish between the two protons. Although the level of asymmetric induction (31 % ee) was not particularly high, this reaction paved the way for subsequent developments in the methodology.

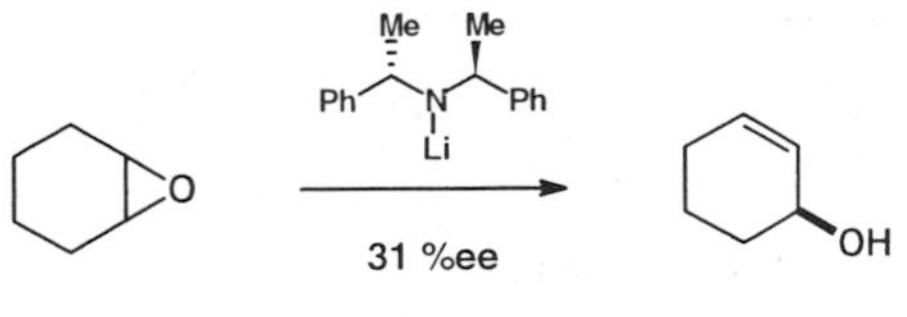

31 %ee

Scheme 2.5

Koga has developed applications of enantiotopically selective enolization reactions. A chiral amide base capable of tight chelation with lithium (Scheme 2.6) provides exceptionally high bias in the enolization of cyclic ketones (up to 97 % ee).[7] The bicyclo[3.3.0]octanone derivative (Scheme 2.7) was used in the synthesis of carbacyclin, a prostacyclin analogue.[8]

Two further examples illustrate the usefulness of enantiotopic selection. In another seminal example, Whitesell showed that the symmetric bicyclo[3.3.0]octadiene (Scheme 2.8) undergoes a clean ene-type reaction with the 8-phenylmenthyl ester of glyoxylic acid to give the product as a single diastereomer in good yield.[9]

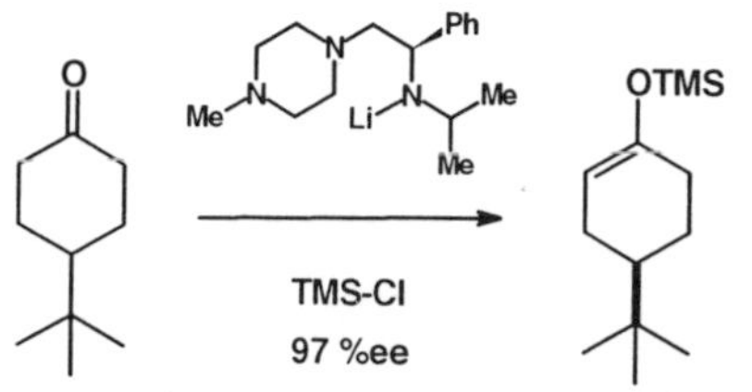

Scheme 2.6

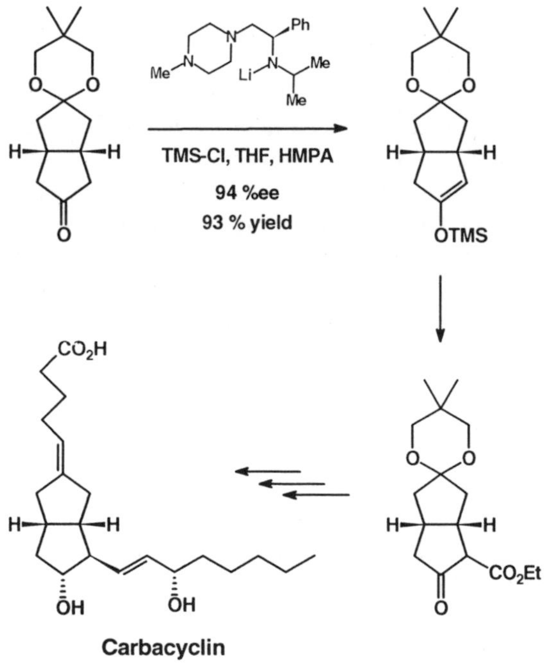

Carbacyclin

Scheme 2.7

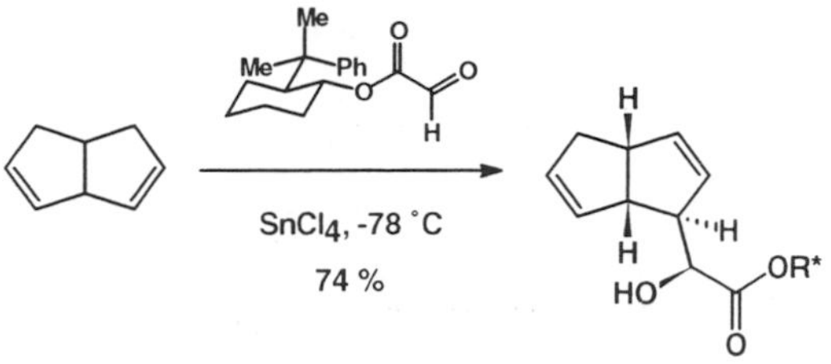

Scheme 2.8

The last example bears some relevance to the question concerning the possibility of distinguishing the two faces of an alkene upon hydroboration. In connection with the synthesis of a prostaglandin $F_{2\alpha}$ intermediate, the cyclopentadieneacetic acid derivative (Scheme 2.9) was hydroborated with a chiral borane reagent, (+)-diisopinocampheylborane, derived from (-)-α-pinene, to give the alcohol in high selectivity (>92 % ee by NMR analysis).[10] Similarly high (95–96 % ee) enantioselectivities were obtained in the hydroboration of 5-methylcyclopentadiene followed by oxidation. The product was used in the asymmetric synthesis of loganin, a biogenetic precursor of several natural products.[11]

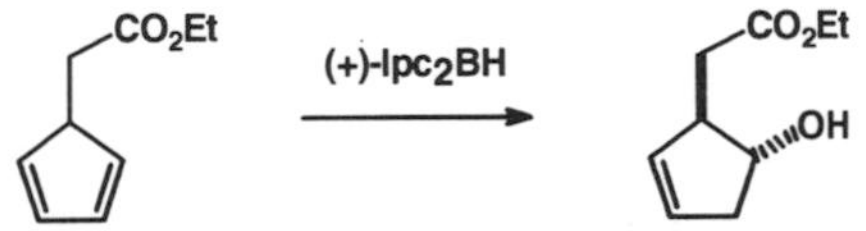

Scheme 2.9

In the example (Scheme 2.9), the attack of the reagent on different faces of the double bonds produces enantiomers, and thus the two faces are enantiotopic. Enantiotopic faces are characterized by the existence of an S_n axis perpendicular to the plane and the absence of a C_2 axis in the plane. The two faces are termed the *Re* and *Si* faces in the following way. The double bond defines a plane with three substituents at the reaction centre. The substituents are given priorities according to the sequence rule. When the double bond is viewed from one face, if the substituent priorities go in a clockwise fashion the face is called *Re*, and if counter-clockwise the face is *Si*. The *Re/Si* nomenclature is exemplified in Figure 2.7 for a carbonyl group, and the substituent R_L is supposed to have higher priority than the substituent R_S.

Preferential attack of a reagent on either face of the double bond is termed *enantiofacial selectivity*. In line with our previous observations on enantiotopic selectivity, enantiofacial selectivity is dependent on the reagent. An achiral reagent gives rise to transition states which are enantiomeric and thus the free energies of the transition states will be similar. There is no difference in the activation energies, and one will obtain an equimolar mixture of the products. Approach of a chiral reagent will give rise to diastereomeric transition states whose energy contents, and thereby the activation energies leading to these transition states, will differ. This will be observed as enantiofacial selectivity. We can now answer our previous question regarding the hydroboration reaction. The two

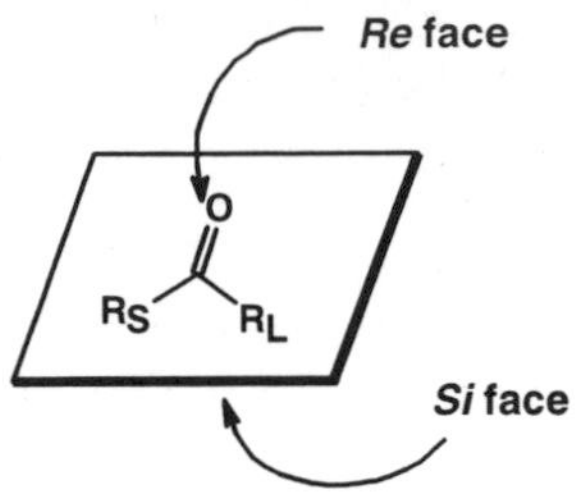

Figure 2.7

faces of the double bond can, at least in principle, be distinguished by the action of a chiral hydroborating reagent. This is clearly manifested in the several examples (including the ones above) where extremely high enantioselectivities have been obtained.[12]

Reduction of carbonyl groups by the action of a chirally modified lithium aluminium hydride reagent is a well established method for achieving enantiofacial differentiation. Scheme 2.10 shows an example of such a process where the reduction of the acetylenic ketone was achieved in high yield and high enantioselectivity.[13] The product was an intermediate in the synthesis of asteriscanolide, a structurally novel sesquiterpene lactone.[14]

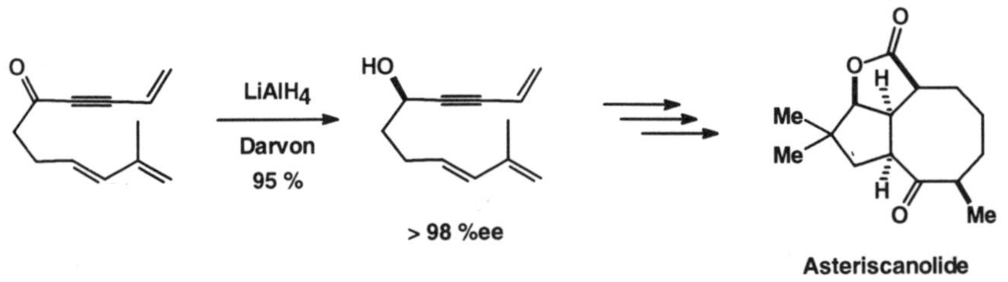

Scheme 2.10

Both enantiotopically and enantiofacially selective reactions are *reagent controlled*, and much effort is currently devoted to the development of asymmetric transformations relying on this concept. Practical realizations of the concepts outlined above will be discussed in more depth in Chapter 3.

A few words need to be said about the specification of the relative stereochemistry in molecules containing two or more chiral centres. As the number of chiral centres increases in a molecule, the description of the relative stereochemistries of the chiral centres can become problematic. Naming stereoisomers has been a prevailing problem since the early days of stereochemistry: dextrorotatory tartaric acid can be correlated chemically with either D- or L-glyceraldehyde, and it was therefore once specified D by the European chemists and L by the Americans. The sequence rules (the CIP, or Cahn–Ingold–Prelog system) were introduced to avoid such ambiguities in compounds containing one chiral centre.[15]

In compounds containing more than one chiral centre, the *threo* and *erythro* descriptors are not unambiguous, either (Figure 2.8). According to the original definition, the relative configurations of two groups on the same side in the Fischer projection are called *erythro*, and those on opposite sides are *threo*. Drawing the main chain in the more convenient zigzag form, the substituents of the *erythro* isomer end up on opposite sides of the plane! The situation got even more confused when inversion of the nomenclature was suggested in 1980.[16] The new system has gained widespread acceptance, and one should bear this in mind when reading the literature.

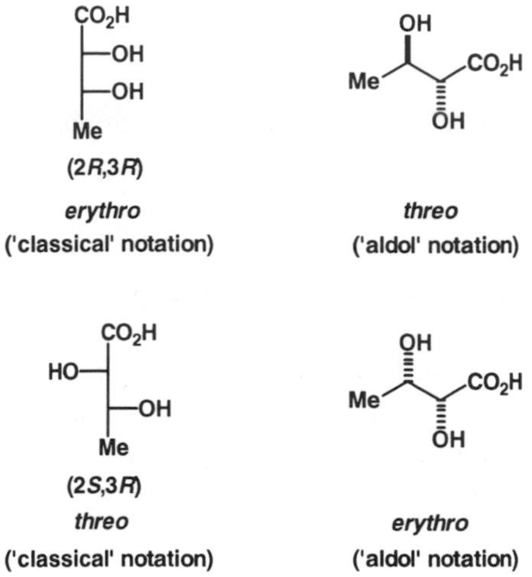

Figure 2.8

Several alternative systems were suggested to avoid the confusion, the most comprehensive of them being the one devised by Seebach and Prelog.[17] This system is based on the sequence rules and application of the *relative topicities* of the reactions. The processes are described simply as being either like (descriptor *lk*) or unlike (descriptor *ul*). The like course of a reaction involves the combination of either *Re,Re* or *Si,Si* faces of two trigonal, stereogenic atoms to give rise to *lk* relative topicity. Similarly, addition from the *Si* face of the *S* enantiomer gives rise to *lk* relative topicity. The relative configurations of the products are similarly described as being like (*l*) for (*R,R/S,S*) or (*R*,R**) and unlike (*u*) for (*R,S/S,R*) or (*R*,S**) (Figure 2.9).

The use of *syn* and *anti* descriptors was proposed by Masamune[18,19] to describe two non-hydrogen substituents being, respectively, either on the same side or the opposite sides of the plane defined by the zigzag main chain. This has the advantage of allowing instant recognition of the relative stereochemistry without necessitating the assignment of the absolute stereochemical descriptors *R* and *S*. With several chiral centres the system can be easily modified, as shown in Figure 2.10. The *syn/anti* descriptors are the ones we shall use throughout this text.

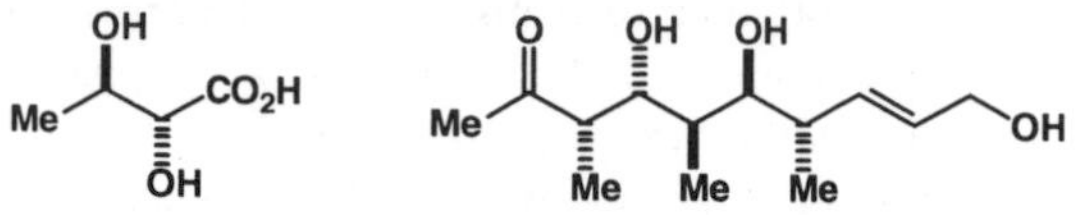

(2*R*,3*R*) *l*-isomer

(3*S*, 4*R*, 5*R*, 6*S*, 7*S*, 8*E*) *u,l,u,l,lk*-isomer

Figure 2.9

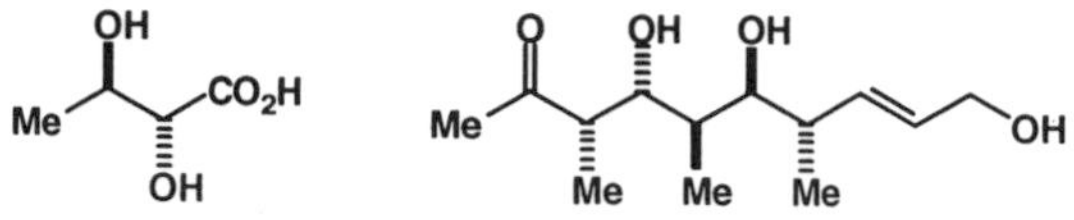

anti-isomer

syn,anti,syn,anti,E-isomer

Figure 2.10

2.1 The need for enantiopure compounds

Chirality is a property which often determines the actions and behaviour of molecules in rather unexpected ways. Lemons and oranges both contain limonene, the different enantiomers giving rise to subtle changes in the aroma properties of these fruits. Similarly, *R*- and *S*-carvone have different tastes: the former tasting of spearmint and the latter of caraway (Figure 2.11).

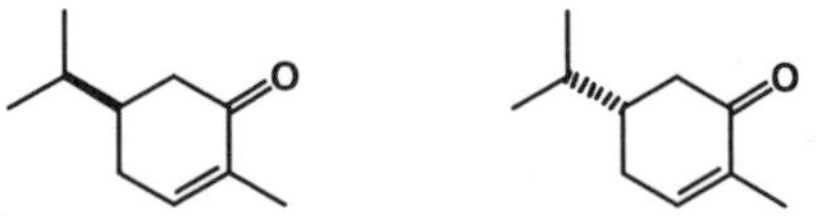

(*R*)-carvone spearmint odor

(*S*)-carvone caraway

Figure 2.11

As natural products and their derivatives and analogues find wide use in our everyday life, from medicines to food additives, it is understandable that for the production of these compounds by synthetic means we need to secure them in enantiopure form.* We simply cannot face risks similar to the infamous thalidomide (Figure 2.12) case of the late 1950s where one enantiomer of the product turned out to be potently teratogenic. Therefore,

* Single enantiomers are often called optically pure compounds. However, as this implies optical activities, which is not necessary for chiral compounds, other terms have been sought. The term homochiral has gained wide use, but the opponents of this term, quite justifiably, claim that there should also be the 'heterochiral' counterpart. 'Enantiomerically pure compound' (or EPC) is another term suggested, but herein we prefer the shorter form 'enantiopure' to describe a single enantiomer of a compound.

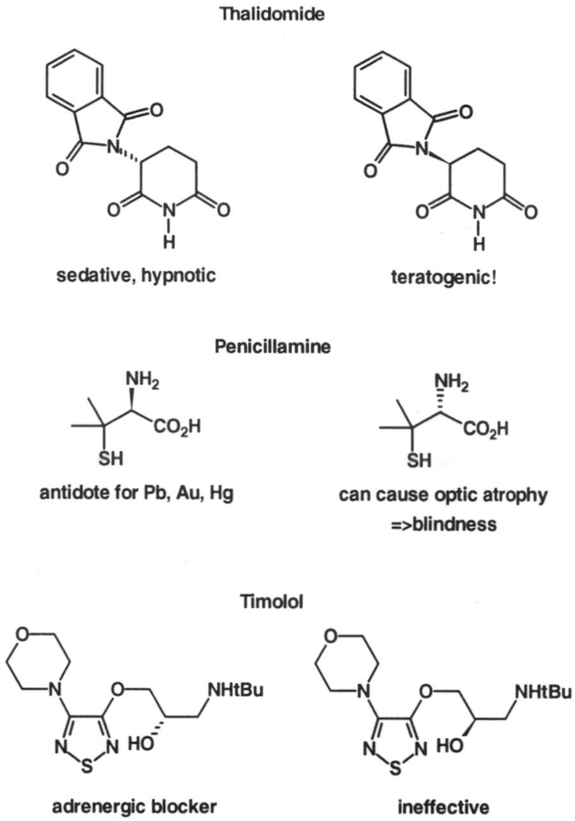

Thalidomide

sedative, hypnotic   teratogenic!

Penicillamine

antidote for Pb, Au, Hg   can cause optic atrophy
                              =>blindness

Timolol

adrenergic blocker   ineffective

Figure 2.12

various drug approval agencies are now strengthening their requirements for the approval of drugs capable of showing optical isomerism. In the drug industry, enantiopure products are playing an increasingly important role, and it is logical to expect that other areas of synthetic chemical activity will follow suit with rapid pace.[20–23]

Natural product synthesis has evolved from the inception of organic chemistry. In the late 19th century the carbochemical industry was born, with its main emphasis on aromatic compounds. By current terminology, the syntheses were based on operational transformations of functional groups, relying most heavily on associative or analogue based planning.

The early part of this century witnessed the events which may be considered as the birth of modern synthesis. Emil Fischer's outstanding achievements in carbohydrate, protein and nucleic acid chemistry, as well as his numerous other synthetic achievements, combined with the efforts of such notable chemists as Gustav Komppa (camphor, 1903), William Perkin (α-terpineol, 1904) and Sir Robert Robinson (tropinone, 1914), sowed the seeds which in the first decade of this century brought around what could be called the first total syntheses of complex natural products (Figure 2.13).

The steady development of physical organic chemistry led to an increased understanding of detailed reaction mechanisms. During the same period, the structural

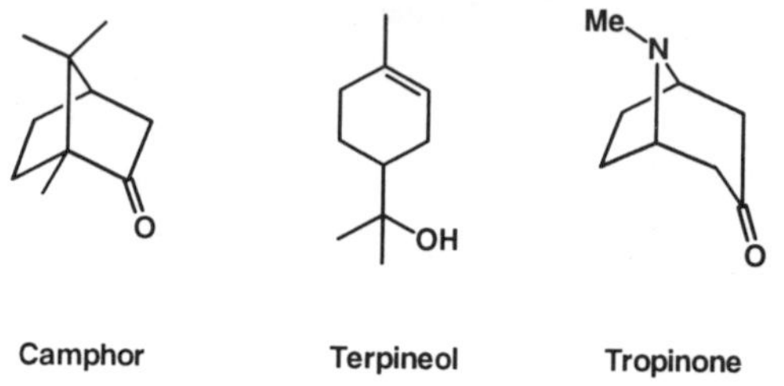

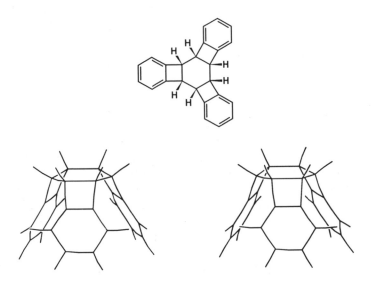

Camphor   Terpineol   Tropinone

Figure 2.13

theory of organic molecules evolved considerably, and this can be seen as the origin of conformational analysis. It is well worth remembering that as late as the 1920s, the chair and boat conformations of cyclohexane were far from being generally accepted. Baeyer's strain theory (from 1885) had taught chemists to consider cyclic carbon frameworks as planar polygons, although Sachse had five years later proposed that cyclohexane could exist in two forms. However, the conformational equilibration of cyclohexane evaded Sachse's analysis, and it took another 50 years before the concept of conformational analysis was established.[24] It is interesting to point out at this stage that planar cyclohexane compounds have now actually been made, as exemplified in Figure 2.14.[25]

Figure 2.14

During the 1940s and 1950s, the development of chromatographic and spectroscopic methods augmented by the development of new selective reagents all added to the armamentarium of the synthetic chemist, who was now able to achieve syntheses undreamed of just a few decades earlier. These included the syntheses of vitamin A (Isler, 1949), penicillin V (Sheehan, 1957), and several alkaloids, including strychnine

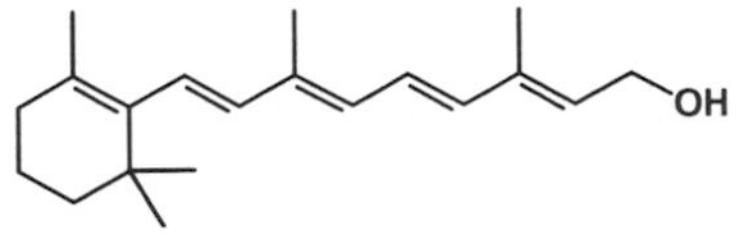

Vitamin A

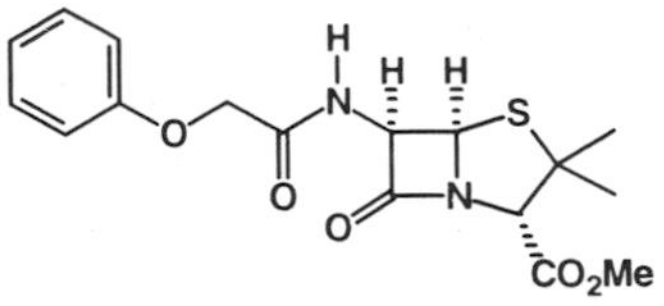

Penicillin V

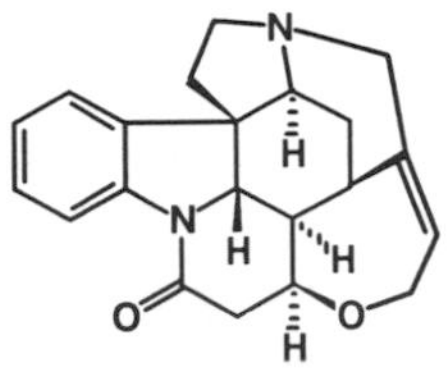

Strychnine

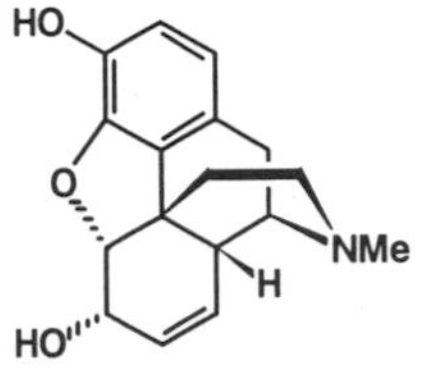

Morphine

Figure 2.15

(Woodward, 1954) and morphine (Gates, 1956) (see Figure 2.15). This was also the era of one of the most brilliant minds in the history of organic chemistry, namely Robert Woodward, whose synthetic masterpieces include such formidable target structures as quinine (1944), patulin (1950), cholesterol and cortisone (1951), lanosterol (1954), reserpine (1956), chlorophyll (1960), colchicine (1963), cephalosporin C (1965), vitamin B_{12} (together with Eschenmoser, 1972), prostaglandin F_{2a} (1973), and erythromycin (finished posthumously, 1981).

The last two or three decades have witnessed another major revolution in the art of synthesis. The value of chirality has been realized, and this has led to spectacular achievements in the application and development of reactions capable of distinguishing enantiomers and diastereomers. Mainly through and as consequence of the efforts of Corey, Stork and Woodward, the design of syntheses has gained much from the application of logical reasoning.[26] Modern day syntheses are highly sophisticated, and the use of techniques unknown in the 1940s and 1950s is quite common today.

The evolution of synthetic techniques and our understanding of the factors governing the structural and reactivity aspects of organic compounds have also led to an overall change in the justifications of synthesis. Whereas only a few decades ago the synthesis of a defined target compound could be a good enough justification for the execution of a multi-step synthesis, currently more and more emphasis is put on the actual design phase of a target compound. Drug molecules, monomers for polymerization, compounds valuable to biochemical or physicochemical studies, etc. are already being designed using highly sophisticated, but at the same time readily available, computer facilities. One could say that one of the present focal points in the research in synthesis is the design and execution of syntheses of defined and designed molecules which possess defined functional properties. This also implies that the strict divisions between the different branches of chemistry, as well as the distinction of chemistry from biochemistry and molecular biology, will have to be surpassed.[27–29]

As the need for enantiopure compounds is rapidly increasing, the objective of obtaining the final compound not only in chemically pure form but also in enantiopure state is usually secured already during the planning stages of the synthesis. Before looking at the various ways of achieving this goal, we shall take a brief excursion to the energetics involved in recognizing chirality.

2.2 Determination of Enantiomeric Purity

Two measures of optical or enantiomeric purity are in common use: *optical purity*, which is based on the optical rotation of the compound, and *enantiomeric excess*, which is independent of the optical behaviour of the compound. As we shall shortly see, these two measures often give similar or identical results, but this is not always necessarily the case.

Optical purity is defined in terms of the optical rotation of the compound. The specific optical rotation $[\alpha]\lambda$ is measured using a polarimeter which gives the rotation angle α (in degrees) of plane polarized light of wavelength λ (usually the sodium D-line, 598 nm, is used) in a cell typically 1 (or l) dm long in a sample concentration c (given in g per 100 mL).

$$[\alpha]\lambda = \frac{\alpha}{1c}$$

The optical purity (% op) is defined as the ratio between the observed specific rotation $[\alpha]_{obs}$ and the maximum specific rotation $[\alpha]_{max}$.

$$\%op = \frac{[\alpha]_{obs}}{[\alpha]_{max}} \times 100$$

This method has its drawbacks, however. In many cases the maximum specific rotation is not known, especially in cases when one synthesizes a previously unknown compound. Experimental conditions also cause ambiguities. One should always use a cell with a long path and a large diameter to avoid local concentration gradients which can cause distorted behaviour of the rotation. Particles in the sample easily modify the observed rotation, as do air bubbles, even very small ones, which refract light. In cases when the sample is coloured, the absorption of the light may cause problems.

Enantiomeric excess (% ee) is defined as the excess of one enantiomer over the other, and this definition makes this measurement unambiguous.

$$\%ee = \frac{|\%R - \%S|}{|\%R + \%S|} \times 100 = 100 - 2\,(\%S)\ (for\ R)$$

Comparison between enantiomeric excess and optical purity reveals that the latter is further hampered by the following experimental problems. The solvent, concentration and, to a lesser extent, temperature affect $[\alpha]$. Reproduction of literature concentrations may be difficult. Especially in cases where the rotation has been reported in chloroform or ethanol, one should always also specify the quality of the solvent (i.e. does the chloroform contain ethanol as a stabilizer, if so how much—the American and European standards vary; in determinations performed in ethanol solutions, what is the percentage of ethanol?). The optical rotation is not always necessarily linear with concentration owing to association effects; in other words, a bimolecular complex of the *R–R* forms of a compound does not necessarily have the same rotation as the (unimolecular) *R* compound. Even small amounts of impurities can adversely affect the outcome of the determination of optical rotation. For instance, if the impurity B has a relatively large rotation (e.g. 100°), the measurement of the rotation of compound A (e.g. −1°) is severely affected; even 1 % of impurity will completely abolish the measurement. It is noteworthy that the impurities need not be optically active, owing to association effects. A case in point is the effect of acetophenone in a solution of phenethanol (Scheme 2.11), where the presence of unreacted acetophenone significantly increases the rotation of the product (from +43.1° with no PhCOMe to +58.3° with a four-fold excess of PhCOMe).[30]

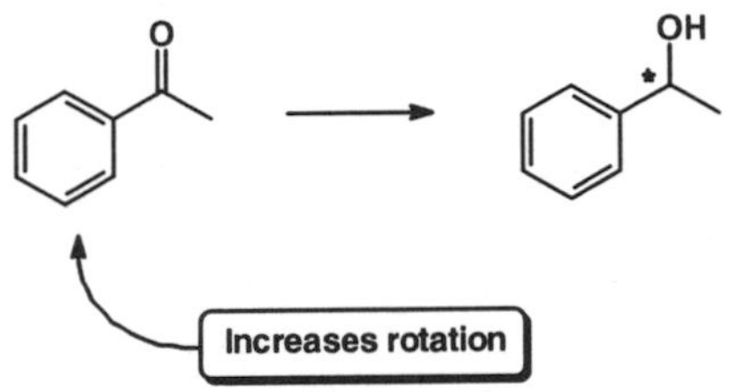

Increases rotation

Scheme 2.11

The determination of enantiomeric excess is in practice performed through the determination of the relative amounts of the two enantiomers. During the course of method development, one should always bear in mind that every transformation subsequent to the chirality forming reaction can also effect kinetic resolution, and thus any further treatment should be avoided (especially any purifications or separations). The determination should be performed with a crude sample directly from the reaction mixture if at all possible. The most direct method is based on the utilization of chiral chromatographic media, and several types of chiral columns are currently available. Quantitation is simple, and no further operations are needed. In cases when this is not possible, one can revert to a number of other methods, bearing in mind the precautions concerned with kinetic resolution. Derivatization with a chiral agent gives rise to a mixture of diastereomers, which can usually be easily separated either by HPLC or NMR methods. Any derivatizing agent capable of achieving separation is acceptable, but the most commonly used ones include α-methoxy-α-trifluoromethyl phenylacetic acid (MTPA, Mosher's acid),[31,32] mandelic acid, phenethyl amines, amino acids and amino alcohols and their derivatives. The derivatization need not be based on covalent bond formation. In the NMR methods, one can also use chiral solvents or chiral shift reagents.

Whatever the strategy one decides to use to establish the enantiomeric purity of the compound, one should always secure the detection limits in each particular case. A narrow peak and a broad peak will give markedly different threshold behaviours with all detection methods.

2.3 Chirality and Thermodynamic Principles of Asymmetric Induction

It is well known that enantiomers do not differ in their physical properties, except when subjected to a chiral environment. For example, the R and S (D and L, respectively) forms of N-*tert*-butoxycarbonylphenylalanine methyl ester (Figure 2.16) give identical UV, IR, NMR and mass spectra, and their chromatographic mobilities on normal phase and (achiral) reversed phase chromatography are also indistinguishable.

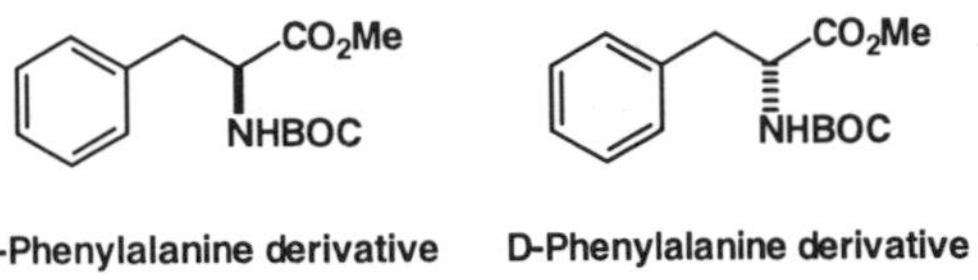

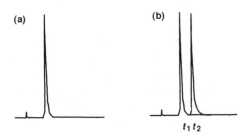

L-Phenylalanine derivative   D-Phenylalanine derivative

Figure 2.16

However, if the two enantiomers are subjected to chromatography on a chiral medium, the two compounds experience the environment differently; they are recognized by the chiral medium as being different. In Figure 2.17(a) the trace shows a typical chromatogram for a racemic mixture on an achiral medium. Figure 2.17(b) represents a chromatogram obtained from chromatography on a chiral medium. Similarly, diastereomeric derivatives (e.g. amides made from (R)- or (S)-phenylethylamine) exhibit dissimilar physical properties, such as chromatographic mobilities. Thus, Figure 2.17(b) could also represent the separation of diastereomers on an achiral medium.

(a)   (b)

$t_1\,t_2$

Figure 2.17 Chromatograms obtained from: (a) racemic compound on an achiral medium; (b) racemic compound or a mixture of diastereomers, on a chiral medium.

The x-axis respresents time, and thereby also the amount of eluent (solvent volume) used to elute the compound out of the column. In calculating the time, one takes into account the void column volume (the time needed to bring the solvent front through the column, shown as the small peak at the left of each chromatogram). The difference in the elution volumes of the two compounds is based on their affinities for the medium, being higher for the compound eluted last out of the column. The efficiency of the separation can be described by the ratio of the retention time indices t_1 and t_2 (e.g. the retention times or retention volumes).

$$\alpha = \frac{t_2}{t_1}$$

This ratio gives an estimate of the 'ease' of separation, and can be used as a guide for selecting, for example, preparative separation methods. In terms of energy, the free energy difference in the retention of the two species interacting with the solid support can be estimated from this ratio of retention times using the Gibbs free energy.

$$\Delta G = -RT \ln \alpha$$

As the ratios α are usually quite small (typically of the order of 1.05–1.5), it is quite obvious that even small energy differences in interaction can be very efficiently distinguished by chromatography.

Let us now turn our attention to the energetics of asymmetric reactions. Not unlike the chromatography example, a reaction producing two enantiomers from a single achiral starting material must be able to distinguish between the two emerging products. The chiral information must come from somewhere, and we shall return to the various methods of asymmetric induction in Section 2.4. For our present discussion on the energetics, we shall assume that the chiral information is mediated by a chiral reagent, although the same analysis applies to all the cases of simple enantioselection. As the starting material is a single compound, the distinction cannot be made at this stage—even a chiral reagent would form two enantiomeric complexes with the same energies. The products are enantiomeric with each other and, by the same principle, are equienergetic. The distinction is possible at the stage of the transition states, where the two transition states TS(1) and TS(2) are diastereomeric, as shown in Scheme 2.12. The energy profiles for the two alternative reactions are shown in Figure 2.18.

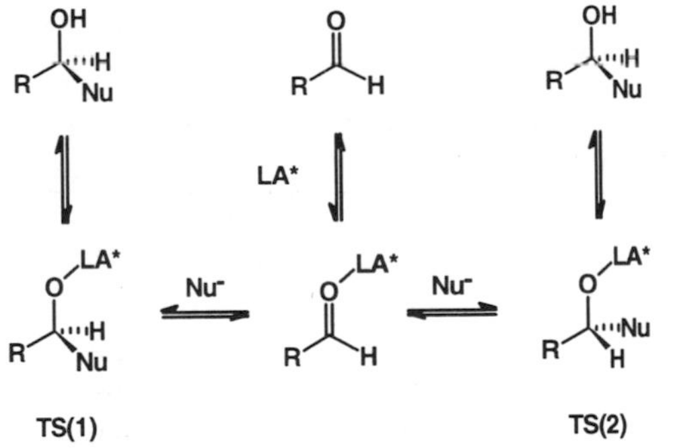

Scheme 2.12

Since the two transition states are diastereomeric, the substituents of the starting material occupy different positions in space. For some reactions involving cyclic, six membered transition states, the distinction between the two alternative transition states can be simply deduced from the general principles of conformational analysis, such as the favoured orientation of a substituent in an equatorial rather than an axial position. We shall return to such cases in Chapter 3 in connection with the individual reactions.

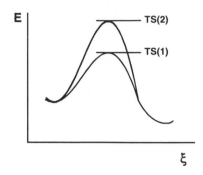

Figure 2.18   Energy diagram for the reaction of an achiral starting material with the enantiomers of a reagent to produce the enantiomeric products.

In the absence of thermodynamically driven equilibration processes, the formation of a single chiral centre is typically under kinetic control, where the energy difference between the transition states determines the relative rates of the two competing reactions, and thereby also the ratio of the products formed. The formation of a second and further chiral centres is more complicated, as this can often also be thermodynamically controlled, i.e. the products are of different energies and equilibration can alter the initial kinetic ratio. For a kinetically controlled process, the product ratio can be calculated from the free energy differences between the transition states or, vice versa, the free energy difference can be calculated from the observed product ratio. Table 2.1 shows the variation of the difference in the activation energies ($\Delta\Delta G$) leading to the major and minor products in an enantioselective reaction as a function of temperature and the observed % ee. As is clearly obvious, the temperature effects are quite remarkable (the energy difference nearly doubles within the common temperature range $-78°C$ to rt). Another point to note is the fact that useful enantioselectivities (>95 % ee) require a substantial energy

Table 2.1 Energy differences (in kJ mol^{-1}) between the two isomers at different temperatures as a function of enantiomeric excess and temperature.

% ee	Temperature (°C)						Ratio
	-100	-78	-40	-23	0	23	
99.9	10.9	12.4	14.7	15.8	17.3	17.7	99.95:0.05
99.5	8.63	9.72	11.6	12.5	13.6	14.7	99.75:0.25
99	7.63	8.59	9.39	11.0	12.0	13.0	99.5:0.5
98	6.62	7.46	8.92	9.55	10.4	11.3	99:1
95	5.28	5.95	7.12	7.63	8.34	9.05	97.5:2.5
90	4.23	4.78	5.70	6.12	6.70	7.25	95:5
80	3.18	3.56	4.27	4.57	4.99	5.41	90:10
70	2.51	2.81	3.35	3.60	3.94	4.27	85:15
60	2.01	2.26	2.68	2.93	3.14	3.39	80:20
50	1.59	1.80	2.14	2.30	2.51	2.72	75:25
40	1.22	1.38	1.63	1.76	1.93	2.10	70:30
30	0.88	1.01	1.22	1.30	1.42	1.51	65:35
20	0.59	0.67	0.80	0.84	0.92	1.01	60:40
10	0.29	0.34	0.38	0.42	0.46	0.50	55:45

difference between the two (diastereomeric) transition states leading to the products. An energy difference of 8 kJ mol^{-1} roughly matches a hydrogen bond in a peptide, or the energy difference between an axial and an equatorial methyl group in a cyclohexane (approximately 7.5 kJ mol^{-1} repulsion for the axial conformer).

2.4 Methods for Obtaining Chiral Compounds

Since the attainment of chiral compounds in enantiopure form is desired, we need to consider the various ways available for achieving this goal. If the desired compound is commercially available, then simply purchasing it may prove a trivial solution to the problem. This is true for a growing number of compounds, as many large and small companies now provide a wide selection of optically active and enantiopure compounds. Another method, which is often the more tedious one, is to isolate the compound from natural sources. This can be a competitive route in those cases where the synthesis has not been achieved and the compound is desperately needed (e.g. the promising anticancer compound taxol, Figure 2.19, belongs to this group). A variation of this method is the development of fermentation processes capable of producing large quantities of the material. Several antibiotics are produced in such a fashion, and widespread efforts are being directed at developing new fermentation avenues for compounds which are currently produced by total synthesis via a long and expensive sequence.

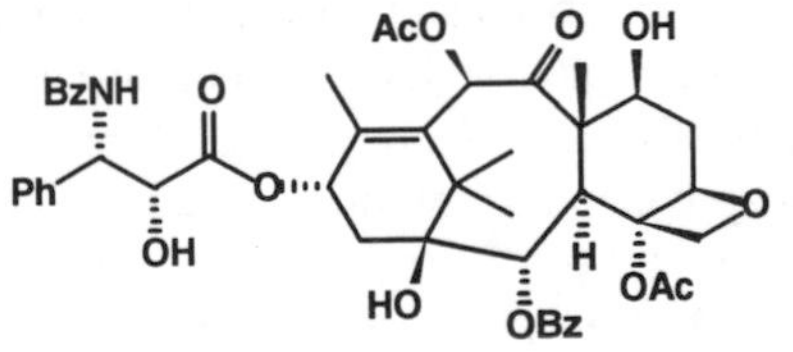

Taxol

Figure 2.19

An economically somewhat less satisfactory method relies on resolution. One synthesizes the target compound in racemic form, and then breaks the racemate to obtain the desired enantiomer. The attractiveness of this method is diminished by the fact that a maximum of 50 % yield can be obtained in what often is the final step in the synthesis. A further disadvantage is that the other enantiomer often ends up in waste. In some cases the unwanted isomer can be epimerized and thus used nearly quantitatively in the synthesis. Resolution can sometimes be achieved by direct crystallization, although this is mainly used for improving the optical purity of many crystalline compounds.[33] Crystallization via diastereomeric derivatives (e.g. phenethylamine or mandelate salts) is common practice even in industrial scale processes. In chromatography, one can use either direct or indirect methods. In the direct methods, one uses one of a large number of chiral columns[34,35] to achieve resolution. This is particularly attractive for analytical separations, and is widely used in connection with method development. In the indirect

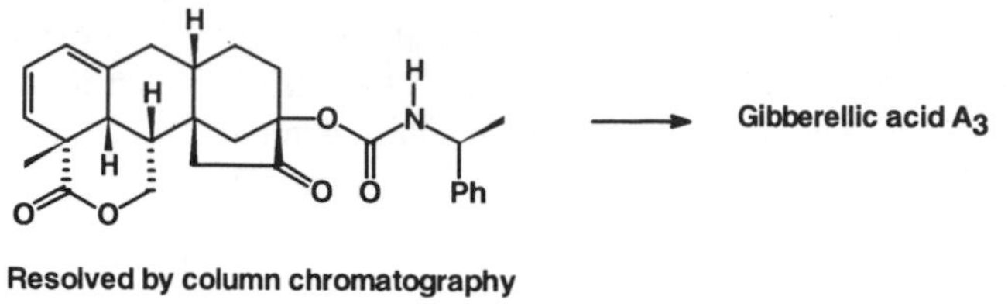

Resolved by column chromatography

Scheme 2.13

methods, the compound to be resolved is derivatized with a suitable chiral agent, and the derivatizing agent is again cleaved after chromatographic separation. The synthesis of gibberellic acid A_3 by Corey is an early and successful example of such a process (Scheme 2.13).[36]

Asymmetric transformations, albeit still rarely predictable, and thus of limited utility, could provide an interesting method for obtaining enantiopure compounds. Distinction must be made between *first-order* asymmetric transformations, in which the equilibrium is shifted to favour one enantiomer in solution, and *second-order* asymmetric transformations, in which one of the enantiomers crystallizes from the solution, and thus drives the equilibrium towards this side. Second-order asymmetric transformations have the unique property that the rate of crystal formation increases as the temperature is increased.[37]

Specific examples of second-order asymmetric transformations are provided by the transformations of phenylglycine derivatives.[38] The transformations proceed via the formation of a Schiff base and subsequent protonation (and concomitant hydrolysis of the imine) to precipitate the D-phenylglycinate (+)-tartrate salt. The optical purities approach 100 % with practically all ketones tested. A similar process is used in the preparation of the following benzodiazepinone (Scheme 2.14) in nearly enantiopure form.[39] The benzodiazepinone is a useful intermediate for the synthesis of a number of cholecystokinin receptor ligands.[40] In this case, the crystallization-induced, second-order asymmetric transformation is brought about through the use of 3,5-dichlorobenzaldehyde and camphorsulfonic acid (CSA). The process affords over 90 % yield of practically enantiopure material.

The underlying basis for second-order asymmetric transformations lies in the thermodynamics of the two interconverting species. One of the enantiomers must be removed from equilibrium, thus giving the driving force to shift the equilibrium to that side. The interconversion rate must also be sufficiently fast so that equilibration is

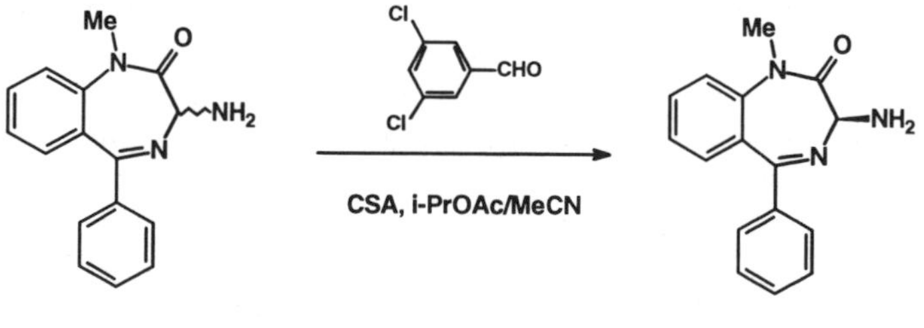

CSA, i-PrOAc/MeCN

Scheme 2.14

feasible. If the energy barrier of the path connecting the two (enantiomeric) compounds is low enough (60–75 kJ mol^{-1} at room temperature), the two forms of the reactants and products readily interconvert. If the reactants and products are enantiomeric, their free energies are equal in an achiral environment, and no separation can be achieved. However, if the products are diastereomeric, with only one of the chiral centres being stereochemically labile, the ground state energies of the products will differ and separation (enrichment) can be achieved. Scheme 2.15 shows an example where dynamic kinetic resolution has been achieved with high selectivity utilizing a chiral catalyst to provide a chiral environment and favour the removal of one of the enantiomers from the equilibrium.[41,42]

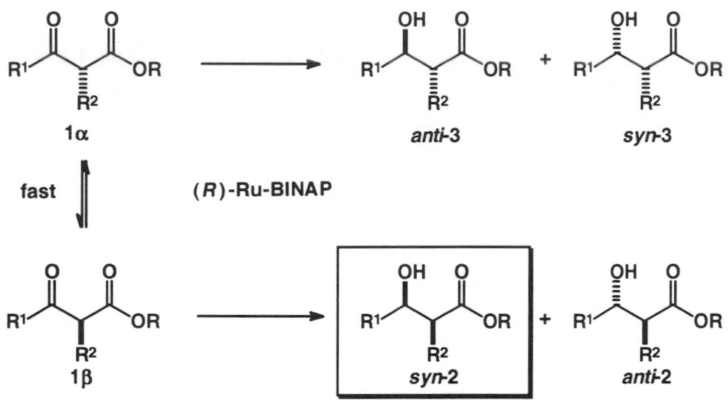

Scheme 2.15

In each case studied, the enantioselectivities were high (90–98 % ee). With the (R)-Ru-BINAP catalyst, the *syn*-**2** products predominated when the R^2 group was an acylamino substituent. In the case of cyclopentan-1-one-2-carboxylate (R^1, R^2 = CH$_2$CH$_2$CH$_2$), the *anti*-**3** product was formed in 99:1 diastereomeric ratio with opposite absolute stereochemistry at the labile C-2. These observations were rationalized as follows. The absolute stereochemistry at C-3 is determined by the BINAP catalyst (R or S). The *anti* selectivity of the hydrogenation of the cycloalkyl substrate was rationalized in terms of a constrained, tricyclic transition state model (Figure 2.20). In the case of the amide substrates, the amido NH is capable of participating in intramolecular hydrogen bond formation (Figure 2.20) which directs the delivery of the hydrogen with high *syn* selectivity. This rationalization was further supported by the strong solvent dependence of the reduction: the diastereoselectivity dropped to 7:3 on changing the solvent from dichloromethane to methanol.

For second-order asymmetric transformations to be of practical utility, the reactants must be in rapid equilibrium, making the removal of one enantiomer the rate determining step. The structure of the starting material must allow for clear distinction between the two transition states in terms of the stabilities of the forming *syn* and *anti* products. Finally, the catalyst must be strongly biased towards selecting one of the enantiomers of the starting material.

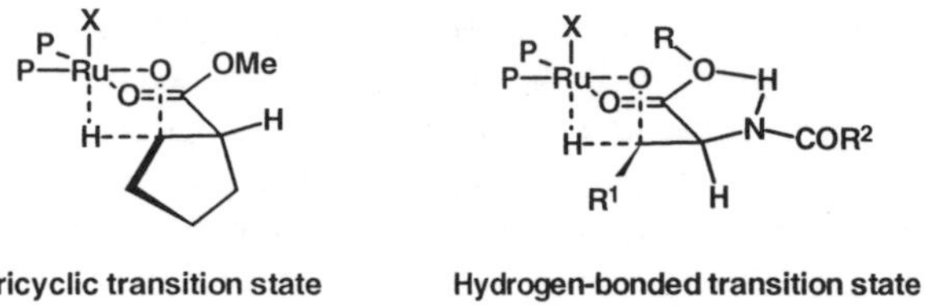

Tricyclic transition state   Hydrogen-bonded transition state

Figure 2.20

Meso compounds potentially give simple access to chiral compounds if one has a way to break the internal mirror symmetry of the compound. This can be achieved by kinetic resolution in several cases, and the most common ones rely on enzymatic reactions. Thus, one can for instance take the cyclic diester derived from benzene-1,2-diol (microbial oxidation product of benzene) and achieve kinetic resolution by the action of porcine liver esterase (PLE) (Scheme 2.16). The optical yields are often quite high.[43,44]

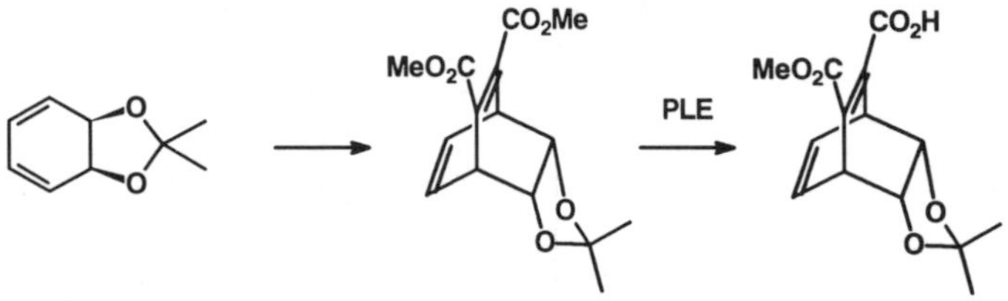

Scheme 2.16

A very useful example of kinetic resolution is based on the asymmetric epoxidation reaction developed by Sharpless.[45–47] In the kinetic resolution, a racemic allyl alcohol is converted to a mixture of the corresponding enantiomerically enhanced epoxy alcohol (glycidol) and the starting material (Scheme 2.17). As we shall see later (Section 3.3.1.1), this transformation is highly influenced by the reagent, and one can often achieve kinetic resolution with very high enantiomeric excess obtained for either product.

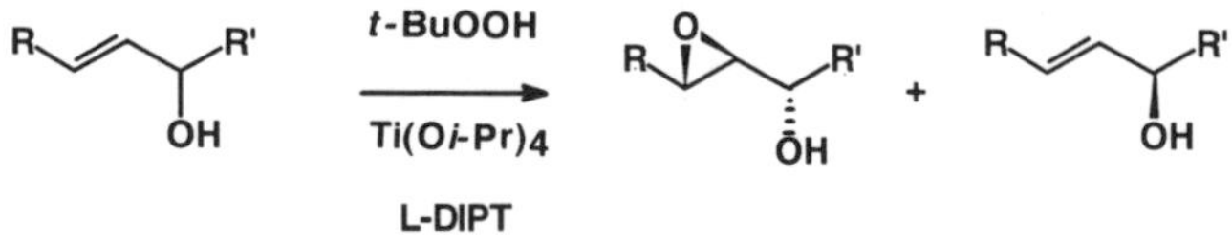

Scheme 2.17

Asymmetric induction based methods give the last category of transformation for obtaining optically pure products. Asymmetric synthesis is a term which has been used to describe stereocontrolled synthetic processes in many differing ways. The concept of asymmetric synthesis was originally introduced by Emil Fischer,[48,49] based on his synthesis of carbohydrates. Marckwald's classical definition of asymmetric syntheses embraced 'those reactions which produce optically active substances from symmetrically constituted compounds with the intermediate use of optically active materials but with the exclusion of all analytical processes'.[50] Strictly speaking, according to this original definition, asymmetric synthesis was intended to mean only processes during which asymmetry is introduced into an achiral starting material. However, such processes are only slowly emerging in practical synthetic chemistry, and other definitions have become prevalent. A common usage is to describe any synthetic operation that produces a new chiral centre in enantiomerically enriched form as being asymmetric. This is the definition we shall also use in this text. Because of the tremendous developments in the level of asymmetric induction achieved over the past two or three decades, we shall place the main emphasis on those processes which will give high levels of asymmetric induction (>90 % ee). We shall further divide the asymmetric processes into three major subclasses—internal, external and relayed asymmetric induction—based on the origin of the chiral information.

The asymmetric information can originate from either the starting material or the reagent. Since the starting material can contain a chiral center as an integral part of the structure, or it can be temporarily introduced into it, we can think of three principal ways of introducing chirality into the emerging centres in the molecule.

One could think of a way, or route, where one uses chiral starting materials whose intrinsic asymmetric information is transferred into the new chiral centres. In such cases we talk about **internal asymmetric induction**. The original chiral centre is thus an integral part of the final product, and this requires careful selection of starting materials in the planning stages of the synthesis. The utilization of the chiral pool compounds[51–53] (amino acids, carbohydrates and terpenes) as starting materials is a hallmark of this approach, whose extension to include also synthetic chiral materials is widely known as the chiron approach.[54] The chiral centre is bound to the reacting system irreversibly through a covalent bond, and remains connected throughout the sequence. The asymmetric induction in such cases relies heavily on diastereoselective processes, as we shall see later (Section 3.3.2). The synthesis of (+)-vincamine from aspartic acid[55] utilized the chiral α-centre of the starting amino acid as the sole source of chiral information in the construction of the three chiral centres in the target molecule (Scheme 2.18). The α-carboxyl group and thus also the original chiral centre is lost later in the synthesis, but other atoms of the starting amino acid are incorporated into the final product, as shown by the bold lines.

One can also think of a route where the chiral information is introduced into the molecule at a suitable stage before the chirality forming reaction, then used as above,

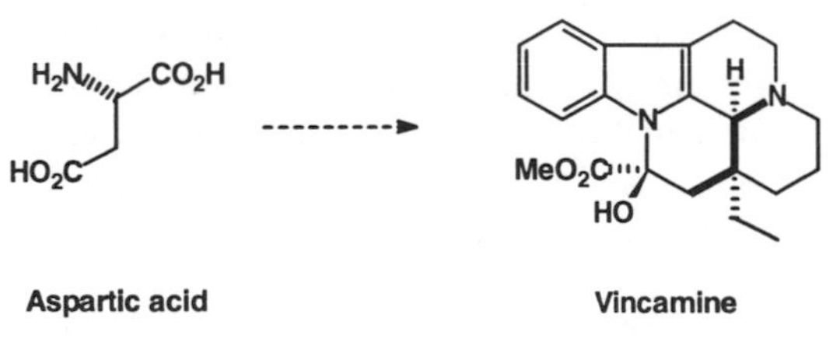

Aspartic acid   **Vincamine**

Scheme 2.18

and the chiral originator then being removed (**relayed asymmetric induction**). One often speaks of chiral auxiliaries in such cases. An inherent drawback of such processes is that one usually needs two extra synthetic operations: one to introduce the chiral auxiliary, and another one to remove the chiral originator. Furthermore, in many cases the chiral auxiliary is completely lost in the following operations, including the chiral information, which is a wasteful process in terms of the economy of information.

Scheme 2.19 illustrates a synthesis of MeBMT, a non-proteinogenic amino acid in cyclosporine, utilizing the chiral auxiliary technique developed by Evans. It is noteworthy that in this synthesis the two oxazolidinone-containing achiral auxiliaries are derived from D- and L-phenylalanine. All of the chiral information is thus derived from the chiral auxiliaries.[56] We shall return to this important class of chiral originators, and we will see that several non-destructive methods have been developed for the removal of the oxazolidinone moiety (see e.g. Section 3.2.2.2).

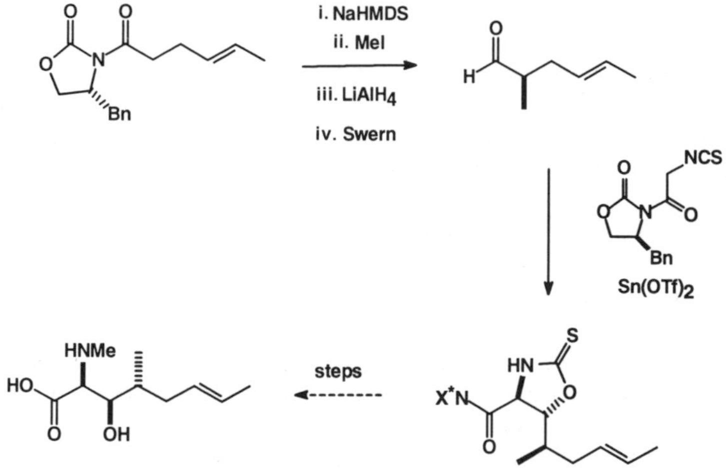

Scheme 2.19

Finally, the third group of reactions involves processes where the chiral information is brought transiently into the reacting system, usually into the transition state, through reversible and/or weak bond formation, used to assemble the reacting partners appropriately, and finally released in its original form at the end of the reaction. Such **external asymmetric induction** processes are catalytic in the chiral moiety. External asymmetric inductions thus correspond to Marckwald's classical definition of asymmetric synthesis. These processes are economically the most desirable ones, since we only need a catalytic quantity of the chiral originator which is later recovered in its original form and thus reusable in other experiments. These are also the ideal goals for the development of asymmetric processes, and, as we shall see later in Chapter 3, many reactions have already succumbed to such catalyst development.

Scheme 2.20 shows a particularly intriguing example of using a catalyst for the enantioselective alkylation of aldehydes to produce optically active secondary alcohols.

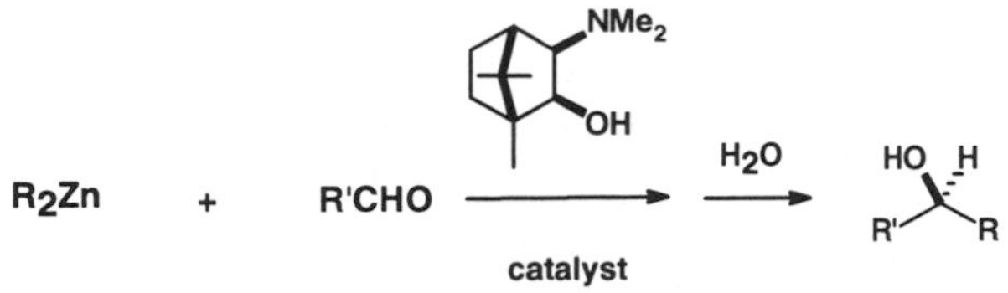

Scheme 2.20

We shall return to this process in more detail later (Section 3.2.1.2), but it is noteworthy that this reaction proceeds with excellent enantioselectivity even when the chiral catalyst is only of 15 % ee (i.e. 57.5:42.5 ratio of enantiomers).[57]

This quite remarkable chirality amplification has its origins in the sterically induced stability and instability of the complexes formed from the ligand and the organozinc reagent.[58,59] The (−)-3-*exo*-(dimethylamino)isoborneol (DAIB) ligand accelerates the addition of the organozinc reagent onto the carbonyl acceptor only as a monomeric species. However, the dimeric species are much more stable, and owing to steric reasons the heteromeric (i.e. from (+)- and (−)-DAIB) meso complex is thermodynamically the most stable. In solution, the chiral diastereomer of the complex has a more pronounced tendency to dissociate into the reactive monomeric species than the meso complex. Therefore, the minor enantiomer is transformed into the stable meso complex, while the major enantiomer produces the more easily dissociable chiral dimer (Scheme 2.21).

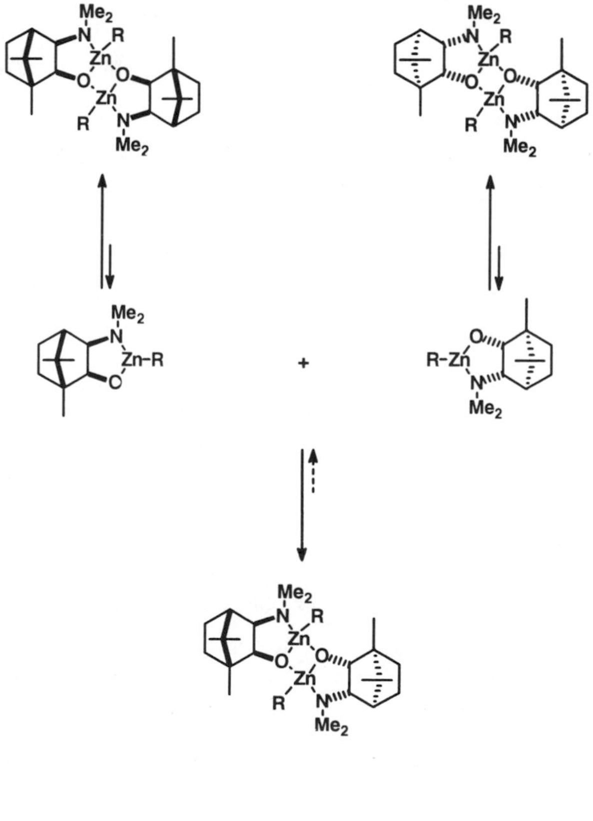

Scheme 2.21

References

1. Huffman, J.W. and Charles, J.T. *J. Am. Chem. Soc.* **90** 6486–6492 (1968).
2. Mislow, K. and Raban, M. in *Top. Stereochem*, vol **1** (Allinger, N.L.: Eliel, E.L., Eds.), John Wiley and Sons: New York, 1967, 1–38.
3. Hanson, K.R. *J. Am. Chem. Soc.* **88**,2731–2742 (1966).
4. Hirschmann, H. and Hanson, K.R. *Tetrahedron* **30**, 3649–3956 (1974).
5. Hill, R.K. and Chan, T.-H. *Tetrahedron* **21**, 2015–2019 (1965).
6. Whitesell, J.K. and Felman, S.W. *J. Org. Chem.* **45**, 755–756 (1980).
7. Shirai, R., Tanaka, M., and Koga, K. *J. Am. Chem. Soc.* **108**, 543–545 (1986).
8. Izawa, H., Shirai, R., Kawasaki, H., Kim, H., and Koga, K. *Tetrahedron Lett.* **30**, 7221–7224 (1989).
9. Whitesell, J.K. and Allen, D.E. *J. Org. Chem.* **50**, 3025–3026 (1985).
10. Partridge, J.J., Chadha, N.K., and Uskokovic, M. *J. Am. Chem. Soc.* **95**, 7171–7172 (1973).
11. Partridge, J.J., Chadha, N.K., and Uskokovic, M. *J. Am. Chem. Soc.* **95**, 532–540 (1973).
12. Brown, H.C., Jadhav, P.K. *Asymm. Synth.* (Morrison, J.D., Ed.) vol **2A**, Academic Press: New York, 1–43 (1983).
13. Ihle, N.C., Correia, C.R.D., and Wender, P.A. *J. Am. Chem. Soc.* **110**, 5904–5906 (1988).
14. San Feliciano, A., Barrero, A.F., Medarde, M., de Corral, J.M.M., Aramburu, A., Perales, A., and Fayos, J. *Tetrahedron Lett.* **26**, 2369–2373 (1985).
15. Cahn, R.S., Ingold, C.K., and Prelog, V. *Angew. Chem.* **78**, 413–447 (1966).
16. Heathcock, C.H., Buse, C.T., Kleschick, W.A., Pirrung, M.C., Sohn, J.E., and Lampe, J. *J. Org. Chem.* **45**, 1066–1081 (1980).
17. Seebach, D. and Prelog, V. *Angew. Chem., Int. Ed. Engl.* **21**, 654–660 (1982).
18. Masamune, S., Ali, Sk. A., Anitman, D.L., and Garvey, D.S. *Angew. Chem.* **92**, 573–575 (1980).
19. Masamune, S., Kaiho, T., and Garvey, D.S. *J. Am. Chem. Soc.* **104**, 5521–5523 (1982).
20. Parshall, G.W. and Nugent, W.A. *CHEMTECH* 184–190, 314–320, 376–383 (1988).
21. Kagan, H.B. *Bull. Soc. Chim. Fr.* 846–853 (1988).
22. Scott, J.W. *Top. Stereochem.*, vol **19**.; Eliel, E.L., Wilen, S.H., Eds., John Wiley & Sons: New York, (1990), pp. 209–226.
23. Crosby, J. *Tetrahedron* **47**, 4789–4846 (1991).
24. Eliel, E.L., Allinger, N.L., Angyal, S.J., and Morrison, G.A. *Conformational Analysis* American Chemical Society: New York, (1965).
25. Mohler, D.L., Vollhardt, K.P.C., and Wolff, S. *Angew. Chem., Int. Ed. Engl.* **29**, 1151–1154 (1991).
26. Corey, E.J. and Cheng, X.-M. *The Logic of Chemical Synthesis* John Wiley & Sons: New York, (1989).
27. Hanessian, S., Franco, J., and Larouche, B. *Pure Appl. Chem.* **62**, 1887–1910 (1990).
28. Corey, E.J. *Angew. Chem., Int. Ed. Engl.* **30**, 455–465 (1991).
29. Seebach, D. *Angew. Chem., Int. Ed. Engl.* **29**, 1320–1367 (1990).
30. Yamaguchi, S. and Mosher, H.S. *J. Org. Chem.* **38**, 1870–1877 (1973).
31. Dale, J.A., Lull, D.L., and Mosher, H.S. *J. Org. Chem.* **34**, 2543–2549 (1969).
32. Dale, J.A. and Mosher, H.S. *J. Am. Chem. Soc.* **95**, 512–519 (1973).
33. Jacques, J., Collet, A., and Wilen, S.H. *Enantiomers, Racemates and Resolution* John Wiley & Sons: New York, (1981).
34. Pirkle, W.H. and Finn, J. *Asymm. Synth.* (Morrison, J.D., Ed.) vol **1**, Academic Press: New York, 87–123 (1983).
35. Duncan, J.D. *J. Liq. Chromatogr.* **13**, 2737–2755 (1990).
36. Corey, E.J., Narisada, M., Hiraoka, T., and Elllison, R.A. *J. Am. Chem. Soc.* **92**, 396–397 (1970).
37. Turner, E.E. and Harris, M.M *Q. Rev. Chem. Soc.* **1**, 299–330 (1947).
38. Clark, J.C., Phillips, G.H. and Steer, M.R. *J. Chem. Soc., Perkin 1*, 475–481 (1976).
39. Reider, P.J., Davis, P., Hughes, D.L., and Grabowski, E.J.J. *J. Org. Chem.* **52**, 955–957 (1987).
40. Bock, M.G., DiPardo, R.M., Evans, B.E., Rittle, K.E., Whitter, W.L., Veber, D.F., Anderson, P.S., and Freidinger, R.M. *J. Med. Chem.* **32**, 13–16 (1989).

41. Noyori, R., Ikeda, T., Ohkuma, T., Widhalm, M., Kitamura, M., Takaya, H., Akutagawa, S., Sayo, N., Saito, T., Taketomi, T., and Kumobayashi, H. *J. Am. Chem. Soc.* 111, 9134–9135 (1989).
42. Kitamura, M., Ohkuma, T., Tokunaga, M., and Noyori, R. *Tetrahedron: Asymm.* **1**, 1–4 (1990).
43. Roberts, S.M. *Chem. Ind.* 384 (1988).
44. Pratt, A.J. *Chem. Br.* 282–186 (1989).
45. Finn, M.G. and Sharpless, K.B. *Asymm. Synth.* (Morrison, J.D., Ed.) Academic Press: New York, (1985, Vol. **5**, chapter 8, pp. 247–308.
46. Woodard, S.S., Finn, M.G., and Sharpless, K.B. *J. Am. Chem. Soc.* **113**, 106–113 (1991).
47. Pfenninger, A. *Synthesis* 89–116 (1986).
48. Fischer, E. *Ber. Bunsenges. Phys. Chem.* **27**, 3231 (1894).
49. Freudenberg, K. *Adv. Carbohydr. Chem.* **21**, 1 (1960).
50. Marckwald, W. *Ber. Bunsenges. Phys. Chem.* **37**, 1368 (1904).
51. Coppola, G.M. and Schuster, H.F. *Asymmetric Synthesis: Construction of Chiral Molecules using Amino Acids* John Wiley & Sons: New York, (1987).
52. Ho, T.-L. *Enantioselective Synthesis: Natural Products from Chiral Terpenes* John Wiley & Sons: New York, (1992).
53. Morrison, J.D. and Scott, J.W. (Ed.s) *Asymm. Synth.*, Vol. **5**, Academic Press: New York, (1985).
54. Hanessian, S. *Total Synthesis of Natural Products: The Chiron Approach* Pergamon Press: Oxford, (1986).
55. Gmeiner, P., Feldman, P.L., Chu-Moyer, M.Y., and Rapoport, H. *J. Org. Chem.* **55**, 3068–3074 (1990).
56. Evans, D.A. and Weber, A.E. *J. Am. Chem. Soc.* **106**, 6757–6761 (1986).
57. Kitamura, M., Okada, S., Suga, S., and Noyori, R. *J. Am. Chem. Soc.* **111**, 4028–4036 (1989).
58. Noyori, R. and Kitamura, M. *Angew. Chem., Int. Ed. Engl.* **30**, 49–69 (1991).
59. Noyori, R. *Science* **248**, 1194–1199 (1990).

3 Asymmetric Synthesis

In this chapter we shall examine the asymmetric reactions that give access to functionalized chiral molecules utilizing the principles we have discussed in Chapter 2. The field of asymmetric synthesis is developing rapidly, and a comprehensive survey of all the known reactions would be impossible. We therefore restrict our study to those representative reactions which give a good idea of where we stand and where the future research might take us. We shall restrict our study to only two functional groups, the carbonyl group and the olefinic double bond, as these are by far the best studied and therefore also the most widely used functionalities in novel bond constructions.

To understand better the underlying reasons for stereoselective processes, we must appreciate the effects of even rather small energy differences. As we have already seen

(Table 2.1), a difference between two alternative activation energies of only about 5 kJ mol^{-1} is needed to bring about a 95:5 ratio, or 90 % ee, in the products. It is also important to be aware of the fact that the relationship between energy and ratio is logarithmic. This directly implies that the prediction of large selectivities requires less accuracy!

We must also be able to understand the origins of these subtle differences in the energy contents of the reacting species. Conformational analysis is the tool that can give us valuable insight, but we should be reminded that we are really interested in transition state geometries, not those of ground states.[1] Direct observation of transition states is, of course, impossible, but fortunately computational methods have been advanced to the level that one can rather reliably model most of the reaction types using relatively easily accessible computational facilities.[2]

Before commencing our studies on the individual types of processes amenable to the creation of stereogenic centres, we shall review some aspects of conformational analysis central to understanding the rationalizations presented later.

3.1 Allylic Strain

It is well known that in cyclohexane the substituents tend to adopt an equatorial position in preference over an axial orientation. This bias can be rationalized as arising from steric gauche type repulsions (Figure 3.1) with the axial protons located two bonds away. Each substituent has a different tendency towards equatorial bias, and simple considerations of the various 1,3-interactions usually lead to qualitatively correct analyses of the cyclohexane conformations.

Figure 3.1

The situation in cyclohexenes is slightly different, especially when the olefinic carbon atoms are also substituted. Such systems have been analysed by Johnson,[3] and he has proposed the concept of allylic strain which has found widespread utility not only in the chemistry of cyclohexene but also recently in the rationalizations of acyclic stereocontrolled reactions. We shall briefly inspect the salient points of allylic strain.

In 2,3-dialkylcyclohexenes where the alkyl groups are relatively large, the dihedral angle between C(3)—R' and C(2)—R is considerably smaller (roughly 35°) than the ideal value of 60°. This is manifested in a steric interaction between the substituents which forces the allylic substituent to adopt an axial position. This type of strain is called A^{1,2} strain to designate its allylic nature and more specifically that it arises from 1,2-interactions of substituents (Figure 3.2).

The facial selectivity (below or above the plane of the ring) on the attack of a reagent (e.g. epoxidizing agent) is controlled by two factors: (i) the stereoelectronic effects demanding as much continuous π-orbital overlap as possible in the transition state, and (ii) the steric hindrance caused by the substituents (R'). In conformer **A** (Figure 3.2), the

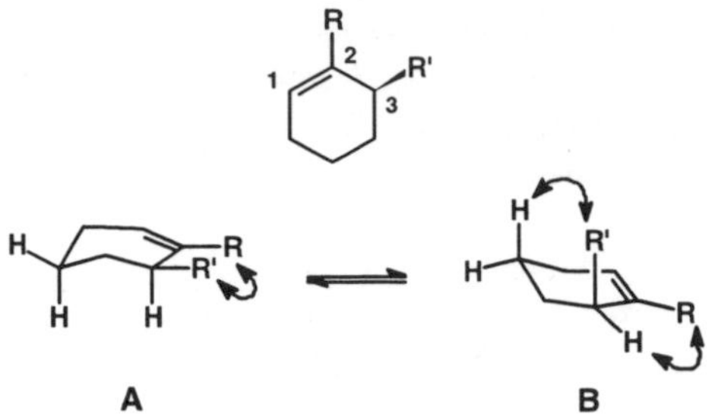

Figure 3.2

substituent R' causes little steric hindrance to the axial approach of the reagent from the lower face of the molecule. In conformer **B**, the large substituent R' effectively blocks the upper face (which would be favoured by the stereoelectronics). Attack from the lower face would necessitate the adoption of a boat-like conformation along the reaction coordinate, which implies a high energy of activation.

The second type of allylic strain, the $A^{1,3}$ strain, is concerned with alkyl substituents at the termini of an allyl system. In methylenecyclohexane conformation **C** (Figure 3.3), the substituent R' lies nearly in the plane of the double bond (in cyclohexanone the dihedral angle $O===C—C—H_{eq}$ is only 4.3°), and can thus experience severe repulsion with the substituent R on the opposite terminus of the double bond. Thus, in the case of large substituents R and R', the conformational equilibrium would favour the conformation **D** with R' pseudoaxial.

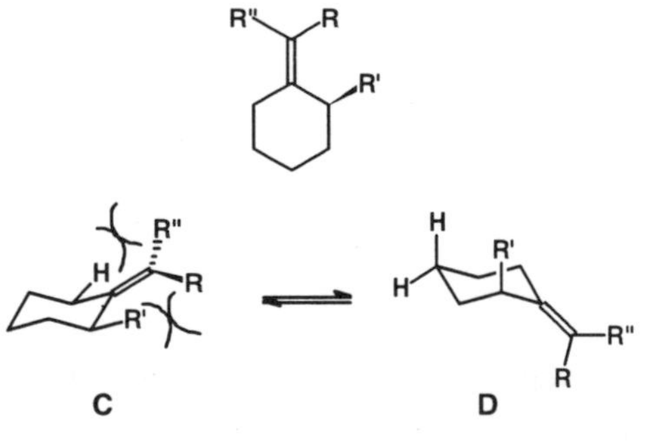

Figure 3.3

Allylic 1,3-strain has proved to be one of the most powerful tools in understanding acyclic stereocontrol.[4] In acyclic systems the conformational bias is amenable to *ab initio* calculations which have shown that the conformational bias can be quite large indeed. In 3-methylbutene, rotation around bond **a** (Figure 3.4) gives two possible minima (the staggered conformation is actually not an energy minimum), of which the one with the allylic hydrogen eclipsed is favoured by 3.1 kJmol^{-1}. Addition of a methyl group at the terminus of the alkene to form a Z alkene raises the energy of the eclipsed methyl so much that it actually represents the maximum energy conformer! The staggered conformation is in this case an energy minimum, but 14.4 kJmol^{-1} above the lowest energy conformer. Thus, for all practical purposes the ground state is correctly represented by the eclipsed hydrogen conformation.

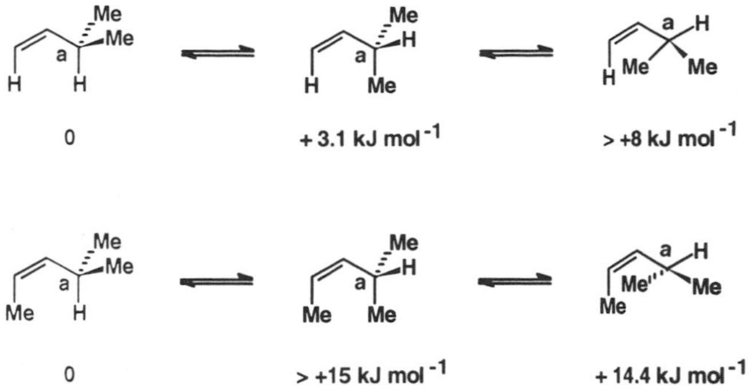

Figure 3.4

Although the energy differences are quite large in the latter case, we should bear in mind that these calculations represent the ground state conformations. Rotation of the dihedral angle is relatively free of penalty in energy up to *ca* 30° rotation, and particularly in reactions which proceed with stereoelectronic control these small energy changes can be easily overcome. Typically the directions of electrophilic attack[5] and nucleophilic attack[6] are under such control. In order to maximize the overlap with either the HOMO (electrophilic attack) or LUMO (nucleophilic attack), the substituent X must be aligned so that the σ-bond to the substituent is parallel to the π-orbital of the double bond. The two alternative conformations can be distinguished by the developing allylic 1,3-strain, and generally good levels of asymmetric induction will be observed.

In electrophilic attack, the interaction of the LUMO of the electrophile with the HOMO of the alkene gives a transition state LUMO which is stabilized by the arrangement of the proximal substituent in such a fashion that the most electropositive allylic group favours an *anti* orientation to maximize donation from a high lying σ-orbital to the transition state LUMO (Figure 3.5).

In nucleophilic attack, the most electronegative group favours an *anti* conformation to maximize electron withdrawal from the reacting π-system, and the most electropositive group prefers an outside position to minimize electron donation to the π-system of the already electron rich transition state. Electropositive groups prefer the outside or inside positions because the interactions of the σ_{C---D} orbital are destabilizing. This destabilization is maximal when the σ-bond is *anti* to the carbonyl plane and minimal when the σ-bond lies in the plane of the carbonyl, which is best accommodated in the conformation where the electropositive substituent is outside (Figure 3.6).

3.2 Reactions of the Carbonyl Group

Carbonyl groups play a central role in synthesis owing to their much developed chemistry. It is no surprise that the asymmetric reactions of the carbonyl group provide perhaps the largest set of reactions capable of giving access to enantioselectivity. The multitude of reactions that carbonyl functionalized compounds undergo include direct attack at the carbonyl carbon, either via reduction (hydride addition) or nucleophilic addition of

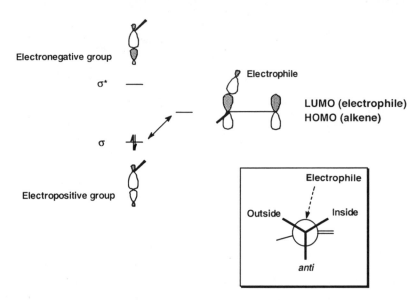

Figure 3.5

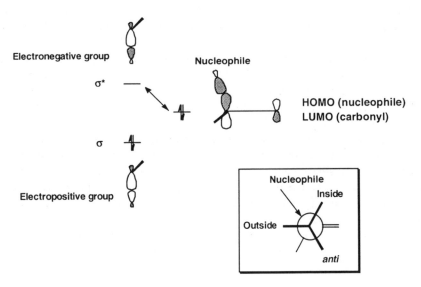

Figure 3.6

alkyl, allyl and propargyl groups, etc. The α-centre can be induced to function as a nucleophile, giving access to alkylation reactions and aldol type addition reactions with a second carbonyl functionality. In α,β-unsaturated carbonyl compounds the β-carbon is electrophilic, and 1,4-additions (Michael) give rise to a further class of substituted carbonyl groups.

3.2.1 Nucleophilic Additions on the Carbonyl Carbon

The stereochemical aspects of the addition of a nucleophile onto a carbonyl group featured in the earliest examples of studies in stereoselectivity. In order to generate a (secondary) chiral centre from a carbonyl carbon atom one has in principle two alternative possibilities. It would seem to be justified to expect that if reduction of a ketone with a suitable chiral hydride reducing agent gives access to one enantiomeric form of the product, C—C bond formation through a chirally mediated delivery of the nucleophile R'$^-$ should give access to the other enantiomer. For most processes, this is what is observed.[7]

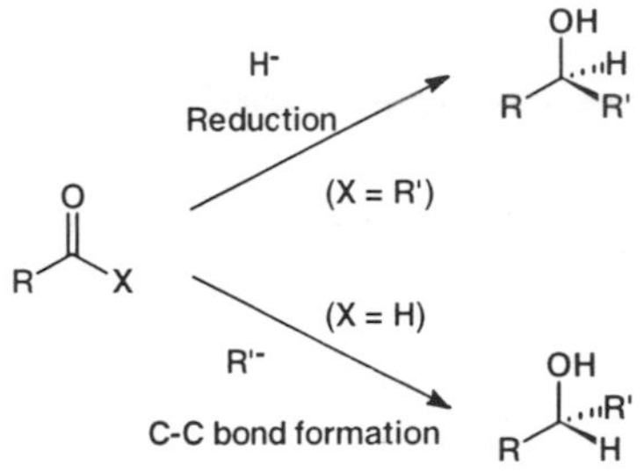

Figure 3.7

The difference between the sizes of R and X is the principal factor affecting the level of asymmetric induction: the larger the difference, the better the selectivity one would expect to obtain. In the light of recent findings, it is almost certain that electronic effects also play a significant role, but alas these effects are not yet fully understood.

The addition of a hydride nucleophile onto the carbonyl carbon atom is a process whose geometrical features have been studied both experimentally and computationally. However, C—C bond construction with either enantio- or diastereocontrol is a much less well developed area, save for cyclization reactions and manipulations of carbocyclic compounds. It is only during the last three or four decades that these important areas of synthesis have been attacked. We shall start our study of carbonyl group reactions by first inspecting some general rationalizations on the geometrical principles of addition reactions.

Starting from the presumption that the nucleophile adds along the plane containing the C≡≡≡O bond and begins at a right angle to the plane of the carbonyl group (the normal plane), one can define an approach vector to define the steric trajectory that the incoming nucleophile would follow.[8] Qualitative considerations,[9,10] model calculations[11,12] and X-ray crystallographic studies[13] have led to the formulation that the nucleophile follows a trajectory along the normal plane, making an angle with the C≡≡≡O line near the tetrahedral angle in magnitude (approximately 107°; however, subtle changes in the steric environment can lead to variations in this angle).[14] An ingenious use of crystallographic data enabled Burgi and Dunitz to propose an estimate for this angle, which is known as the Burgi–Dunitz angle (α_{BD}; Figure 3.8).[15]

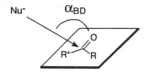

Figure 3.8

The observed angle can be rationalized also in terms of frontier molecular orbital (FMO) theory.[16-18] The HOMO–HOMO interaction is destabilizing since both orbitals are occupied. Destabilization is maximized when the attack angle is acute, and correspondingly destabilization is minimized with an obtuse approach angle. The stabilizing interaction between the LUMO of the electrophilic component (carbonyl acceptor) and the HOMO of the nucleophile (Nu^-) is also maximized with an obtuse angle, since the overlap integrals between the HOMO and the p-orbitals of the carbonyl LUMO are of opposite sign (Figure 3.9). Attack angles on carbonyl groups are smaller than those on alkenes, and further polarization of the carbonyl by coordination with a Lewis acid further reduces the magnitude of the angle.

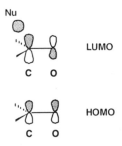

Figure 3.9

If in this case the group R* is chiral (Figure 3.8), the reaction will proceed through diastereomeric transition states, and the two products will be formed in unequal amounts. The rationalization for the effect of the R* group was first put forth by Cram, who suggested that the carbonyl compound will adopt a conformation where the carbonyl group is flanked by the two smaller substituents (S and M), and the largest substituent is eclipsed with the alkyl group R (Figure 3.10). The nucleophile then approaches from the side of the smallest group S.[19] However, this model does not correspond to either the ground state or the transition state structure.

Several alternative mechanistic rationalizations have been advanced, and the best agreement is given by the Felkin model,[6] which can be derived in the following manner. Because of steric interactions not unlike the allylic strain effects, in the ground state conformation of an α-chiral carbonyl compound the chiral side chain tends to be oriented in such a manner that the largest group eclipses the carbonyl group. The transition state for the reaction occurs early along the reaction coordinate, and therefore resembles the stucture of the reactant. Rotation of the C(=)O—C_α bond gives the arrangement shown in Figure 3.11 with the largest group perpendicular to the carbonyl group axis.

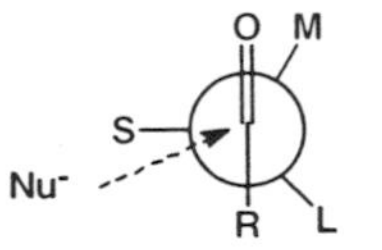

Cram open-chain

Figure 3.10

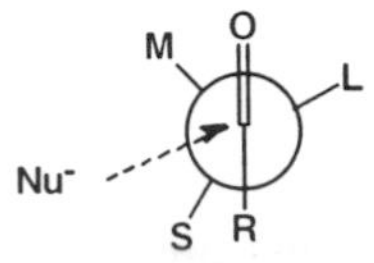

Felkin-Anh

Figure 3.11

Attack of the nucleophile from the less hindered side, opposite to the largest substituent, is now observed. In this arrangement, the antibonding orbital σ^* of the C_a—L bond is *anti* to the forming C—Nu bond.

This model also agrees with *ab initio* calculations.[11] In this model L represents the largest group, or the group whose bond to the α-carbon (L—C_α) provides the greatest $\sigma^*-\pi^*$ overlap with the antibonding π^*-orbital of the carbonyl group.

In connection with the above studies Cherest et al. also observed that lithium aluminium hydride reductions of a series of α-chiral dialkyl ketones gave increasing diastereoselectivities on increasing the size of the alkyl group on the side opposite to the stereogenic chiral centre (Scheme 3.1).[6]

Heathcock and coworkers have formulated a clarification on the above rationalizations.[20] According to this explanation, if the two alkyl substituents on the carbonyl carbon are of similar size, the incoming nucleophile will follow the approach along the Burgi–Dunitz angle. However, as the size difference between the two

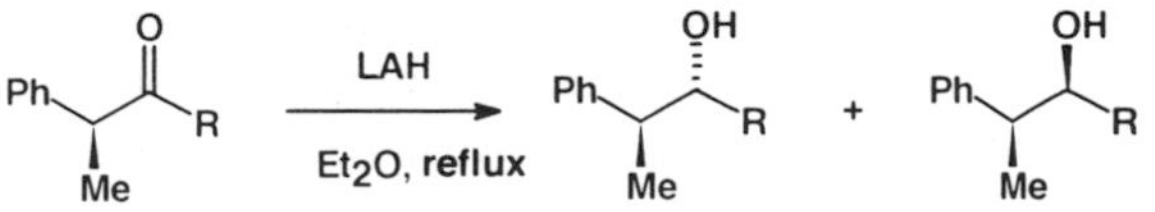

R	A	:	B
Me	3		1
Et	3		1
i-Pr	5		1
t-Bu	50		1

Scheme 3.1

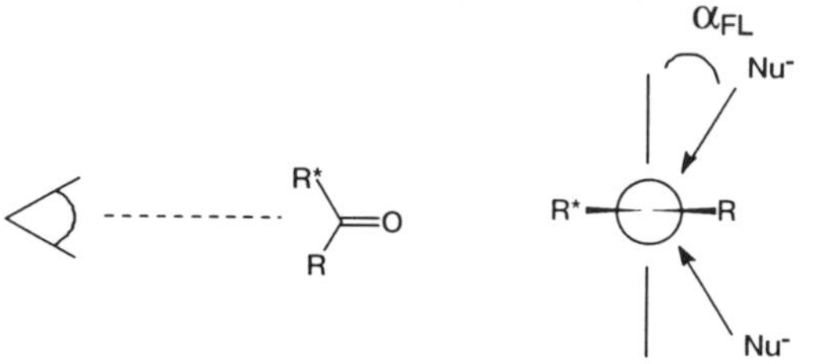

Figure 3.12

substituents increases, the incoming nucleophile should try to avoid the larger one, and the approach vector would thus be tilted away from the normal plane. They call this angle the Flippin–Lodge angle α_{FL}, which is related to the difference in steric bulk on the two sides of the normal plane (Figure 3.12).

In the reduction of the above series of ketones (Scheme 3.1), one should therefore expect that the larger the R group is, the further the Flippin–Lodge angle will be tilted towards the chiral centre, and therefore the diastereoselectivity should increase. The observed results are in full accordance with this rationalization.

3.2.1.1 Asymmetric reduction of carbonyl compounds

The two faces of the plane of an unsymmetrically substituted carbonyl compound RCOR' are enantiotopic, and thereby the reactions of such compounds with chiral reducing agents give rise to asymmetric induction. The development of efficient chiral versions of the common hydride reducing agents based on boron and aluminium has occupied a central position in the method development over the past few decades.[21,22]

The first claim to asymmetric reduction of a ketone with a chirally modified lithium aluminium hydride is based on simple alcohol modifiers.[23,24] Reduction of aliphatic ketones with a reagent prepared by mixing lithium aluminium hydride and camphor (which is presumably reduced to mostly isoborneol) was claimed to give optically active alcohols. Nearly two decades later the observation could be reproduced, but giving only 2 % ee.[25] The strategy, based on monohydroxy modifiers, is hampered by facile disproportionation of the chiral alkoxyaluminium hydride species. Efforts to overcome this problem have led to significant developments of the reagents. Two of the most successful strategies are based on diol modifiers and chiral 1,3-amino alcohol modifiers.

The first succesful diol modifiers were based on monosaccharide derived diols.[26,27] It was found that one of the two active hydrogens in the aluminium hydride complex reacted much faster than the other one to give S stereochemistry in the product alcohol. Treatment of the reagent with one equivalent of ethanol led to inversion of the observed sense of asymmetric induction to give predominantly the R alcohol. This can be explained by preferential removal of the more reactive hydrogen H_a by the alkoxide and the formation of a trialkoxyaluminium hydride complex (Figure 3.13).

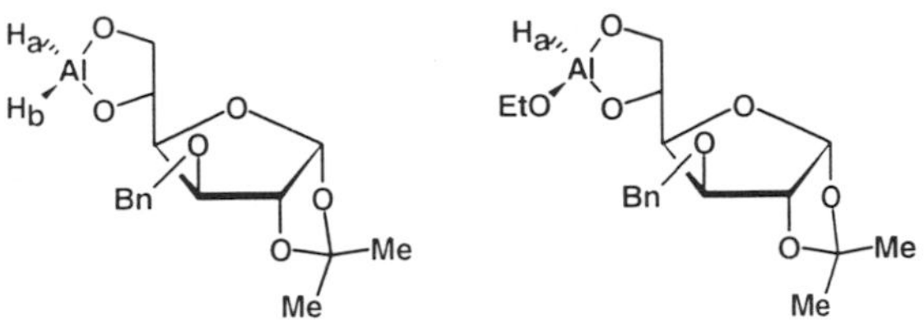

Figure 3.13

By far the best results with diol modifiers are obtained with the reagent derived from axially chiral binaphthol (1,1'-bi-2,2'-naphthol), lithium aluminium hydride and a simple alcohol (methanol or ethanol).[28-30] The reducing agent, BINAL-H (Figure 3.14), is capable of reducing unsaturated carbonyl compounds with exceptionally high enantioselectivities, very often with nearly complete enantioselectivity. The presence of the secondary modifier (methanol or ethanol) is crucial: in the reduction of acetophenone without the ethanol, only 2 % ee was obtained, and utilization of bulkier alcohols led to reversal of the sense of asymmetric induction. The results were rationalized with the diastereomeric six membered structure shown in Figure 3.14. The indicated 1,3-diaxial interaction will favour the placement of the bulkier substituent L equatorial and the less bulky one S axial.

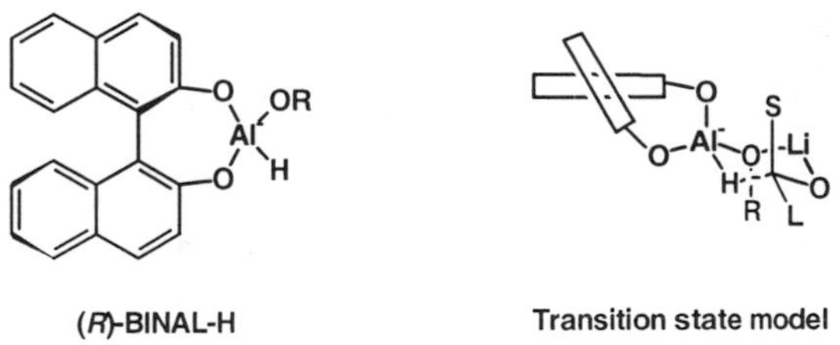

(*R*)-BINAL-H   Transition state model

Figure 3.14

Alkyl acetylenic ketones were reduced in 84–94 % ee, but increasing the size of the alkyl group has a deleterious effect on the enantiofacial differentiation. Thus, ethynyl isopropyl ketone gives only 57 % ee. Olefinic ketones were also reduced in substantially high enantiomeric excess. The dienone β-ionone gave nearly 100 % ee upon reduction with (*S*)-BINAL-H. Reductions of the enones *en route* to prostaglandins with (*S*)-BINAL-H gave nearly quantitative asymmetric induction.[30] Reduction of cyclopentenedione with the same reagent gave 4-hydroxy-2-cyclopenten-1-one in 94 % ee (Figure 3.15).[31] This compound is a key nucleus in the three-component synthesis of prostaglandins.

One of the prevailing problems in the Corey synthesis of prostaglandins has been the generation of the C-15 stereocentre.[32,33] BINAL-H reduction of either the bicyclic lactone intermediate (with either THP or acetyl protection on the C-11 hydroxyl group) or of the monocyclic prostaglandin $F_{2\alpha}$ intermediate gave practically exclusive selectivity for the production of the desired 15*S* isomer (Figure 3.16).[30]

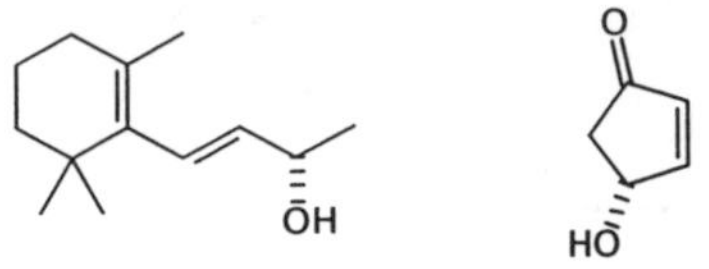

Figure 3.15

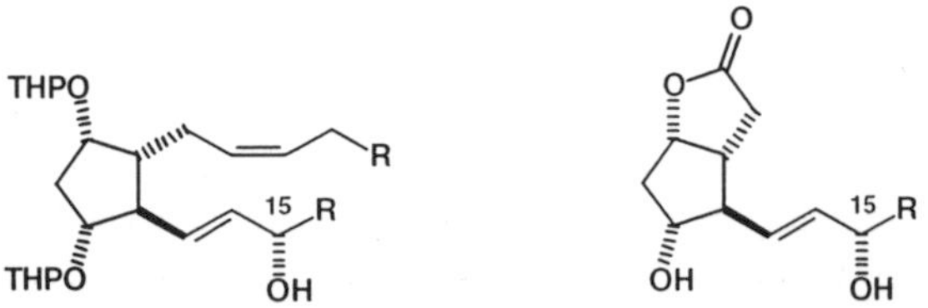

Figure 3.16

In an interesting application, a stannyl ketone is reduced with (*S*)-BINAL-H with 93 % ee. After protection of the secondary carbinol as its methoxymethyl (MOM) ether, the stannane behaves as a chiral α-alkoxyalkyl anion. The final butanolide is obtained with undiminished enantiomeric purity (Scheme 3.2).[34]

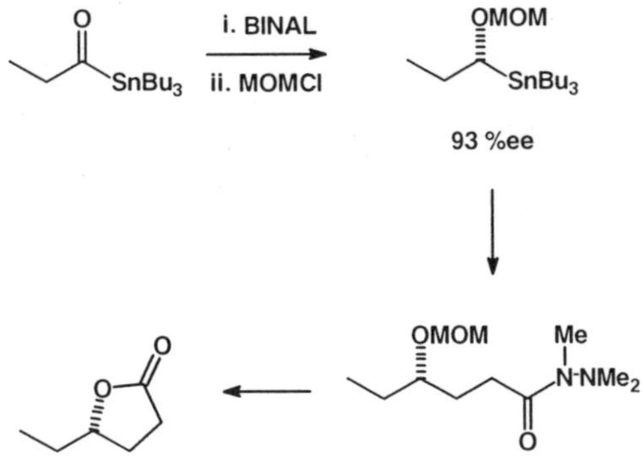

Scheme 3.2

Utilization of amino alcohols led to efficient asymmetric induction even earlier than the diol–based strategy, especially the methods employing (2*S*,3*R*)-4-dimethylamino-3-methyl-1,2-diphenyl-2-butanol (Darvon alcohol) as the chiral modifier (Figure 3.17).[35,36] Cohen has further developed the methodology based on the Darvon reduction.[37] The reduction is highly enantioselective for acetylenic ketones (typically 60–90 % ee).

A number of chiral terpene derived borane reducing reagents have been developed, mainly by the groups of Brown[22] and Midland.[38] The structures of three of the most widely used compounds, all derived from pinene, are shown in Figure 3.18. The proposed

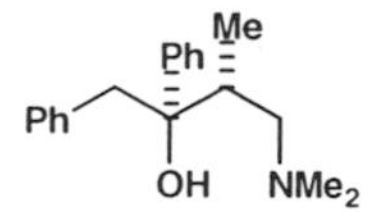

Darvon alcohol

Figure 3.17

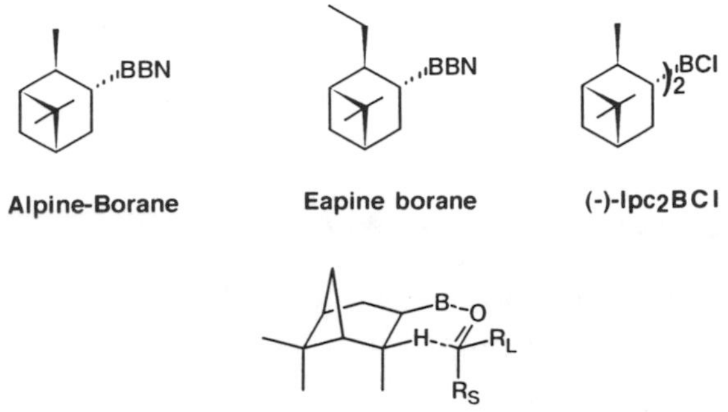

Alpine-Borane   Eapine borane   (-)-Ipc$_2$BCl

Proposed transition state model
for pinanyl borane reductions

Figure 3.18

six membered, boat-like transition state for the reduction is also shown.[39] The large substituent R_L occupies the pseudoequatorial position to avoid steric interference with the methyl group of the reagent. The B—C—C—H arrangement is nearly planar, leading to enhancement of the reaction rate.[40,41]

Alpine-Borane is the trade name of the Aldrich Chemical Company for B-3-pinanyl-9-borabicyclo[3.3.1]nonane (BBN = 9-borabicyclo[3.3.1]nonane). Alpine-Borane reduces aldehydes, 1,2-dicarbonyl compounds and especially α,β-unsaturated ketones and acetylenic ketones rapidly and with high enantioselectivity.[42,43] A most remarkable distinction between the two faces of an alkynyl vinyl ketone has been noted (Scheme 3.3).[44]

The triple bond provides a good functional handle for further elaborations. The triple bond can be isomerized to the terminal position, without affecting the chiral propargylic centre, with potassium 3-aminopropylamide (KAPA).[45] This strategy has been applied in the synthesis of a hexenolide in practically optically pure form.[46] The R isomer, massoilactone, is the defense allomone of the formicine ant (Scheme 3.4).

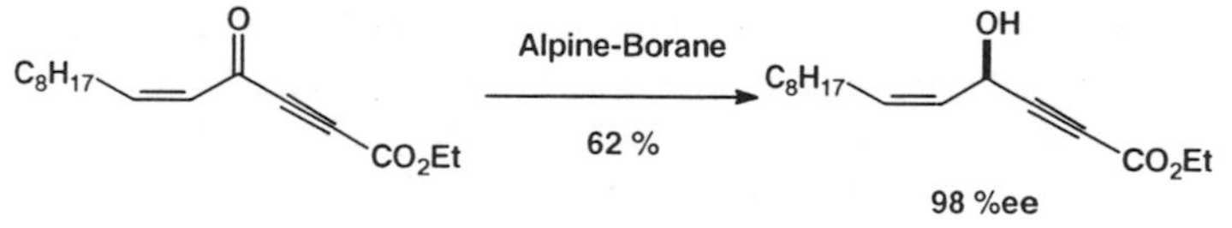

Scheme 3.3

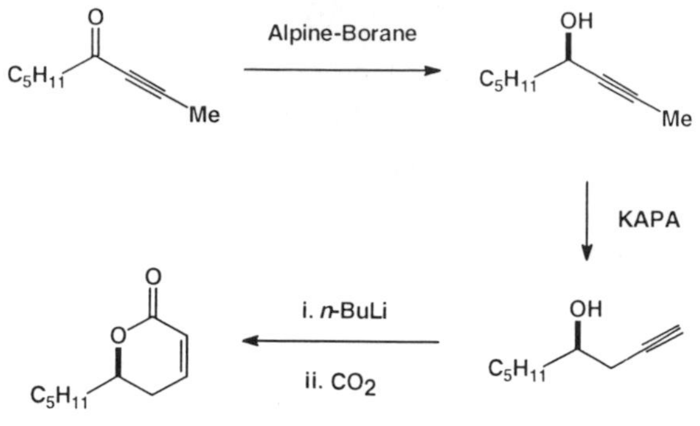

Scheme 3.4

Reduction of alkynyl ketones is feasible in the presence of other ketones, including aromatic ones, but not in the presence of aldehydes, as these are reduced much faster than the ynones. Running the reduction without solvent allows the reduction of α-keto esters and α-halo ketones in high enantioselectivity.[47] Aryl ketones can be reduced under high pressure conditions (2000 atm).[39,48] However, aliphatic ketones and enones give low enantioselectivities even under these conditions. Reduction of chiral ketones was also studied at high pressures, and of the ketones studied carvone provided interesting results. With (*R*)-Alpine-Borane, *d*-carvone was reduced in a 4.6:1 ratio of diastereomers with the diequatorial isomer predominating (Scheme 3.5). With the same reagent, *l*-carvone resisted reduction completely, even after five days at 6000 atm.

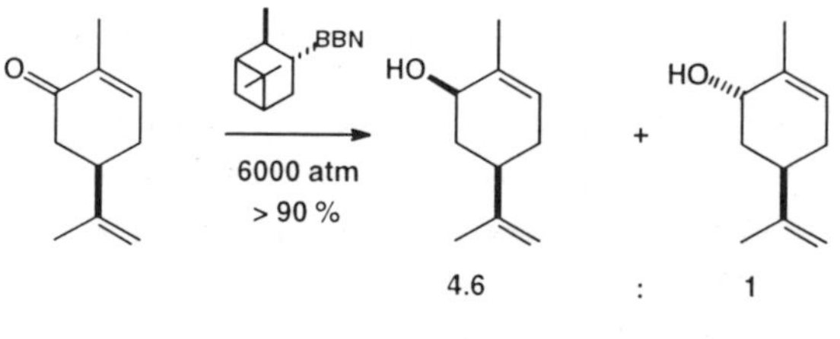

Scheme 3.5

The low reactivity of Alpine-Borane towards ketones can be enhanced by increasing the Lewis acidity of the boron. This has been achieved through the development of chlorodiisopinocampheylborane (Ipc$_2$BCl).[49] Aromatic ketones and α-tertiary alkyl ketones can be reduced with high enantioselectivity, but simple aliphatic ketones still give poor results.

Masamune has developed a C_2 symmetric 2,5-dimethylborolane as a reducing agent, capable of achieving the reduction of relatively symmetrical dialkyl ketones in high enantiomeric purity.[50] 2-Butanone and 2-octanone are reduced in *ca* 80 % ee, and a larger difference between the two ketone substituents leads to practically complete facial discrimination by the reagent. The mechanism of the reduction is believed to be complex. The actual reacting species is formed from a dimer of the borolane and the corresponding

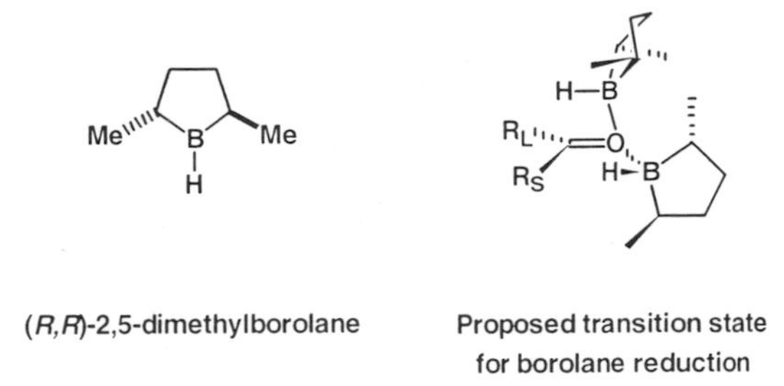

(R,R)-2,5-dimethylborolane

Proposed transition state
for borolane reduction

Figure 3.19

2,5-dimethylborolanyl methanesulfonate, which is formed during the preparation of the reagent. The methanesulfonate plays a catalytic role in the reduction, coordinating to the carbonyl group of the ketone through the lone pair *syn* to the smaller alkyl group (Figure 3.19).[51]

One of the singularly important developments in asymmetric reduction is based on Itsuno's observation that a reducing agent derived from (S)-2-amino-3-methyl-1,1-diphenylbutan-1-ol, itself derived from leucine, and borane gives high enantioselectivity on reduction of ketones (95 % ee for the reduction of acetophenone).[52-56] Corey et al. have extended this method and developed a more efficient reagent from proline.[57] They were able to isolate the active chiral reductant, the oxazaborolidine reagent. The higher selectivity observed with this Corey–Bakshi–Shibata (CBS) reagent is presumably due to the higher bias exerted by the concave/convex discrimination of the bicyclic oxazaborolidine reagent (Figure 3.20).

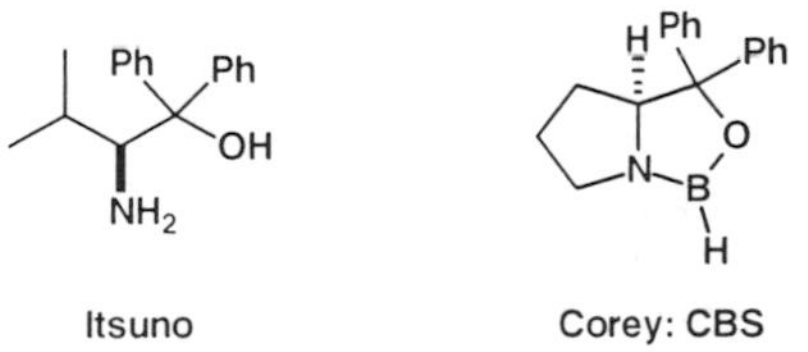

Itsuno

Corey: CBS

Figure 3.20

The major breakthrough in the development of the Corey oxazaborolidine reagent came with the observation that the isolated oxazaborolidines function as efficient chiral catalysts in the reduction reactions. Usually only *ca* 5–10 mol % of the catalyst is needed to effect the reduction in high enantioselectivity (typically >95 % ee). Among the hydride donors, borane–tetrahydrofuran complex, borane–dimethyl sulfide complex (BMS) and catecholborane are the favoured ones. The structure of the oxazaborolidine has been the subject of further development, and variations include changing the substituents (*β*-naphthyl instead of phenyl,[58-59] boron substituents, etc). However, the original diphenylprolinol derived compound is rather general and is a good choice as the starting point for more detailed optimization studies.[60,61]

The B—H complex is air and moisture sensitive, whereas the B–alkyl catalysts can be stored and handled in air.[62] This has obvious practical advantages, and the reagent is also more readily prepared than the hydrogen analogue. The observed enantioselectivities retained the high values. This reducing agent was applied in the synthesis of a prostaglandin intermediate where a diastereoselectivity of 9:1 was observed with the *S* catalyst (Scheme 3.6).

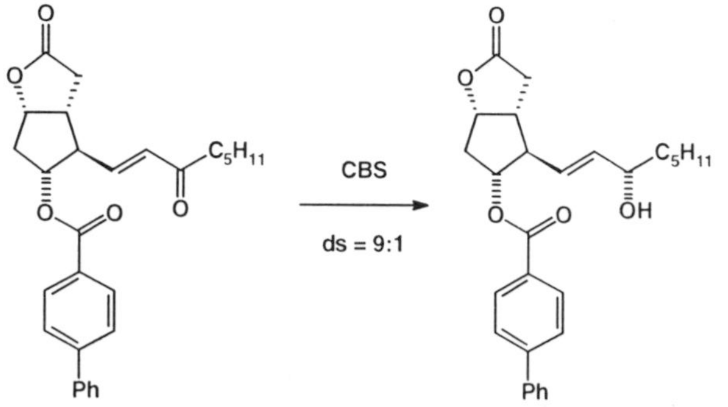

Scheme 3.6

In the synthesis of platelet activating factor (PAF) antagonist *trans*-2,5-diaryltetrahydrofuran, the same catalyst gave 95 % ee on reduction. This was further elaborated through DIBAL reduction to the lactone, conversion to the corresponding α-bromo ether and reaction with an aryl Grignard reagent to the potential PAF agent (Scheme 3.7).

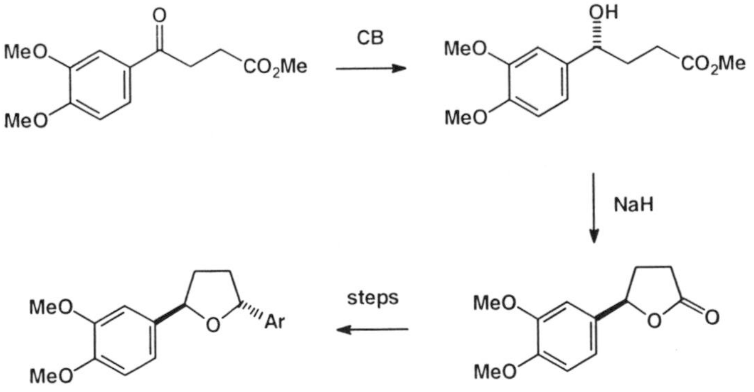

Scheme 3.7

Whereas the B—H and B—Me catalysts work rapidly at 0 °C or room temperature but fail at lower temperatures, the B–butyl catalyst is capable of effecting reduction at low temperatures. Especially with catecholborane as the stoichiometric hydride donor, the non-catalyzed reaction is suppressed, and enones and α,α,α-trihalomethyl ketones undergo rapid reduction to give the corresponding alcohols in high enantiopurity.[63] In the case of the trifluoroacetylmesitylene (Scheme 3.8), the reduction with the S-catalyst was observed to give 100 % ee of the R-product.[64,65] The sense of enantioselectivity was inverted with the acetylmesitylene (99.7 % ee). This result implies that the catalyst coordinates to the lone pair *anti* to the CF$_3$. The extremely high enantioselectivity observed cannot be simply explained by the mere steric size difference between the CF$_3$ and the mesityl groups, but substantial electronic repulsion between the electron rich fluorine atoms and the negatively charged boron atom may be involved.

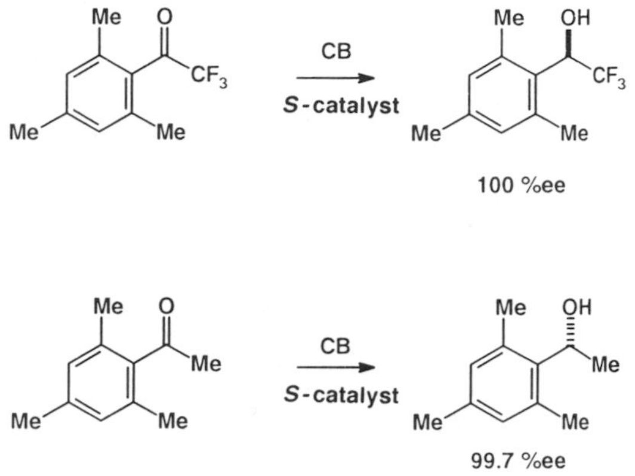

Scheme 3.8

The mechanistic rationale for the CBS reduction has been presented,[66] and the catalysts have been modelled using quantum chemical calculations.[67-69] The reaction is described as involving a six membered transition state involving one molecule each of the catalyst, the borane and the substrate. The borane and the carbonyl compound are complexed *cis* relative to each other on the oxazaborolidine, and hydrogen is transferred via a boat-form, six membered transition state (Scheme 3.9).

3.2.1.2 Asymmetric alkylation of the carbonyl group

The direct nucleophilic alkylation of a prochiral carbonyl group with facial discrimination has been a long-standing goal for synthesis. Early success was achieved by the use of a steroidal chiral auxiliary in the synthesis of atrolactic acid from phenyl glyoxylate with a methyl Grignard reagent (Scheme 3.10).[70] The enantiomeric excess was 69 % ee, which must be considered extremely high for those days.

Some two decades later, Whitesell demonstrated the utility of 8-phenylmenthol (PM) as a powerful chiral auxiliary. Grignard alkylations of PM esters of α-keto acids gave up to 98 % ee (Scheme 3.11).[71,72]

As we have already discussed, addition of organometallic reagents to aldehydes provides a powerful tactic for the generation of secondary alcohols (Scheme 3.12). One can simultaneously construct a new chiral centre and a new carbon–carbon bond. Although development of an asymmetric version of this reaction has met with limited success so far, the addition of unsaturated nucleophiles (most notably allyl and propargyl) has the added advantage of introducing a further functional handle which

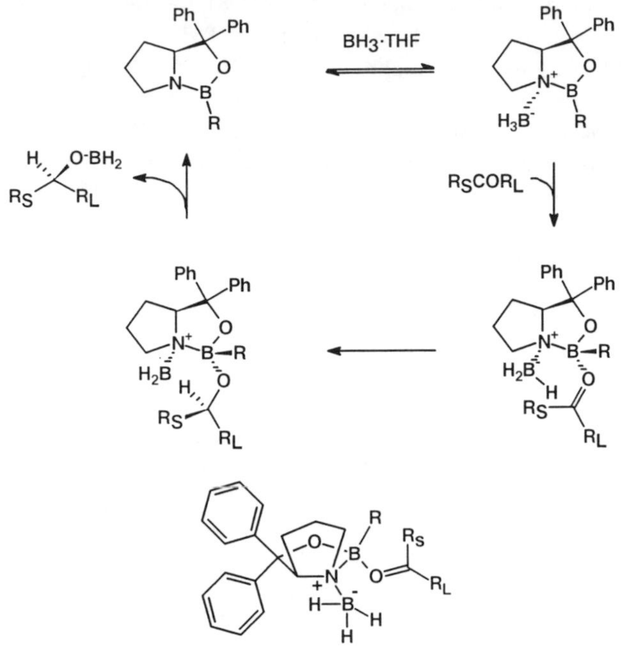

Scheme 3.9

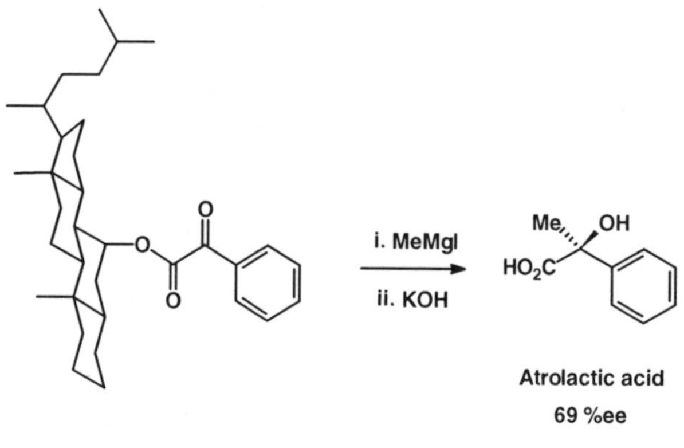

Atrolactic acid

69 %ee

Scheme 3.10

can provide access to more elaborate structures. Coordination of chiral diamine ligands with the organometallic species gives rise to chiral complexes capable of enantiofacial differentiation on the aldehyde targets. Both the nucleophilicity and basicity of the organometals are enhanced by coordination.[73] Alkyllithium[74,75] and Grignard reagents[76] have been successfully coupled with aromatic aldehydes in the presence of a proline derived diamine ligand to give access to enantiomerically enriched products. Although

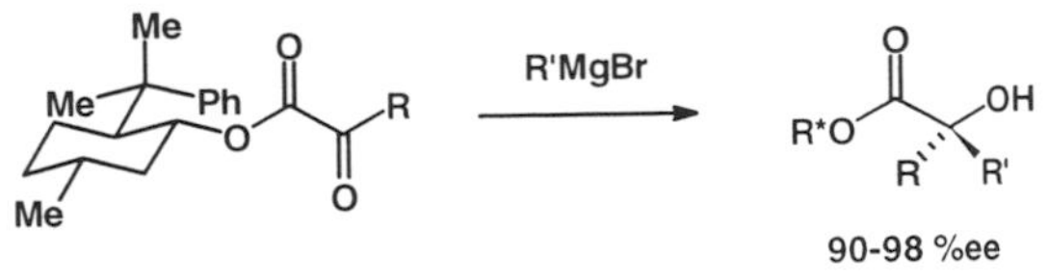

Scheme 3.11

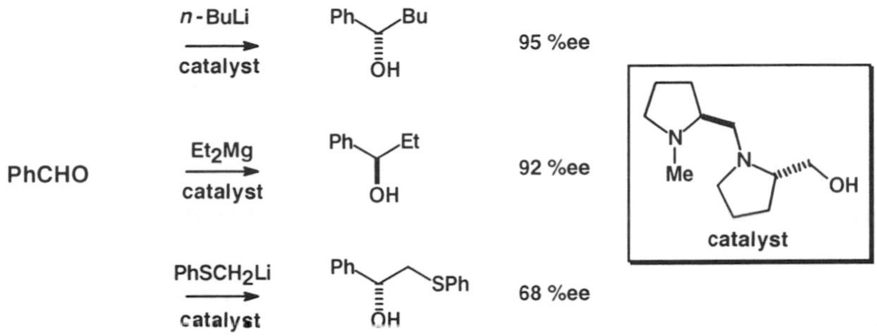

Scheme 3.12

the ligands can usually be recovered and recycled, at least a stoichiometric amount of the ligand is needed because of the competing uncatalysed reactions which produce racemic material.

Organozinc reagents are among the oldest known organometallic species, in fact diethylzinc was the first organometallic to be synthesized.[77] Organozinc reagents react relatively sluggishly with aldehydes even at room temperature in non-donor solvents. This reactivity can be accelerated by coordination with, for example, chiral amino alcohols, which provides accelerated asymmetric catalysis. We have already seen in Chapter 2 the case of the chirality multiplication in the alkylation reaction of benzaldehyde with dialkylzinc reagents in the presence of a partially racemic, camphor derived amino alcohol. A number of chiral catalysts have been devised for this purpose, including the ones shown in Figure 3.21 (enantioselectivities are given for the addition of diethylzinc to benzaldehyde).[78–85]

The sulfonamide reagent[79] is used together with titanium tetraisopropoxide and the organozinc reagent, which can also lead to the formation of a titanium organometallic species acting as the catalytically active chiral nucleophile.

The amino alcohol mediated asymmetric induction is satisfactorily explained by a six membered transition state,[83–85] as shown in Scheme 3.13. A similar model was also suggested for the prolinamine derived ligand. The aldehyde and the diethylzinc both coordinate on the same, sterically less hindered face of the ligand, and the alkyl group is transferred to the *Si* face of the carbonyl group.

Titanium complexes have recently found wide application as chiral catalysts for nucleophilic additions to carbonyl compounds.[86] Several structurally quite varied catalyst systems have been developed, including the tartrate derived, seven membered ring titanium complexes in Figure 3.22. These are efficient ligands for alkylzinc, alkyllithium and alkyl Grignard reagents, achieving enantioselectivities of up to 98 % ee.

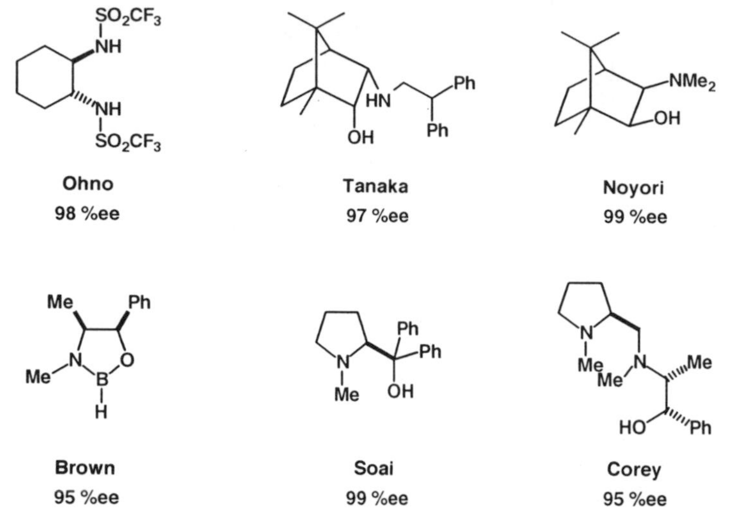

Figure 3.21

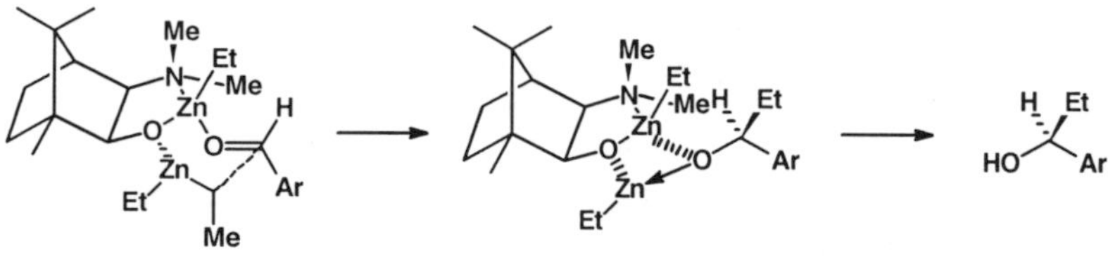

Scheme 3.13

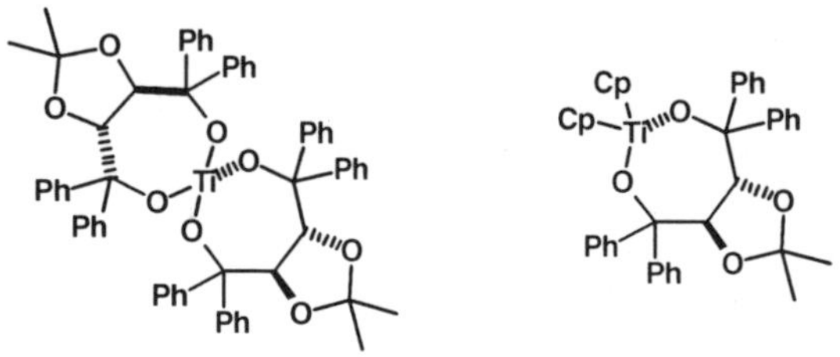

Figure 3.22

The same tartaric acid derived ligand, $\alpha,\alpha,\alpha',\alpha'$-tetraaryl-2,2-dimethyl-1,3-dioxolane-4,5-dimethanol (TADDOL), has also been used for enantioselective Grignard alkylation with a wide variety of carbonyl substrates (Scheme 3.14).[87] Addition of alkyl Grignard reagents to acetophenone occurs from the *Re* face. Aryl, heteroaryl and α,β-unsaturated alkyl ketones give the highest enantioselectivities, whereas alkyl ketones give lower

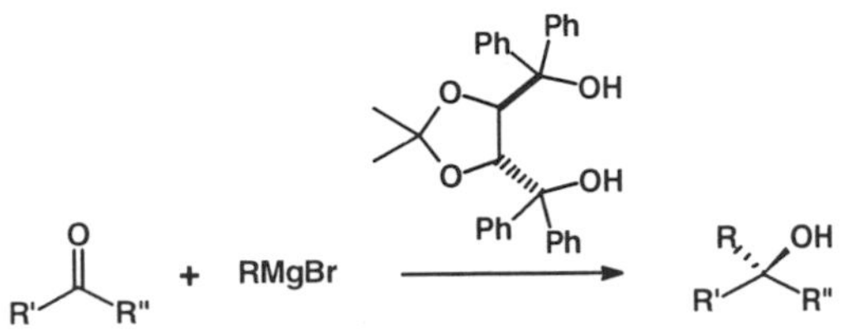

Scheme 3.14

selectivities coupled with lower yields. The reaction is sensitive to solvent effects: when the solvent is changed from THF to Et_2O, the selectivity is lost. The reaction mixture is heterogeneous, and the asymmetric induction may be mediated either by chiral aggregate formation [(metal-R)$_m$(metal–X$^*_n$)], or by chiral Lewis acid (R*OMgBr) activation of the ketone substrate in the rate determining step.

3.2.1.3 Asymmetric allylation/propargylation of the carbonyl group

Many of the above catalyst systems also handle allyl and propargyl transfers quite satisfactorily, but some more specific systems have also been developed. These reactions bear a striking resemblance to the aldol reaction, and many of the generalizations presented in that context will also hold true for the allylmetal reactions with aldehydes (Scheme 3.15)

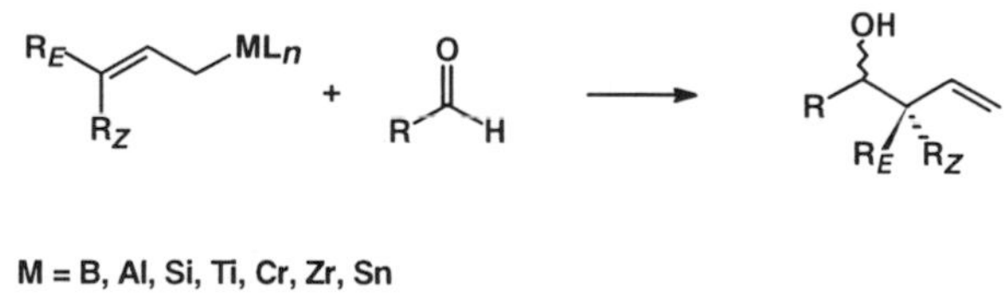

M = B, Al, Si, Ti, Cr, Zr, Sn

Scheme 3.15

The reaction is successful with boron and a wide variety of metals (Al, Si, Ti, Cr, Zr, Sn).[88] The stereochemical outcome can be classified into one of three groups: (i) type 1, in which the *syn/anti* ratio of the products reflects the *Z* or *E* geometry of the allyl moiety (B, Al, Sn); (ii) type 2 reactions which are *syn* selective irrespective of the olefin geometry (Sn, Si); and (iii) type 3 reactions which are *anti* selective irrespective of the olefin geometry (Ti, Cr, Zr).[89]

Crotylation with stereochemically defined crotylmetal reagents (R_E or R_Z = Me) gives rise to masked aldol products which can be converted to 3-hydroxyaldehydes simply by ozonolysis. Powerful asymmetric reagents, based on crotylboranes[90] and crotylboronates,[91,92] have been described.

The proposed origin of the asymmetric induction in the case of allylboronates is shown in Figure 3.23. Rotation of the B—O bond clockwise moves the aldehyde non-bonding

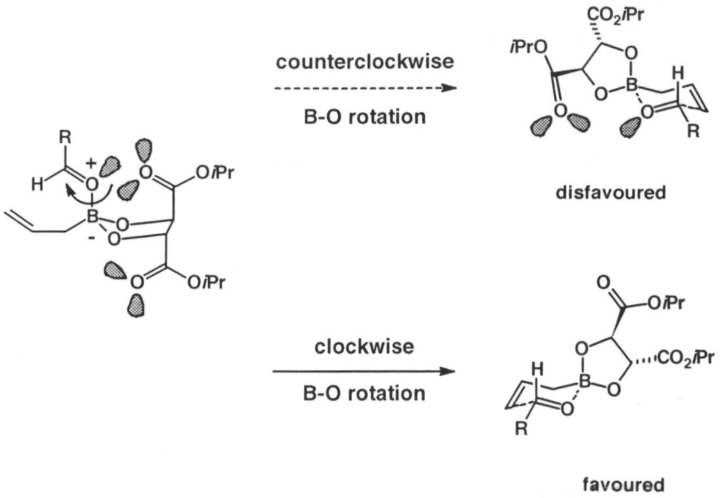

Figure 3.23

lone pair away from the proximate ester, which is pseudoaxial in the five membered dioxaborolane ring. The alternative rotation in a counter-clockwise sense would give an assembly with the ester carbonyl and the aldehyde non-bonding lone pair close to each other.

Chiral allyltitanates have also been successfully utilized in crotylation reactions.[93,94] Reactions with various chiral and achiral aldehydes give excellent enantio- and diastereoselectivities, as exemplified by the allylation of the serine derived aldehyde (Scheme 3.16).

In Figure 3.24 some commonly used chiral boron reagents are collected, along with typical enantioselectivities for benzaldehyde.[95-100]

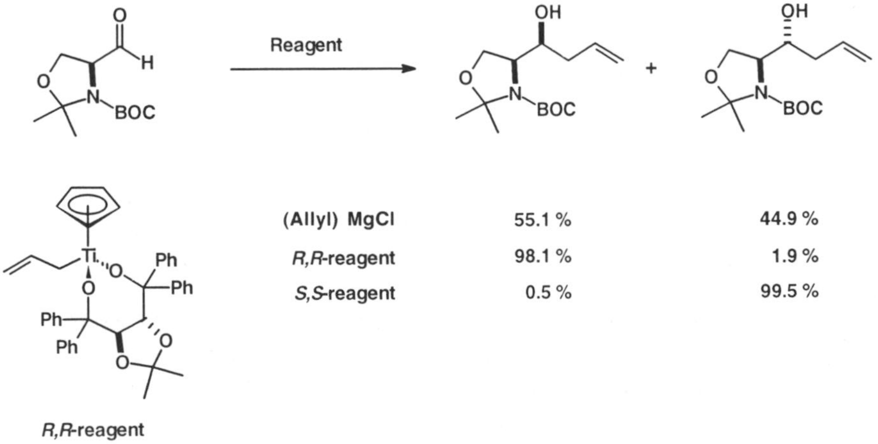

(Allyl) MgCl	55.1 %	44.9 %
R,R-reagent	98.1 %	1.9 %
S,S-reagent	0.5 %	99.5 %

R,R-reagent

Scheme 3.16

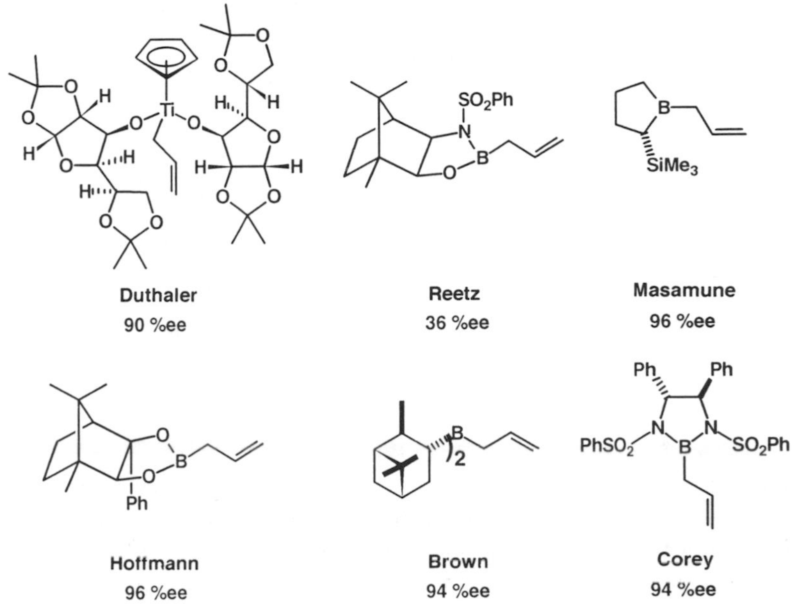

Figure 3.24

3.2.2 Reactions at the α-Carbon (Enolate Chemistry)

The α-centre of a carbonyl compound provides an efficient entry point for the introduction of a new stereogenic centre in a molecule. Enolate chemistry being one of the most extensively studied fields in organic synthesis, it is no surprise that methods for asymmetric α-alkylation of carbonyl compounds have also been well developed. Two factors need to be considered in discussing asymmetric induction in enolate alkylations, namely the E/Z geometry of the enolate and the source of the asymmetric information, i.e. enantiofacial selectivity.

To define a unique system of nomenclature for the enolate geometry, we shall in this text distinguish the two stereoisomeric forms of the enolate (or enol ether) as E and Z with the descriptors for the atoms defining the relationship in parentheses. We shall use as the reference groups the main carbon chain (R) and the enolized oxygen (O). Figure 3.25 shows the E(O,R) and Z(O,R) enolates for a ketone, ester, thioester, and alkoxyketone enolate.

The geometry of the enolate is of prime importance for the stereochemical outcome of any reaction occurring at the α-carbon. It is therefore important to remind ourselves of the factors governing the stereochemistry of enolization. Herein, we shall describe the process in terms of a simplified molecular orbital theory and conformational analysis. From conformational analysis, we know that the most stable (ground state) geometry of a ketone is the one having an α-hydrogen eclipsed with the carbonyl oxygen. This is shown in Figure 3.26 as conformation **A**. As this is the ground state conformation, and

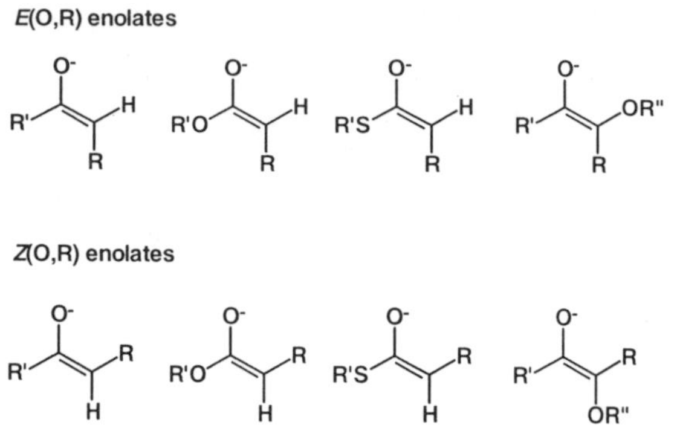

E(O,R) enolates

Z(O,R) enolates

Figure 3.25

as the ground state can hardly be expected to be reactive enough to drive the reaction, we must consider what happens as we rotate the C_α—C===O bond. Rotation of the bond by $30°$ (conformation **B**) or by $90°$ (conformation **C**) brings the C_α—H bond perpendicular to the plane of the carbonyl group, or expressed in other words, parallel to the plane of the p-orbitals of the carbonyl group. This leads to maximum overlap of the σ^*-orbital of the C_α—H bond with the (electron attracting) π^*-orbital. Conformation **C** suffers from strain between the R' group and the carbonyl compared to conformation **B**, which is thus to be expected to be the favoured one leading to the *kinetic* product.

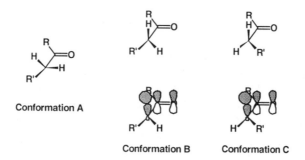

Conformation A   Conformation B   Conformation C

Figure 3.26

The practical value of this analysis is the notion that the hydrogen which is (nearly) perpendicular to the plane of the carbon atom is the most acidic one.[101] In cyclic systems, such as cyclohexanone, this is manifested in the higher acidity of the axial α-protons compared with the equatorial ones. In acyclic ketones, this leads to stereoselectivity of enolate formation as follows. Conformation **B** is the favoured pathway leading to the *E*(O,R) enolate which is the kinetic product. However, as one increases the size of the R' group the steric repulsion between the two alkyl substituents (R and R') becomes dominant (allylic A^{1,2} strain), and the order of the stabilities of the two pathways leading to the *E*(O,R) and *Z*(O,R) enolates will be reversed. Similar arguments also hold for the protonation of enolate anions.[102]

3.2.2.1 α-Alkylation of enolate anions

In the asymmetric α-alkylation of the enolate anions, the methods thus far developed utilize either internal or relayed asymmetric induction. Thus, the chiral information is either covalently connected to the starting material, leading to diastereotopic selection in the transition states, or brought into the transition state with the aid of a chiral auxiliary which must be cleaved later. We shall not discuss the first case in this chapter, and a few comments are sufficient for the latter case. The commonest method of utilizing the chiral derivatizing agents in enolate α-alkylation depends on the use of labile enamine type derivatization. Ketone hydrazones have been widely used in this context,[103] but we shall take our example from enamine chemistry. The asymmetric formation of the CD ring system of steroids is a prevailing problem (Section 8.3.2). The most straightforward way to achieve this is to utilize the classical Robinson annulation, provided that the reaction can be harnessed in an enantioselective form. Catalysis by enamines is known for these reactions, and utilization of a chiral amine for the formation of the enamine nucleophile accomplishes the desired function.[104, 105] Thus, proline catalysis gives rise to the nearly enantiopure product aldol, a result of enantiofacial selection of the two enantiotopic carbonyl groups (Scheme 3.17).

Scheme 3.17

We shall see some further examples of the utilization of chiral auxiliaries in the asymmetric synthesis of specific natural products in later chapters (e.g. 5.6 and 8.3.2).

3.2.2.2 Aldol reaction

Two- and three- carbon units, acetate and propionate, play a central role in the biosynthesis of most of the natural products. These (related) relatively simple structural units are used in a most versatile way to construct structures of tremendous stereochemical complexity from only a limited number of distinctly different reaction types. The reaction of an enolate with an aldehyde, the aldol reaction, provides a rapid means of setting up several stereogenic relationships in a single step (Scheme 3.18). As the aldol reaction is also one of the oldest systematically studied reactions in organic chemistry, it is no wonder that the recent advances in understanding the stereochemical aspects of the aldol reaction have been intensively studied, and one can presently predict the outcome of such a process or design a specific reaction for a given stereochemical problem.

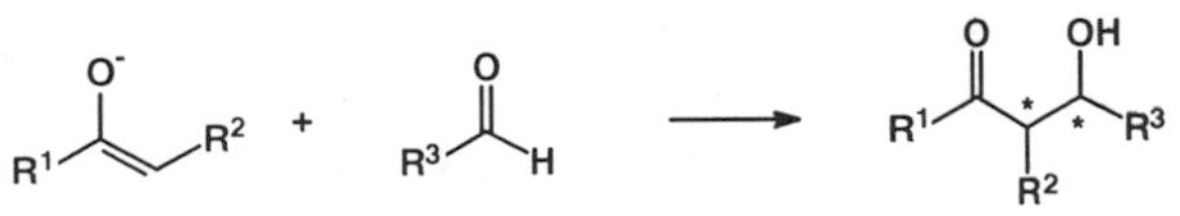

Scheme 3.18

In this section, we will inspect the factors affecting the stereochemistry of the product aldol in the light of currently accepted rationalizations on the mechanism of the aldol addition. The reaction has been studied by several groups, and numerous reviews have been written that highlight different aspects of the reaction.[106-110]

A number of transition state geometries have been suggested to explain the general trends in the diastereoselectivity of the aldol reaction. The earliest was the pericyclic model proposed by Zimmerman and Traxler,[111] which consisted of a six membered, chelated, chair-like transition state. This chelation imposes a gauche arrangement of the reacting enolate double bond and the accepting carbonyl double bond. To avoid steric repulsion with the enolate R^1 substituent, the aldehyde substituent R^3 prefers to assume an equatorial position (Figure 3.27).

Figure 3.27

When this model is applied to the aldol reaction, one can draw the general conclusion that $E(O,R)$ enolates will preferentially give *anti* aldol products, whereas $Z(O,R)$ enolates will lead to *syn* products. This generalization holds true in broad terms, but several further refinements need to be made. Inspection of the transition structures in Scheme 3.19 leads to the following conclusions. The correlation is dependent on the relative sizes of the substituents. Bulky R^3 groups will lead to higher selectivity. Especially in the case of $Z(O,R)$ enolates, the correlation is enhanced if both R^1 and R^3 are sterically demanding. If R^2 is very large, the normal trend (Z gives *syn*, E gives *anti*) is reversed. Steric congestion in the transition structure also enhances the selectivity, and thus boron enolates usually give higher selectivities than the more loose lithium enolates (the O—Li bond is 1.9–2.0 Å whereas the O—B bond is 1.4–1.6 Å).[112,113] Similarly, the increase in bond length[114-116] and decrease in solvation enthalpy[117] lead to looser transition structures and breakdown of the selectivity suggested by the Zimmerman–Traxler model.

Not all the aldol reactions follow the generalizations given above, and thus further classification of the reactions has emerged. The reactions following the Zimmerman–Traxler cyclic transition state and yielding mainly Z–*syn* and E–*anti* selectivity are known as type 1 aldol processes. Type 2 reactions are *syn* selective regardless of the enolate geometry, and are common to a wide range of enol derivatives, including enol silanes, enol stannanes, zirconium enolates and enol borates. Both open and cyclic transition state geometries have been proposed for these reactions.

Ketene acetals and ketene thioacetals tend to give *anti* products regardless of the enolate geometry. These *anti* selective aldol reactions are known as type 3 processes. As in the case of type 2 processes, both open and closed transition state geometries have been proposed, but much disagreement still remains.

The open transition state proposed by Yamamoto is shown in Scheme 3.20 for the *syn* selective addition of tin or zirconium enolates to aldehydes.[118] The enolate and carbonyl moieties are aligned in an antiperiplanar fashion. For both the $E(O,R)$ and $Z(O,R)$ enolates, the transition state leading to the *anti* aldol product is destabilized through the steric interaction between the R^2 and R^3 substituents.

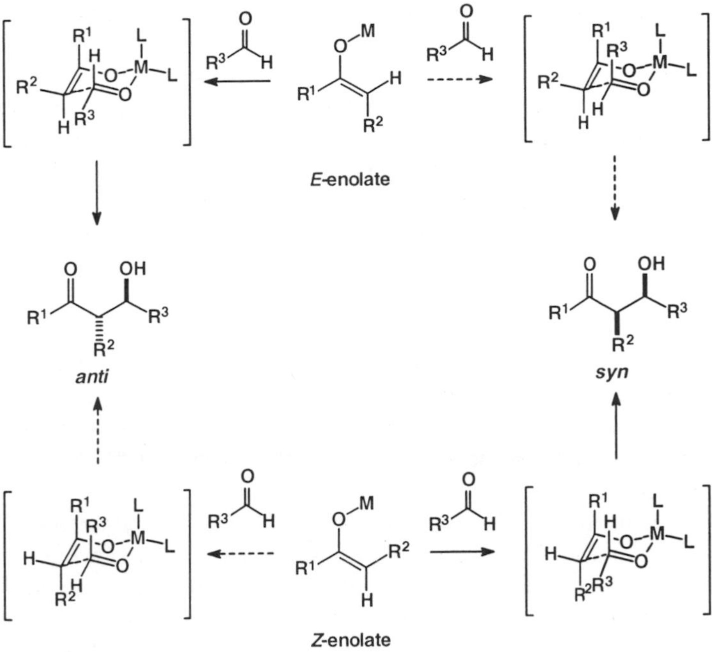

Scheme 3.19

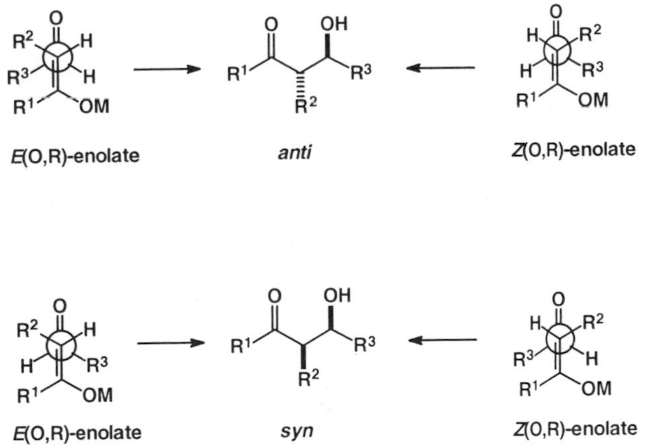

Scheme 3.20

Recent *ab initio* calculations suggest that: (i) the open transition state is favoured for metal-free enolates; (ii) cyclic transition states best account for the case of lithium enolates with distorted Burgi–Dunitz attack angles due to short bonds to lithium; and (iii) a close balance between chair and twist-boat cyclic transition structures exists for enol borinates and enol borates, depending on the substitution pattern.[119,120]

The aldol reactions show an increasing preference for the participation of the cyclic transition state and thereby *syn* product preference (synclinal orientation of the enolate and the aldehyde) with increasing cation coordinating ability ($K^+ < Na^+ < Li^+ < MgBr^+$). Reactions with a 'naked' enolate, where the cation has been sequestered, show a strong preference for an open transition state in which the dipole moments of the enolate and aldehyde tend to cancel each other (antiperiplanar orientation). Solvent effects are negligible, especially for the strongly coordinating cations (e.g. Li). The use of additives, such as HMPA, effectively leads to attenuation of the coordinating power of the cation. As a result, the tendency towards an open transition state is increased.[121]

Introduction of asymmetry into the aldol reaction can be achieved using any of the general methods discussed in Chapter 2. Here we shall examine two of the generally useful strategies, namely the utilization of a chiral auxiliary (relayed asymmetric induction) and the use of chiral reagents (or catalysts, external asymmetric induction).

Of the chiral auxiliaries, the Evans chiral oxazolidinones have gained widespread use in many laboratories, thanks to their generally high and predictable level and sense of asymmetric induction.

The normal mode for the Evans aldol process produces mainly the *syn* product. This is due to the chelated, six membered transition state, which favours the observed products.[122] Of the two possible *syn* aldol products, the one forming through β-attack (see Scheme 3.21) is favoured on steric grounds. Rotation of the enolate carbonyl–nitrogen bond can give rise to two different enolates, whose reactions from either the α- or β-face will result in diastereomeric *syn* aldol products.

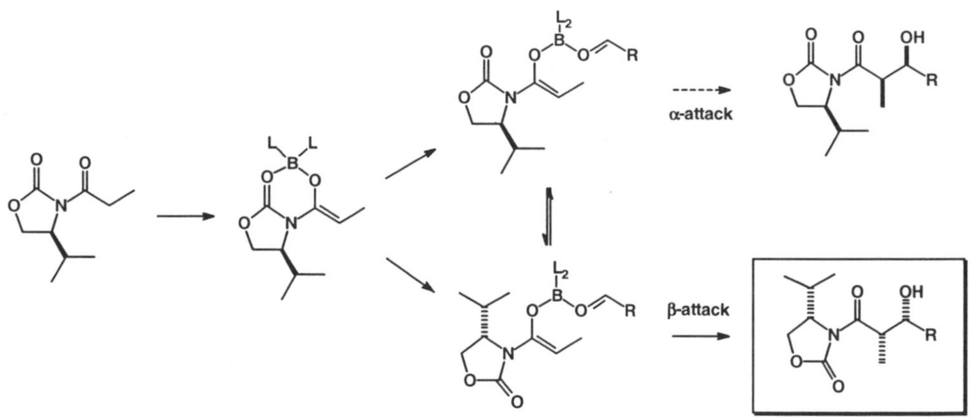

Scheme 3.21

In the addition, approach of the electrophile from the β-face is favoured and this is explained as being the result of stabilization of the enolate geometry through dipolar organization (Figure 3.28).

If the metal is not coordinated with the oxazolidone carbonyl oxygen, the attack of an electrophile occurs from the *Re* face of the enolate double bond.[122] This is explained as being the result of equilibration of the oxazolidinyl enolate to the conformation which

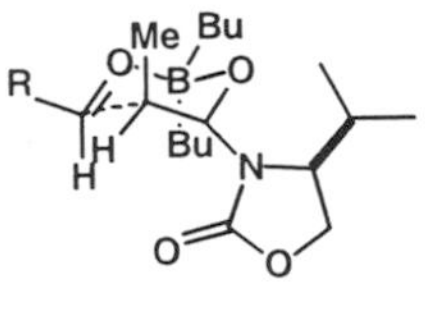

Figure 3.28

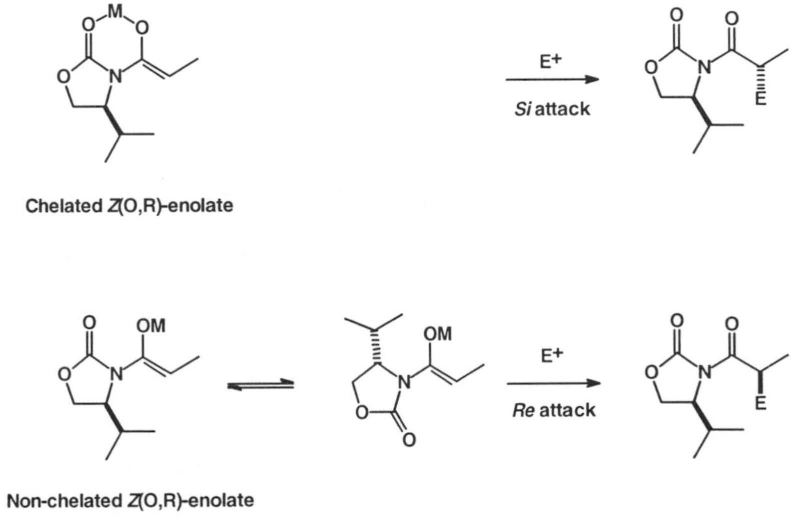

Chelated Z(O,R)-enolate

Non-chelated Z(O,R)-enolate

Scheme 3.22

places the dipoles opposite to each other. If the metal is coordinated, the attack occurs from the *Si* face to give the diastereomeric product, as shown in Scheme 3.22.[123]

The chelated oxazolidinyl enolates allow control of the stereochemistry of the aldol products at the α-carbon if one can control whether or not the metal is chelated at the time of the aldol reaction. Titanium enolates can be employed to obtain the non-Evans *syn* aldol products.[124] The chelated complex is slightly looser (Ti—O bond length 1.62–1.73 Å) and the titanium is hexacoordinate (Scheme 3.23), allowing an approximately octahedral coordination around titanium, whereas boron is approximately tetrahedrally coordinated.

Also the *syn/anti* ratio in the Evans aldol reaction can be affected very easily through the choice of the Lewis acid employed.[125–127] Assuming an open transition state, the form **A** (Scheme 3.24) is favoured if the Lewis acid employed is small (e.g. SnCl$_4$, TiCl$_4$) because it minimizes the gauche interactions about the forming bond. The product will therefore be the non-Evans *syn* aldol. However, if one uses a large Lewis acid (e.g. Et$_2$AlCl) transition state **B** becomes more favourable because of the methyl–Lewis acid interaction. The product will therefore be the *anti* aldol. The observed *anti/syn* selectivities range from 74:26 (PhCHO) to 95:5 (*i*-PrCHO and *t*-BuCHO).

Heathcock has also recently presented a protocol for the preparation of all the four possible stereoisomeric aldol products from a single aldol reagent. The *Z* lithium and *Z*

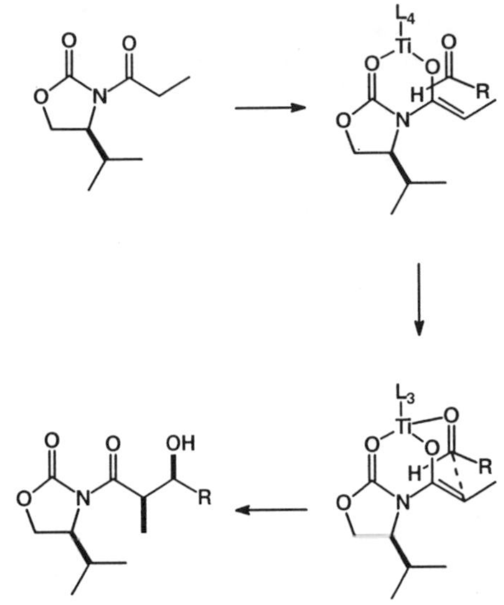

Scheme 3.23

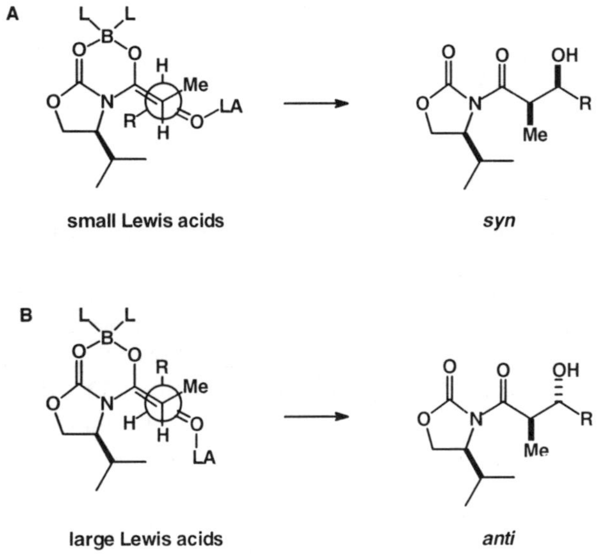

Scheme 3.24

boron enolates give access to the two diastereomeric *syn* aldol products. Metallation with bromomagnesium tetramethylpiperidide gives an *E* magnesium enolate which leads to one *anti* aldol. Transmetallation of the *E* magnesium enolate with (triisopropoxy)titanium chloride gives an *E* titanium enolate which gives rise to the diastereomeric *anti* aldol (Scheme 3.25).[128,129]

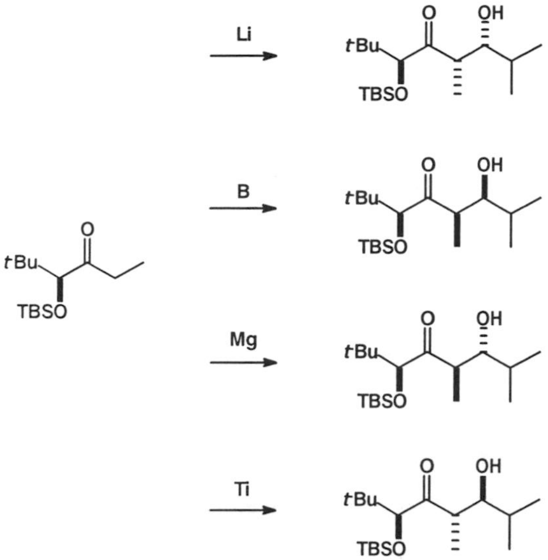

Scheme 3.25

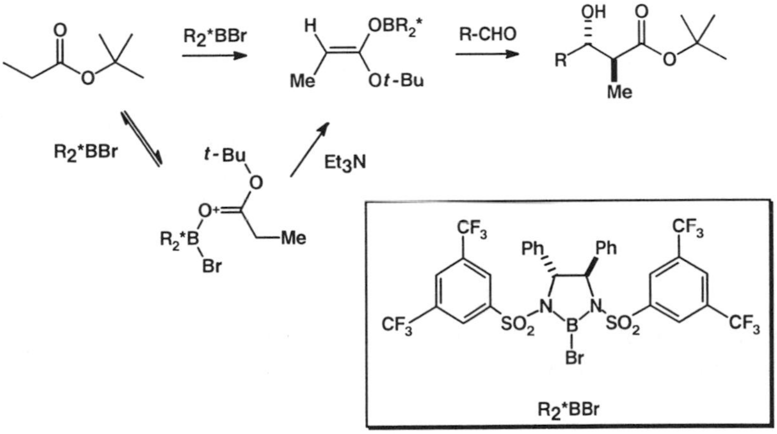

Scheme 3.26

Chiral catalysts based on diazaborolidines[130] and acyloxyboranes[131] have also been developed for aldol reactions. In the example in Scheme 3.26, the *syn/anti* selectivity can be controlled simply by the choice of the aldol substrate. Ester enolates give predominantly *anti* products, whereas thioester enolates give rise to *syn* aldol products.[130]

Chiral acyloxyboranes (CAB) can be used in less than stoichiometric quantities.[131] Thus, treatment of a silyl enol ether and aldehyde with 20 mol % of the tartaric acid derived CAB gives the *syn* aldol products (regardless of the geometry of the starting enol ether) in high % ee (Scheme 3.27).

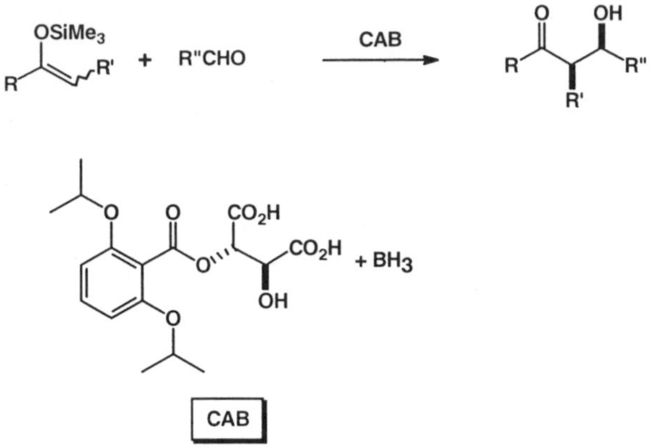

Scheme 3.27

3.2.3 Reactions at the β-Carbon of an Enone

Camphor derived *N*-enoyl sultams undergo highly diastereoselective addition with a wide range of organometallic reagents (Scheme 3.28). Phosphine stabilized alkyl- and alkenylcuprates give diastereoselectivities typically higher than 85 % de, but as the products are highly crystalline they can be crystallized to practically 100 % diastereopurity.[132] Alkyl Grignard reagents undergo an equally successful conjugate addition.[133]

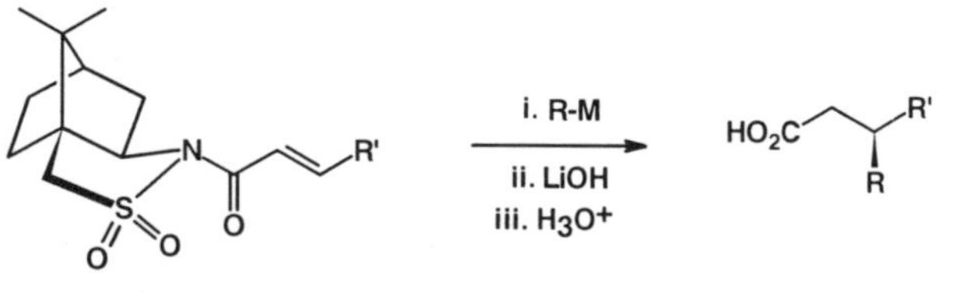

Scheme 3.28

The steric course of the reaction can be rationalized as involving internal chelation of the organomagnesium species with the sultam oxygen (Figure 3.29). Formation of a dimeric type reagent with another molecule of the organomagnesium reagent places the nucleophilic R group in such a position that the delivery of the nucleophile occurs from the front face of the olefin.

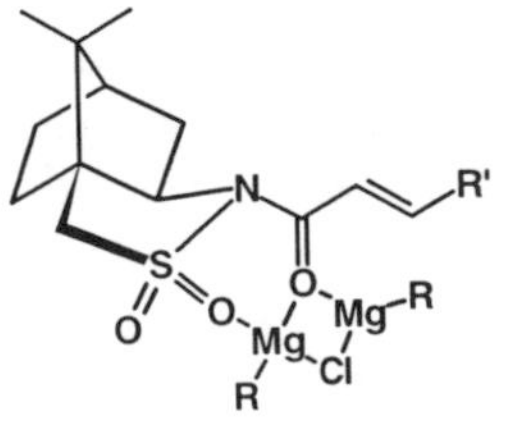

Figure 3.29

The conjugate addition of Grignard reagents on α-alkyl crotonoyl sultams provides an interesting route to 2,3-disubstituted carboxylic acid derivatives with high diastereocontrol. The intermediate enolate can be protonated to give rise to products with two new stereogenic centres. The change in both the sense of absolute stereocontrol and diastereoselectivity upon changing the nucleophile merits some comments. With Grignard reagents, the expected product **A** (Scheme 3.29) is the predominant one (97–99 % of product mixture).[133] Dialkyllithium organocuprates R$_2$CuLi react with opposite enantiofacial selectivity to give **B** as the major product (85–90 %).[134] Finally, Grignard reagents with copper(I) catalysis give the *syn* product **C** with high selectivity (84–97 %).[135]

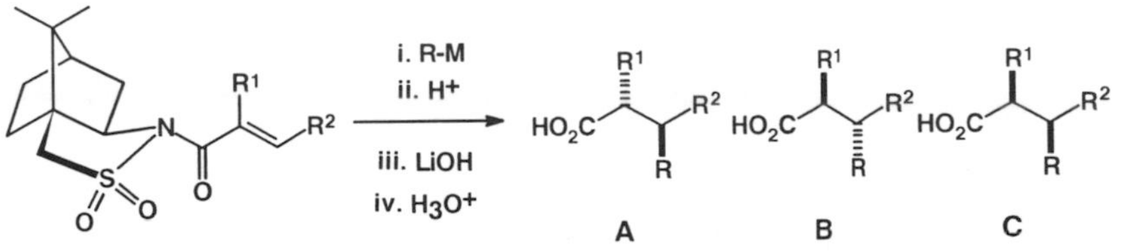

Scheme 3.29

The structurally related bicyclic dicyclohexyl sulfonamide derivative (Scheme 3.30) also functions as an efficient chiral auxiliary in the alkylation of enoates.[136–138] The sense of asymmetric induction is opposite to that observed for the tricyclic camphorsultams. This is owing to the fact that for steric (and electronic) reasons, the enoate adopts an s-*trans* conformation preferentially. The s-*trans* conformer is *ca* 1.3 kJ mol^{-1} more stable in the ground state, but coordination with a Lewis acid further stabilizes this conformation. The nucleophile then attacks the double bond from the *Si* face.

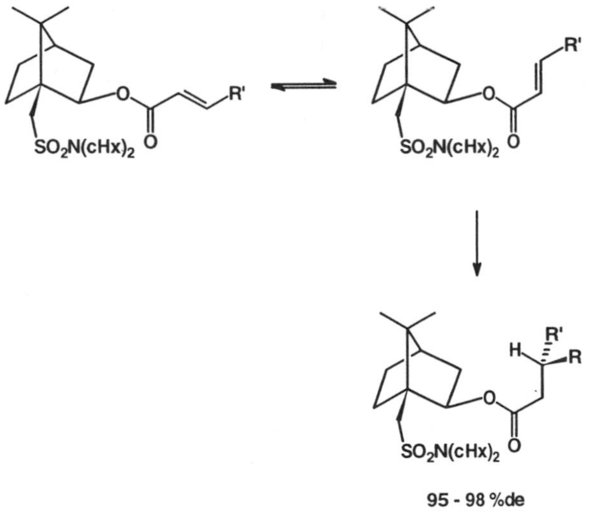

95 - 98 %de

Scheme 3.30

Other ester chiral auxiliaries have been used with moderate to good diastereoselectivity. The *E* crotonate esters derived from 8-phenylmenthol have given good diastereoselectivities (up to 99 % de) on reaction with the Mukaiyama cuprates ($RCu.BF_3$). The corresponding *Z* isomers as well as tri- and tetrasubstituted enoates gave only modest selectivity.[139]

Fumaric acid semialdehydes have been converted to the corresponding hemiaminals and aminals with a number of chiral aminols and diamines, respectively. Particularly promising results have been obtained by the Mukaiyama and Scolastico groups. Mukaiyama has utilized the aminals derived from (*S*)-prolinamine, whose reaction with Grignard reagents under copper(I) catalysis gives rise to the *R* absolute stereochemistry at the new stereogenic centre.[140] Complexation of the magnesium to the more basic bridgehead nitrogen is suggested to be the directing factor (Scheme 3.31).

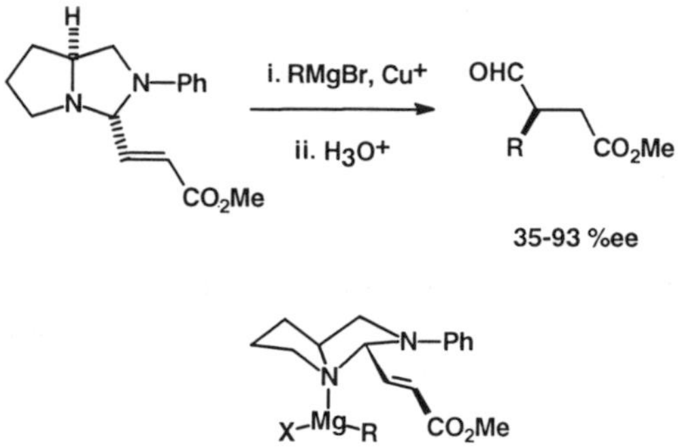

Scheme 3.31

Scolastico has employed hemiaminals derived from norephedrine (Figure 3.30). α, β-Unsaturated aldehydes, ketones and esters with the oxazolidine auxiliary reacted with lithium dialkylcuprates to give the alkylation products with 80–90 % de.[141,142]

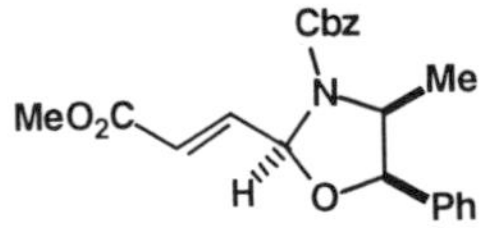

Figure 3.30

3.3 Reactions of Olefins

In this section, we shall discuss the enantioselective refunctionalization of olefinic bonds. Catalytic asymmetric hydrogenation, although of wide practical utility, will be omitted, but some examples will be included, for example, in connection with the asymmetric synthesis of amino acids. The 1,2- and 1,3-dihydroxylated unit is a very common structural unit in natural products. It is no wonder that these functional systems have become two of the most studied difunctional systems in the recent past. In this section, we shall discuss some recent advances in the stereocontrolled synthesis of such functional arrays, especially the methods relying on aldol type transformations, oxidations and stereocontrolled reductions.

3.3.1 Oxidation

Setting up a 1,2-dihydroxy system is most often carried out without formal change of the oxidation level of the two participating carbon atoms. Thus, one usually starts with an olefin, and simply adds the equivalent of hydroperoxide onto it (Scheme 3.32).

Scheme 3.32

Depending on whether one wants to introduce the two hydroxyl groups *syn* or *anti* with respect to the emerging single bond, one has several alternative routes to follow. The simplest one is to rely on *cis* hydroxylation, usually carried out through osmylation (catalytic OsO$_4$, with *N*-methylmorpholine *N*-oxide (NMMO) or trimethylamine *N*-oxide (TMANO) as the stoichiometric oxidant). However, this requires the pre-existence of the desired *cis* or *trans* olefin, which in many cases becomes the limiting factor. The alternative is to use epoxidation followed by epoxide opening. This strategy also requires access to sterechemically homogeneous olefins, and also introduces another factor— regioselective opening of the epoxide. During the last two decades or so, powerful methodology has been developed for both direct asymmetric epoxidation (AE) and asymmetric dihydroxylation (AD) for all of the six structurally distinct classes of olefins (Figure 3.31).

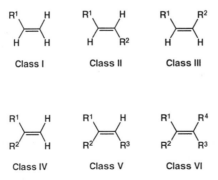

Class I   Class II   Class III

Class IV   Class V   Class VI

Figure 3.31

3.3.1.1 Asymmetric epoxidation

Epoxidation of an olefin leads to the addition of the epoxide oxygen on one face of the molecule. The regioselectivity of the epoxide opening can usually be predicted quite simply by classical means (inductive effects, or polarity control),[143] and necessarily invokes an inversion of stereochemistry at the carbon being attacked by the nucleophile—in the case of a hydroxyl nucleophile or its surrogate, effectively leading to *trans* dihydroxylation (in the case of Nu⁻ being HO⁻) of the original olefin functionality (Scheme 3.33).

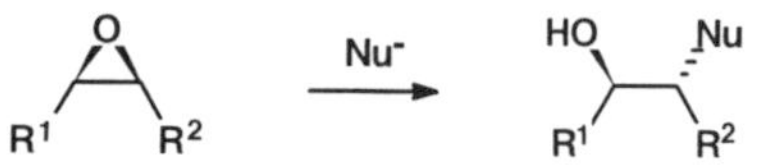

Scheme 3.33

In the case of allyl alcohols, one can utilize the existing chirality either as a directing group (active volume) or as a blocking group (inactive volume).[144] Careful choice of the oxidant gives one the option to choose either an intramolecular delivery of the oxygen (Henbest oxidation) to give the *syn* addition product,[145] or pre-blocking the allylic hydroxyl with a suitable (bulky) protecting group to protect this same face from the attack of the oxidant, giving rise to the *anti* product (Scheme 3.34).

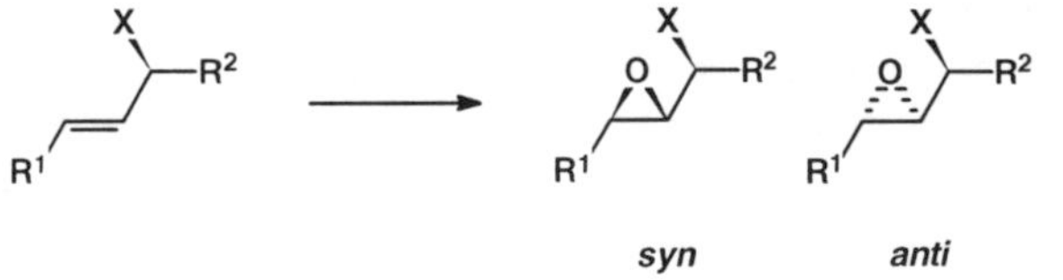

syn   *anti*

Scheme 3.34

This rationale is based on the application of the 1,3-allylic strain which explains the conformation of the allyl alcohol system, and accordingly the method is most powerful with *Z* olefins. The minimum energy conformation corresponds to the conformer where the carbinol hydrogen is eclipsed with the olefinic linkage.[146] In this conformer, the two faces of the olefin are now clearly distinct, and, in the case of an oxidant capable of complexation with the hydroxyl function, amenable to specific means of epoxidation (Scheme 3.35).

Scheme 3.35

If one were to single out one method that has had the widest impact on asymmetric synthesis over the past decade or so, one would probably nominate the Sharpless asymmetric epoxidation. This reaction has been developed to allow industrial scale operations, and it has found wide application in the synthesis of innumerable natural products and medicinally important agents. The Sharpless epoxidation is suitable for most allylic alcohols, which themselves are easily produced by classical means from simpler starting materials.

A complex formed from a tartrate ester (usually diisopropyl or diethyl tartrate) and titanium tetraisopropoxide is used as the chiral catalyst. The catalyst combines the olefin to be oxidized and the oxidant (*tert*-butyl hydroperoxide or cumene hydroperoxide) in such a manner that the delivery of the oxygen occurs principally from one face of the alkene. The detailed mechanism of the reaction has been studied,[147-150] and a reasonable model has been advanced to explain all the experimental observations.

The epoxidation is typically run in an aprotic, non-polar solvent (dichloromethane or even isooctane). The reaction temperature depends on the substrate, as does the eventual enantioselection. The enantioselectivities are high in most cases, and this fact has led to the development of a legion of methods for the utilization of glycidols in the synthesis of other functionalized systems.

The epoxide rings in glycidols can be opened regioselectively in a variety of ways. Treatment with Red-Al (sodium bis(methoxyethoxy)aluminium hydride, earlier also known as Vitride) gives the 1,3-diol selectively.[151] Treatment of the glycidol with an alkyl isocyanate or alkyl chloroformate followed with acid leads to 1,2,3-triols with the two terminal hydroxyl functions protected as a carbonate (Scheme 3.36).[152] Similar treatment with an isocyanate followed by base leads to 2-amino-1,3-diols, suitable intermediates in the synthesis of sphingosine derivatives and hydroxy amino acids.

Although the Sharpless epoxidation has proven to be of wide utility, it still has some limitations. Firstly, only allylic alcohols seem to work well with regard to asymmetric

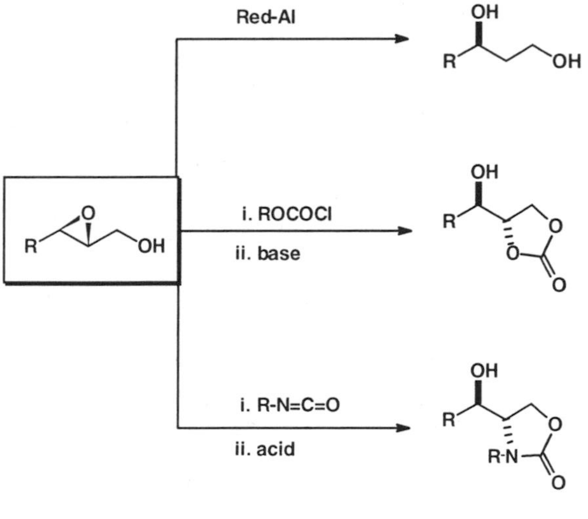

Scheme 3.36

induction. In fact, adding just one carbon, to give a homoallylic alcohol, degrades the induction down to *ca* 50 % ee, hardly meeting the desired criteria any more. This seems to be owing to the structure of the catalyst and the transition state, as inversion of the sense of absolute stereochemistry at the emerging chiral centre has also been observed.

    The second major problem lies in the fact that the substrate needs to be an alcohol. Without the hydroxyl group the olefin does not bind to the titanium species, and is thus not amenable to asymmetric induction. Much effort has been directed at overcoming this shortfall, and some new catalytic systems seem to be at least on their way to being universal epoxidation catalysts, e.g the manganese catalysed epoxidation by Jacobsen (Scheme 3.37).[153]

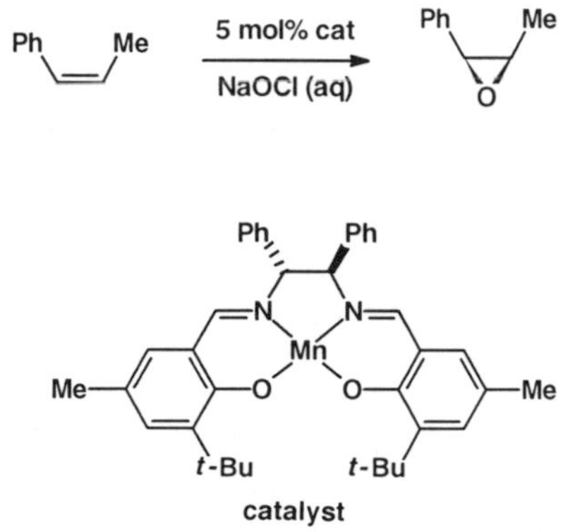

Scheme 3.37

3.3.1.2 Asymmetric dihydroxylation

The stereochemical questions regarding the introduction of asymmetry from the existing chiral centres in the molecule have been studied by experimental and theoretical methods for allyl alcohol substrates. 1,3-Allylic strain has been proposed to be the key factor governing the stereoselectivity.[3,4] The lowest energy ground state conformation of allyl alcohols places the carbinol hydrogen synperiplanar with the olefin to avoid steric repulsion with the group R_Z on the distal atom of the olefin (Figure 3.32).

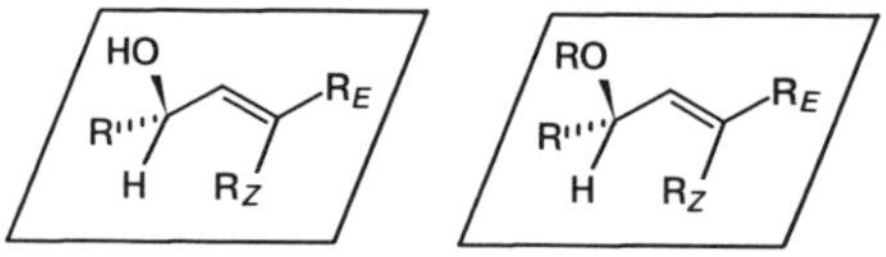

Figure 3.32

On this structure, one would expect the delivery of the two oxygens either from the same face where the hydroxyl group resides (by its participation through a chelation controlled process) or that the face selectivity would be simply governed by steric effects, i.e. the difference in the sizes of the HO and R groups. Kishi has contributed significantly on the study of osmium mediated *cis* hydroxylations, and his group has shown that the allyl alcohols are, in fact, poorer in exerting diastereocontrol than the corresponding allyl ethers.[154] Allyl esters show similarly poor selectivity. Chelation plays a minimal role, and it was originally suggested that the selectivity is size controlled. Houk has further developed the model for the asymmetric induction, and according to this model the ethereal oxygen does participate in the direction of the hydroxylation by donating electron density from the $\sigma*$ orbital to the olefin's π orbital, effectively increasing the electron density on the face opposite to the oxygen and thus making it more nucleophilic towards the (electrophilic) oxidizing agent (inside alkoxy effect, see Section 3.1).[5] This would also explain why the allyl esters are poorer in directing the stereoselectivity, as they are also poorer donors of electron density.

Dihydroxylation necessarily leads to the introduction of the two new hydroxyl groups on the same face of the existing olefin. External asymmetric induction is possible, and recently a number of efficient catalytic methods have been reported. Several chiral catalysts have been designed and tested, and most of them rely on C_2 symmetric ligands for the osmium.[155-160] As shown in Figure 3.33, the ligands can lead to effectively complete asymmetric induction in favourable cases.

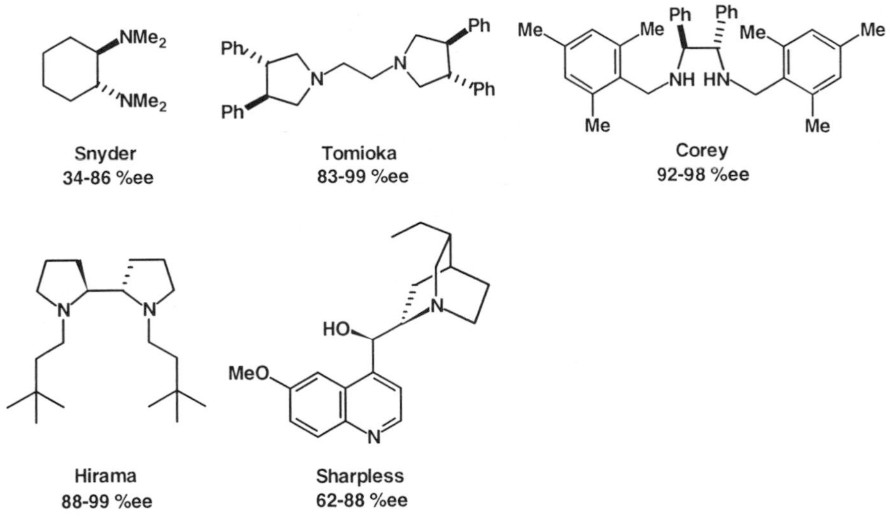

Figure 3.33

Sharpless has been active also in the *cis* hydroxylation of olefins, and he has devised catalytic systems capable of delivering the asymmetric information very powerfully. The chirality can be most conveniently derived from a derivative of quinine, as shown by Sharpless and co-workers.[161,162] Quite recently they have published a widely applicable method depending on chiral phthalazine ligands made from either dihydroquinidine or

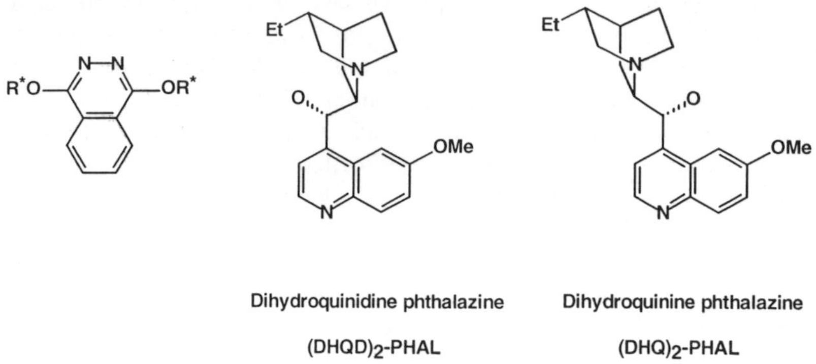

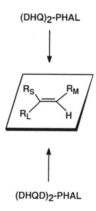

Dihydroquinidine phthalazine   Dihydroquinine phthalazine

(DHQD)$_2$-PHAL   (DHQ)$_2$-PHAL

Figure 3.34

dihydroquinine (Figure 3.34).[163] Instead of osmium tetroxide, dipotassium osmate is used as the oxidant, with potassium ferricyanide functioning as the stoichiometric oxygen source.

The transition state for osmylation has been modelled,[150,164,165] and Sharpless has also devised a simple way of predicting the product based on the heuristic model shown in Figure 3.35.

(DHQ)$_2$-PHAL

$$R_S \diagdown\quad R_M$$
$$R_L \diagup\quad H$$

(DHQD)$_2$-PHAL

Figure 3.35

3.3.2 Reduction

In this section, we shall briefly look at two further ways of constructing an aldol unit, by way of diastereocontrolled reduction of a 3-hydroxy ketone, and by way of catalytic asymmetric hydroboration.

3.3.2.1 1,3-Dioxygenated systems

As previously discussed in Section 3.2.2.2, the aldol reaction and its modifications provide a powerful tool for setting up 3-hydroxycarbonyl systems with excellent stereocontrol.

Stereospecific reduction of the carbonyl function would then provide a plausible access to *syn*- and *anti*-1,3-diols.[166] Methods have been devised for the utilization of the existing chirality in directing the facial selectivity in hydride reductions, and in this section we shall briefly examine the most promising ones.

Use of the β-chiral centre as an active volume necessitates pre-coordination of the hydride reducing agent with the free hydroxyl group (Scheme 3.38), thus setting the stage for intramolecular delivery of the hydride through, usually, a six membered, cyclic transition state. This kind of strategy was first realized with cyclic substrates.[167]

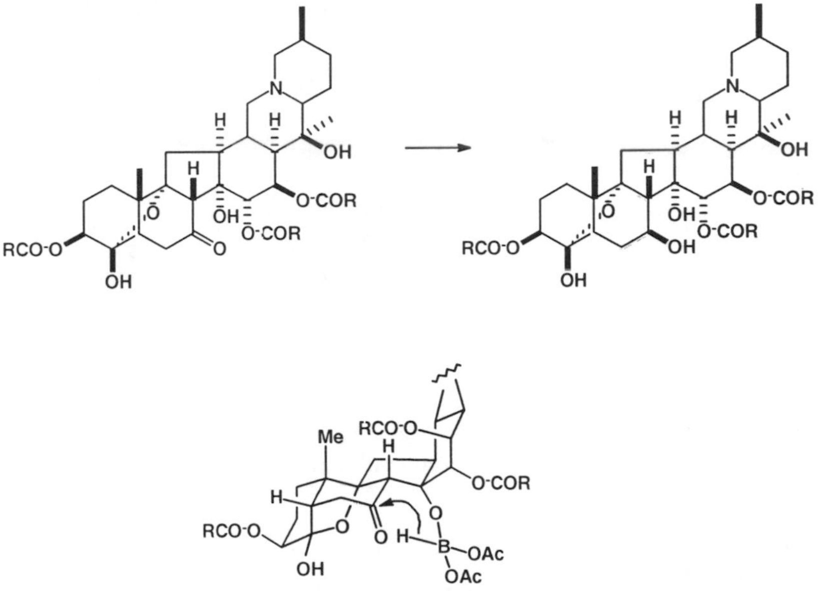

Scheme 3.38

Sodium triacetoxyborohydride, easily prepared from sodium borohydride and acetic acid,[168,169] functions as a selective reducing agent. This is based on the realization that the triacetoxyborohydride is a weak reducing agent, incapable of reducing the carbonyl function, whereas the derived alkoxydiacetoxyborohydride is more powerful as a reducing agent.[167] Thus, the pre-complexation is a precondition for the reduction, giving the desired distinction between intramolecular and intermolecular reduction, besides activating the reducing agent. This method has later found numerous applications in the synthesis of several natural products and pharmacologically interesting compounds, as exemplified in Scheme 3.39 for a key step in the synthesis of a portion of milbemycin β_1.[170]

Evans has extended this strategy into acyclic systems, where tetraalkylammonium triacetoxyborohydride (Scheme 3.40) is used to reduce the 3-hydroxy ketone to the *anti*-1,3-diol with high diastereoselectivity.[171] This protocol has again found wide use in the synthesis of many natural polyketides.

Complementary *syn* selective reduction is also available. This is based on pre-complexation of the hydroxy-ketone with either an alkoxydialkylborane[172] or lithium iodide (Scheme 3.40).[173] In the latter case, lithium aluminium hydride can be used

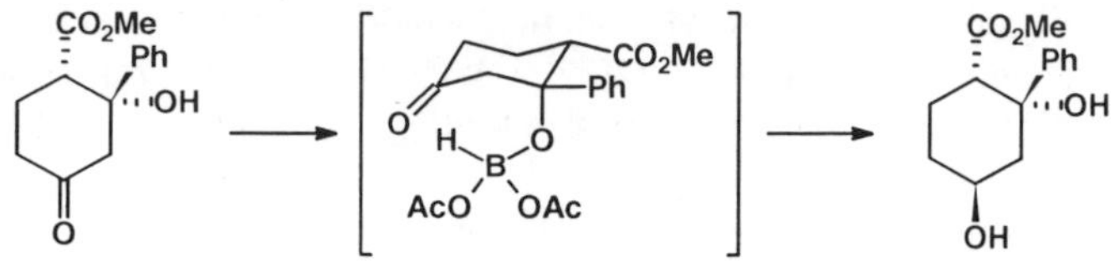

Scheme 3.39

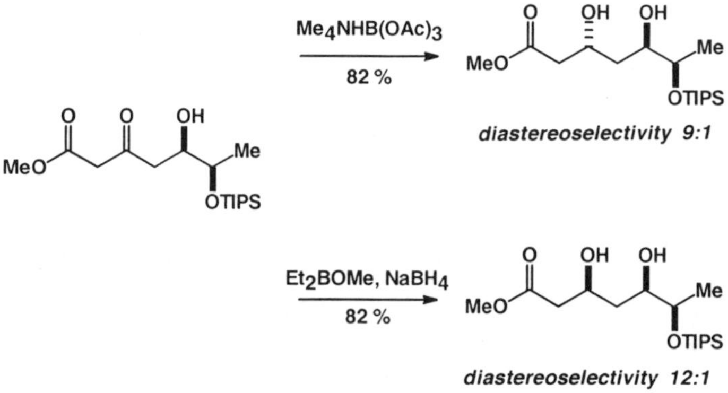

Scheme 3.40

as the reducing agent. Again, Evans has developed a powerful modification of the alkoxydialkylborane mediated reduction protocol.[174]

3.3.2.2 Hydroboration

One of the early examples of high substrate-controlled diastereoselectivity was achieved in the hydroboration of an acyclic, secondary allylic alcohol (Scheme 3.41).[175] This finding soon led to the discovery of catalysed, reagent diastereocontrolled hydroborations.[176,177]

Finally, enantioselective hydroborations have also been achieved using the same principles and chiral phosphine ligands as the source of asymmetric information.[178,179] The mechanism of this reaction has been studied.[180–183]

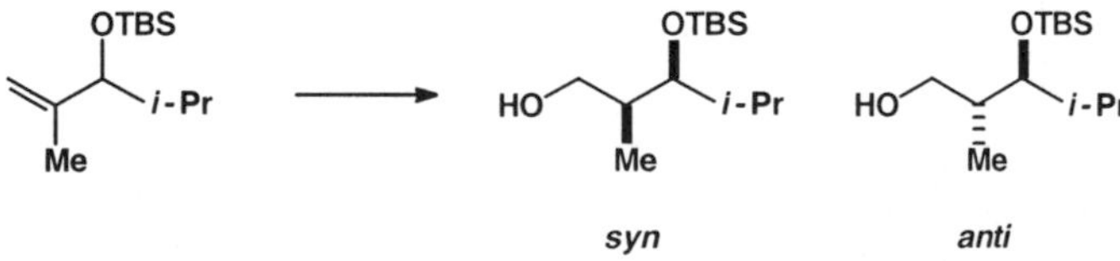

Reagent	syn	anti
9-BBN	5	95
CB, Rh(PPh3)3Cl	97	3

Scheme 3.41

The seminal Still paper introduced the concept of two-directional chain elongation, which has proven to be very powerful in the synthesis of several skipped polyols often encountered in natural products.[184] Hydroboration,[175] *cis* hydroxylation[185] and asymmetric epoxidation[184] methods have been used, as exemplified by Schemes 3.42 and 3.43.

Scheme 3.42

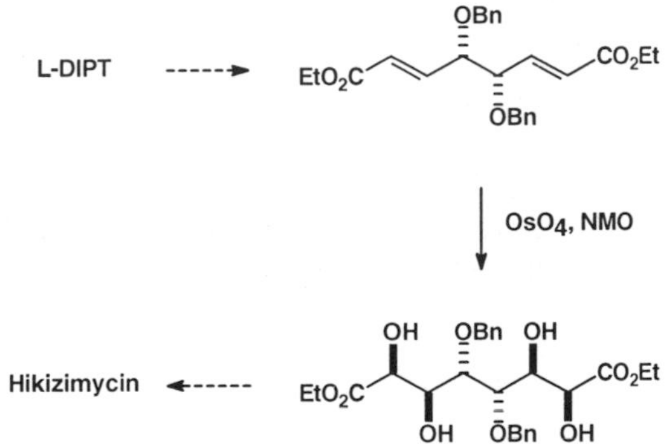

Scheme 3.43

3.3.3 Asymmetric Diels–Alder Reactions

The Diels–Alder reaction is one of the most powerful methods of generating stereochemically controlled products. The fact that the reaction proceeds through a tight cyclic transition state secures mechanistically controlled stereochemistry, potentially at all newly developing stereogenic centres. This is exactly where the power of the reaction lies; besides building two new carbon–carbon bonds, the Diels–Alder reaction also constructs four chiral centres out of sp^2 atoms. The relative stereochemistry of substituents R^2 and R^3 (see Scheme 3.44) in the emerging cyclohexene ring is secured by the stereochemistry of the starting diene. Similarly, the stereochemistry of the dienophile determines the relative stereochemistry of the substituents R^1 and R^4. This leads to two possible diastereomeric products with the R^1, R^2 substituents either *cis* or *trans* to each other. Steric effects and secondary orbital control are the determining factors in promoting either the *endo* or *exo* pathway.

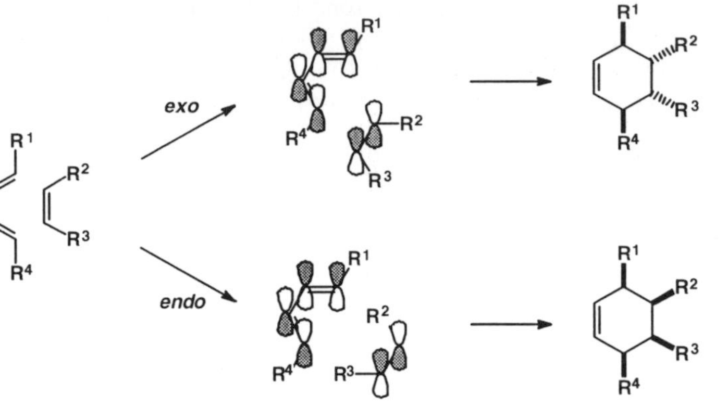

Scheme 3.44

Induction of absolute stereochemical control in the Diels–Alder reaction thus gives access to a functional compound possessing four consecutive chiral centres in an enantiopure form, obviously a very attractive possibility and therefore one of the general strategies which has received wide attention.[186,187] The earliest successes were based on the use of chiral auxiliaries, but in the mid-1970s the first catalytic, asymmetric versions of the reaction were reported,[188] and current developments in external asymmetric induction are rapidly taking over the relayed induction methods. As the chiral auxiliary based asymmetric induction can be explained through typical stereoelectronic control elements (see Scheme 3.45),[189] we shall here only briefly look at the newer catalysts.

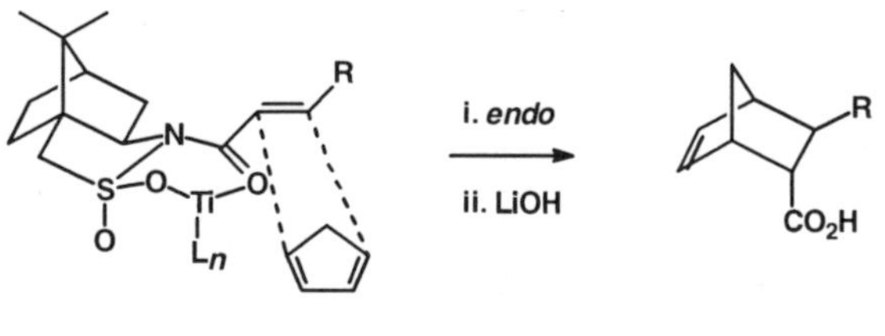

Scheme 3.45

The Danishefsky hetero-Diels–Alder reaction is an example which provides rapid access to enantiopure carbohydrate derivatives. In this reaction (Scheme 3.46), the catalyst is the chiral shift reagent tris[3-(heptafluoropropylhydroxymethylene)-(+)-camphorato]eur opium(III) [Eu(hfc)$_3$].[190]

Ti, Al and B have been successfully utilized in the Lewis acid catalyzed Diels–Alder reactions, and generally the highest enantioselectivities are achieved when using the powerful diol ligands such as BINOL[191] and TADDOL[192] and also the bis(oxazolines).[193] Recent additions to the catalysts also include the chiral acyloxyboranes (CAB)[194] and the oxazaborolidines (Figure 3.36).[69]

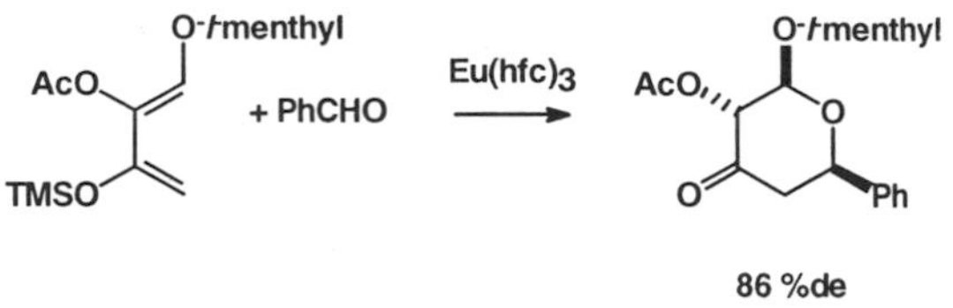

Scheme 3.46

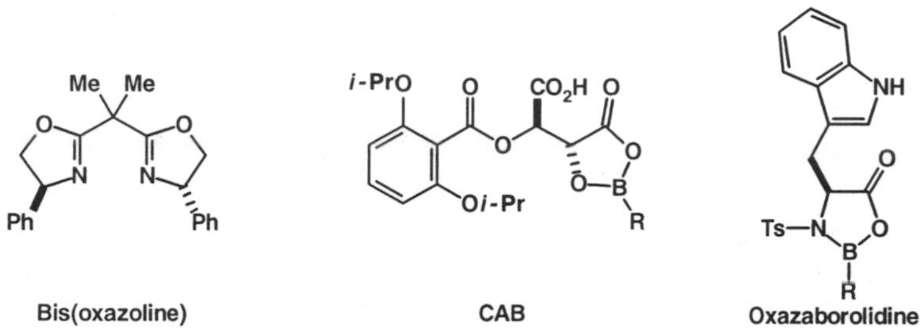

Bis(oxazoline)   CAB   Oxazaborolidine

Figure 3.36

References

1. Barton, D.H.R. *Experientia* **6**, 316–329 (1950).
2. Houk, K.N., Paddon-Row, M.N., Rondan, N.G., Wu, Y.-D., Brown, F.K., Spellmeyer, D.C., Metz, J.T., Li, Y., and Loncharich, R.J. *Science* **231**, 1108–1117 (1986).
3. Johnson, F. *Chem. Rev.* **68**, 375–413 (1968).
4. Hoffmann, R.W. *Chem. Rev.* **89**, 1841–1860 (1989).
5. Houk, K.N., Tucker, J.A., and Dorigo, A.E. *Acc. Chem. Res.* **23**, 107–113 (1990).
6. Cherest, M., Felkin, H., Prudent, N. *Tetrahedron Lett* 2199–2204 (1968).
7. Whitesell, J.K., Deyo, D., and Bhattacharya, A. *J. Chem. Soc., Chem. Commun.* 802 (1983).
8. Baldwin, J.E. *J. Chem. Soc., Chem. Comm.* 738–741 (1976).
9. Menger, F.M. *Tetrahedron* **39**, 1013–1040 (1983).
10. Liotta, C.L., Burgess, E.M., and Eberhardt, W.H. *J. Am. Chem. Soc.* **106**, 4849–4852 (1984).
11. Anh, N.T. and Eisenstein, O. *Nouv. J. Chim.* **1**, 61–70 (1977).
12. Scheiner, S., Lipscomb, W.M., and Kleier, D.A. *J. Am. Chem. Soc.* **98**, 4770–4777 (1976).
13. Burgi, H.B. and Dunitz, J.D. *Acc. Chem. Res.* **16**, 153–161 (1983).
14. Burgi, H.B., Dunitz, J.D., and Shefter, E. *J. Am. Chem. Soc.* **95**, 5065–5067 (1973).
15. Burgi, H.B. and Dunitz, J.D. *Acc. Chem. Res.* **16**, 153–161 (1983).
16. Fukui, K. *Acc. Chem. Res.* **4**, 57–64 (1971).
17. Houk, K.N. *Acc. Chem. Res.* **8**, 361–369 (1975).
18. Fleming, I. *Frontier Orbitals and Organic Chemical Reactions*, John Wiley & Sons, London, 1976.
19. Cram, D.J. and Abd Elhafes, F.A. *J. Am. Chem. Soc.* **74**, 5828–5835 (1952).
20. Heathcock, C.H. *Aldrichimica Acta* **23**, 99–111 (1990).
21. Singh, V.K. *Synthesis* 605–617 (1992).
22. Morrison, J.D. *Asymmetric Synthesis.* Vol. **2** (Ed.) Academic Press: New York, 1985.
23. Bothner-By, A.A. *J. Am. Chem. Soc.* **73**, 846 (1951).
24. Portoghese, P.S. *J. Org. Chem.* **27**, 3359–3360 (1962).
25. Minoura, Y. and Yamaguchi, H. *J. Polym. Sci. A-1* **6**, 2013–2021 (1968).

26. Landor, S.R., Miller, B.J., and Tatchell, A.R. *Proc. Chem. Soc.* 227; (1964) *Chem. Abstr.* **61**:14518e (1964).
27. Landor, S.R., Miller, B.J., and Tatchell, A.R. *J. Chem. Soc.* 197–201 (1967).
28. Noyori, R., Tomino, I., and Tanimoto, Y. *J. Am. Chem. Soc.* **101**, 3129–3131 (1979).
29. Noyori, R., Tomino, I., Tanimoto, Y., and Nishizawa, M. *J. Am. Chem. Soc.* **106**, 6709–6716 (1984).
30. Noyori, R., Tomino, I., Yamada, M., and Nishizawa, M. *J. Am. Chem. Soc.* **106**, 6717–6725 (1984).
31. Suzuki, M., Morita, Y., Koyano, H., Koga, M., and Noyori, R. *Tetrahedron* **46**, 4809–4822 (1990).
32. Corey, E.J., Schaaf, T.K., Huber, W., Koelliker, U., and Weinshenker, N.M. *J. Am. Chem. Soc.* **92**, 397–398 (1970).
33. Corey, E.J., Becker, K.B., and Varma, R.K. *J. Am. Chem. Soc.* **94**, 8616–8618 (1972).
34. Chong, J.M. and Mar, E.K. *Tetrahedron Lett.* **31**, 1981–1984 (1990).
35. Yamaguchi, S. and Mosher, H.S. *J. Org. Chem.* **38**, 1870–1877 (1973).
36. Reich, C.J., Sullivan, G.R., and Mosher, H.S. *Tetrahedron Lett.* 1505–1508 (1973).
37. Cohen, N., Lopresti, R.J., Neukom, C., and Saucy, G. *J. Org. Chem.* **45**, 582–588 (1980).
38. Midland, M.M. *Chem. Rev.* **89**, 1553–1561 (1989).
39. Midland, M.M. and McLoughlin, J.I. *J. Org. Chem.* **49**, 1316–1317 (1984).
40. Midland, M.M., Tramontano, A., and Zderic, S.A. *J. Organomet. Chem.* **156**, 203–211 (1978).
41. Midland, M.M. and Zderic, S.A. *J. Am. Chem. Soc.* **104**, 525–528 (1982).
42. Midland, M.M., Greer, S., Tramontano, A., and Zderic, S.A. *J. Am. Chem. Soc.* **101**, 2352–2355 (1979).
43. Midland, M.M., McDowell, D.C., Hatch, R.L., and Tramontano, A. *J. Am. Chem. Soc.* **102**, 867–869 (1980).
44. Midland, M.M. and Graham, R.S. *Org. Synth.* **63**, 57–65 (1984).
45. Midland, M.M., Halterman, R.L., Brown, C.A., and Yamaichi, A. *Tetrahedron Lett.* **22**, 4171–4172 (1981).
46. Midland, M.M., Tramontano, A., Kazubski, A., Graham, R.S., Tsai, D.J.S., and Cardin, D.B. *Tetrahedron* **40**, 1371–1380 (1984).
47. Brown, H.C. and Pai, G.G. *J. Org. Chem.* **50**, 1384–1394 (1985).
48. Midland, M.M., McLoughlin, J.I., and Gabriel, J. *J. Org. Chem.* **54**, 159–165 (1989).
49. Brown, H.C., Chandrasekharan, J., and Ramachandran P.V. *J. Am. Chem. Soc.* **110**, 1539–1546 (1988).
50. Imai, T., Tamura, T., Yamamuro, A., Sato, T., Wollmann, T.A., Kennedy, R.M., and Masamune, S. *J. Am. Chem. Soc.* **108**, 7202–7204 (1986).
51. Masamune, S., Kennedy, R.M., Petersen, J.S., Houk, K.N., and Wu, Y.-D. *J. Am. Chem. Soc.* **108**, 7404–7405 (1986).
52. Itsuno, S., Ito, K., Hirao, A., and Nakahama, S. *J. Chem. Soc., Chem. Commun.* 469–470 (1983).
53. Itsuno, S., Ito, K., Hirao, A., and Nakahama, S. *J. Org. Chem.* **49**, 555–557 (1984).
54. Itsuno, S., Nakano, M., Miyazaki, K., Masuda, H., Ito, K., Hirao, A., and Nakahama, S. *J. Chem. Soc., Perkin Trans. 1*, 2039–2044 (1985).
55. Itsuno, S., Nakano, M., Ito, K., Hirao, A., Owa, M., Kanda, N., and Nakahama, S. *J. Chem. Soc., Perkin I*, 2615–2619 (1985).
56. Itsuno, S., Sakurai, Y., Ito, K., Hirao, A., and Nakahama, S. *Bull. Chem. Soc. Jpn* **60**, 395–396 (1987).
57. Corey, E.J., Bakshi, R.K., and Shibata, S. *J. Am. Chem. Soc.* **109**, 5551–5553 (1987).
58. Corey, E.J. and Link, J.O. *Tetrahedron Lett.* **30**, 6275–6278 (1989).
59. Corey, E.J. and Link, J.O. *J. Org. Chem.* **56**, 442–444 (1991).
60. Corey, E.J., Shibata, S., and Bakshi, R.K. *J. Org. Chem.* **63**, 2862–2864 (1988).
61. Mathre, D.J., Jones, T.K., Xavier, L.C., Blacklock, T.J., Reamer, R.A., Mohan, J.J., Turner Jones E.T., Hoogsteen, K., Baum, M.W., and Grobowksi, E.J.J. *J. Org. Chem.* **56**, 751–762 (1991).
62. Corey, E.J., Bakshi, R.K., Shibata, S., Chen, C.-P., and Singh, V.K. *J. Am. Chem. Soc.* **109**, 7925–7926 (1987).

63. Corey, E.J. and Bakshi, R.K. *Tetrahedron Lett.* **31**, 611–614 (1990).
64. Corey, E.J., Cheng, X.-M., Cimprich, K.A., and Sarshar, S. *Tetrahedron Lett.* **32**, 6835–6838 (1991).
65. Corey, E.J. *Pure Appl. Chem.* **63**, 1209–1216 (1990).
66. Corey, E.J., Bakshi, R.K., and Shibata, S. *J. Am. Chem. Soc.* **109**, 5551–5553 (1987).
67. Nevalainen, V. *Tetrahedron: Asymm* **2**, 63–74 (1991).
68. Nevalainen, V. *Tetrahedron: Asymm* **2**, 1133–1155 (1991).
69. Corey, E.J. and Lok, T.-P. *J. Am. Chem. Soc.* **113**, 8966–8967 (1991).
70. Dauben, W.G., Dickel, D.F., Jeger, O., and Prelog, V. *Helv. Chim. Acta.* **36**, 325–336 (1953).
71. Whitesell, J.K., Bhattacharya, A., and Henke, K. *J. Chem. Soc., Chem. Commun.* 988–989 (1982).
72. Whitesell, J.K., Deyo, D., and Bhattacharya, A. *J. Chem. Soc., Chem. Commun.* 802 (1983).
73. Mazaleyrat, J.-P. and Cram, D.J. *J. Am. Chem. Soc.* **103**, 4585–4586 (1981).
74. Soai, K. and Mukaiyama, T. *Chem. Lett.* 491–492 (1978).
75. Mukaiyama, T., Soai, K., Sato, T., Shimizu, H., and Suzuki, K. *J. Am. Chem. Soc.* **101**, 1455–1460 (1979).
76. Soai, K. and Mukaiyama, T. *Bull. Chem. Soc. Jpn* **52**, 3371–3376 (1979).
77. Frankland, E. *Justus Liebigs Ann. Chem.* **79**, 171 (1849).
78. Soai, K. and Niwa, S. *Chem. Rev.* **92**, 833–856 (1992).
79. Takahashi, H., Kawakita, T., Yoshioka, M., Kobayashi, S., and Ohno, M. *Tetrahedron Lett.* **30**, 7095–7098 (1989).
80. Tanaka, K., Ushio, H., and Suzuki, H. *J. Chem. Soc., Chem. Commun.* 1700–1701 (1989).
81. Joshi, N.N., Srebnik, M., and Brown, H.C. *Tetrahedron Lett.* **30**, 5551–5554 (1989).
82. Kitamura, M., Okada, S., Suga, S., and Noyori, R. *J. Am. Chem. Soc.* **111**, 4028–4036 (1989).
83. Corey, E.J. and Hannon, F.J. *Tetrahedron Lett.* **28**, 5233–5236 (1987).
84. Corey, E.J. and Hannon, F.J. *Tetrahedron Lett.* **28**, 5237–5240 (1987).
85. Corey, E.J., Yuen, P.-W., Hannon, F.J., and Wierda, D.A. *J. Org. Chem.* **55**, 784–786 (1990).
86. Duthaler, R.O. and Hafner, A. *Chem. Rev.* **92**, 807–832 (1992).
87. Weber, B. and Seebach, D. *Angew. Chem., Int. Ed. Engl.* **31**, 84–86 (1992).
88. Hoffmann, R.W. *Angew. Chem.* **94**, 569–590 (1982).
89. Denmark, S.E. and Weber, E.J. *Helv. Chim. Acta* **66**, 1655–1660 (1983).
90. Brown, H.C., Bhat, K.S., and Randad, R.S. *J. Org. Chem.* **54**, 1570–1576 (1989).
91. Roush, W.R., Hoong, L.K., Palmer, M.A.J., and Park, J.C. *J. Org. Chem.* **55**, 4109–4117 (1990).
92. Roush, W.R., Hoong, L.K., Palmer, M.A.J., Straub, J.A., and Palkowitz, A.D. *J. Org. Chem.* **55**, 4117–4126 (1990).
93. Hafner, A., Duthaler, R.O., Marti, R., Rihs, G., Rothe-Streit, P., and Schwarzenbach, F. *J. Am. Chem. Soc.* **114**, 2321–2336 (1992).
94. Oertle, K., Beyeler, H., Duthaler, R.O., Lottenbach, W., Diediker, M. and Steiner, E. *Helv. Chim. Acta* **73**, 353–358 (1990).
95. Riediger, M., Duthaler, R.O. *Angew. Chem., Int. Ed. Engl.* **29**, 494–495 (1989).
96. Reetz, M.T. *Pure Appl. Chem.* **60**, 1607–1614 (1988).
97. Short, R.P. and Masamune, S. *J. Am. Chem. Soc.* **111**, 1892–1984 (1989).
98. Hoffmann, R.W. and Herold, T. *Chem. Ber.* **114**, 375–383 (1981).
99. Brown, H.C. and Jadhav, P.K. *J. Am. Chem. Soc.* **105**, 2092–2093 (1983).
100. Corey, E.J., Yu, C.-M., and Kim, S.S. *J. Am. Chem. Soc.* **111**, 5495–5496 (1989).
101. Corey, E.J. and Sneen, R.A. *J. Am. Chem. Soc.* **78**, 6269–6278 (1956).
102. Zimmerman, H.E. *Acc. Chem. Res.* **20**, 263–268 (1987).
103. Enders, D. In *Asymmetric Synthesis* Vol. 3 (Morrison, J.D., Ed.), Academic Press: New York, Vol. 3., 275–339 (1984).
104. Hajos, Z.G. and Parrish, D.R. *J. Org. Chem.* **39**, 1615–1621 (1974).
105. Eder, U., Sauer, G., and Wiechert, R. *Angew. Chem.* **83**, 492–493 (1971).
106. Evans, D.A., Nelson, J.V., and Taber, T.R. In *Top. Stereochem.* (Eliel, E.L. and Wilen, S.H., Eds.) Wiley Interscience: New York, 1983, Vol. **13**, pp. 1–116.
107. Heathcock, C.H. In *Comprehensive Carbanion Chemistry* (Buncel, E. and Durst, T., Eds.) Elsevier: New York, 1984, vol. 5B, p. 177.

108. Heathcock, C.H. In *Asymm. Synth.* Vol. **3** (Morrison, J.D., Ed.) Academic Press: New York, 213–274 (1984).
109. Mukaiyama, T. *Org. React.* **28**, 203–331 (1982).
110. Heathcock, C.H. *Science* **214**, 395–400 (1981).
111. Zimmerman, H.E. and Traxler, M.D. *J. Am. Chem. Soc.* **79**, 1920–1923 (1957).
112. Seebach, D. *Angew. Chem.* **100**, 1685–1715 (1988).
113. Nevalainen, V. *Tetrahedron: Asymm.* **2**, 63–74 (1991).
114. Amstutz, R., Schweizer, W.B., Seebach, D., and Dunitz, J.E. *Helv. Chim. Acta* **64**, 2617–2621 (1981).
115. Willard, P.G. and Carpenter, G.B. *J. Am. Chem. Soc.* **108**, 462–468 (1986).
116. Willard, P.G. and Carpenter, G.B. *J. Am. Chem. Soc.* **107**, 3345–3346 (1985).
117. House, H.O., Prabhu, A.V., and Phillips, W.V. *J. Org. Chem.* **41**, 1209–1214 (1976).
118. Yamamoto, Y. and Maruyama, K. *Tetrahedron Lett.* **21**, 4607–4610 (1980).
119. Li, Y., Paddon-Row, M.N., and Houk, K.N. *J. Am. Chem. Soc.* **110**, 3684–3686 (1988).
120. Li, Y., Paddon-Row, M.N., and Houk, K.N. *J. Org. Chem.* **55**, 481–493 (1990).
121. Denmark, S.E. and Henke, B.D. *J. Am. Chem. Soc.* **113**, 2177–2194 (1991).
122. Evans, D.A., Bartroli, J., and Shih, T.L. *J. Am. Chem. Soc.* **103**, 2127–2129 (1981).
123. Evans, D.A., Ennis, M.D., and Mathre, D.J. *J. Am. Chem. Soc.* **104**, 1737–1739 (1982).
124. Nerz-Stormes, M. and Thornton, E.R. *J. Org. Chem.* **56**, 2489–2498 (1991).
125. Danda, H., Hansen, M.M., and Heathcock, C.H. *J. Org. Chem.* **55**, 173–181 (1990).
126. Walker, M.A., Heathcock, C.H. *J. Org. Chem.* **56**, 5747–5750 (1991).
127. Hayashi, K., Hamada, Y., and Shioiri, T. *Tetrahedron Lett.* **32**, 7287–7290 (1991).
128. Van Draanen, N.A., Arseniyadin, S., Crimmins, M.T., and Heathcock, C.H. *J. Org. Chem.* **56**, 2499–2506 (1991).
129. Heathcock, C.H. *Aldrichimica Acta* **23**, 99–111 (1990).
130. Corey, E.J. and Kim, S.S. *J. Am. Chem. Soc.* **112**, 4976–4977 (1990).
131. Furuta, K., Maruyama, T., and Yamamoto, H. *J. Am. Chem. Soc.* **113**, 1041–1042 (1991).
132. Oppolzer, W. Mills, R.J., Pachinger, W., and Stevenson, T. *Helv. Chim. Acta* **69**, 1542–1545 (1986).
133. Oppolzer, W., Poli, G., Kingma, A.J., Starkemann, C., and Bernardinelli, G. *Helv. Chim. Acta* **70**, 2201–2214 (1987).
134. Oppolzer, W., Kingma, A.J., and Poli, G. *Tetrahedron* **45**, 479–488 (1989).
135. Oppolzer, W. and Kingma, A.J. *Helv. Chim. Acta* **72**, 1337–1345 (1989).
136. Oppolzer, W. *Angew. Chem.* **96**, 840–854 (1984).
137. Oppolzer, W., Dudfield, P., Stevenson, T., and Godel, T. *Helv. Chim. Acta* **68**, 212–215 (1985).
138. Oppolzer, W. *Tetrahedron* **43**, 1969–2004 (1987).
139. Oppolzer, W., Moretti, R., Godel, T., Meunier, A., and Löher, H. *Tetrahedron Lett.* **24**, 4971–4974 (1983).
140. Asami, M. and Mukaiyama, T. *Chem. Lett.* 569–572 (1979).
141. Scolastico, C. *Pure Appl. Chem.* **60**, 1689–1698 (1988).
142. Bernardi, A., Cardani, S., Pilati, T., Poli, G., Scolastico, C., and Villa, R. *J. Org. Chem.* **53**, 1600–1607 (1988).
143. Ho, T.-L. *Polarity Control for Synthesis* John Wiley & Sons: New York, 1991.
144. Winterfeldt, E. *Prinzipien und Methoden der Stereoselektive Synthese* Vieweg & Sohn: Braunschweig, 1988.
145. Henbest H.B. and Wilson, R.A. *J. Chem. Soc.* 1958–1965 (1957).
146. Hasan, I. and Kishi, Y. *Tetrahedron Lett.* **21**, 4229–4232 (1980).
147. Corey, E.J. *J. Org. Chem.* **55**, 1693–1694 (1990).
148. Woodard, S.S., Finn, M.G., and Sharpless, K.B. *J. Am. Chem. Soc.* **113**, 106–113 (1991).
149. Finn, M.G. and Sharpless, K.B. *J. Am. Chem. Soc.* **113**, 113–126 (1991).
150. Jorgensen, K.A. *Tetrahedron: Asymm.* **2**, 515–532 (1991).
151. Finan, J.M. and Kishi, Y. *Tetrahedron Lett.* **23**, 2719–2722 (1982).
152. Ma, P., Martins, S.V., Masamune, S., Sharpless, K.B., and Viti, S.M. *J. Org. Chem.* **47**, 1378–1380 (1982).

153. Jacobsen, E.N., Zhang, W., Muci, A.R., Exker, J.R., and Deng, L. *J. Am. Chem. Soc.* **113**, 7063–7064 (1991).
154. Cha, J.K., Christ, W.J., and Kishi, Y. *Tetrahedron* **40**, 2247–2255 (1974).
155. Tokles, M. and Snyder, J.K. *Tetrahedron Lett.* **27**, 3951–3954 (1986).
156. Tomioka, K., Nakajima, M., and Koga, K. *J. Am. Chem. Soc.* **109**, 6213–6215 (1987).
157. Tomioka, K., Nakajima, M., Iitaka, Y., and Koga, K. *Tetrahedron Lett.* **29**, 573–576 (1988).
158. Corey, E.J., DaSilva Jardine, P., Virgil,S., Yuen, P.-W., and Connell, R.D. *J. Am. Chem. Soc.* **111**, 9243–9244 (1989).
159. Oishi, T. and Hirama, M. *J. Org. Chem.* **54**, 5834–5835 (1989).
160. Hirama, M., Oishi, T., and Ito, S. *J. Chem. Soc., Chem. Commun.* 665–666 (1989).
161. Jacobsen, E.N., Marko, I., Mungall, W.S., Schröder, G., and Sharpless, K.B. *J. Am. Chem. Soc.* **110**, 1968–1670 (1988).
162. Wai, J.S.M., Marko, I., Svendsen, J.S., Finn, M.G., Jacobsen, E.N., and Sharpless, K.B. *J. Am. Chem. Soc.* **111**, 1123–1125 (1989).
163. Sharpless, K.B., Amberg, W., Bennani, Y.L., Crispino, G.A., Hartung, J., Jeong, K.-S., Kwong, H.-L., Morikawa, K., Wang, Z.-M., Xu, D., and Zhang, X.-L. *J. Org. Chem.* **57**, 2768–2771 (1992).
164. Corey, E.J. and Lotto, G.I. *Tetrahedron Lett.* **31**, 2665–2668 (1990).
165. Wu, Y.-D., Wang, Y., and Houk, K.N. *J. Org. Chem.* **57**, 1362–1369 (1992).
166. Oishi, T. and Nakata, T. *Synthesis* 635–645 (1990).
167. Saksena, A.K. and Mangiaracina, P. *Tetrahedron Lett.* **24**, 273–276 (1983).
168. Gribble, G.W. and Ferguson, D.C. *J. Chem. Soc., Chem. Commun.* 535–536 (1975).
169. Gribble, G.W. and Nutaitis, C.F. *Org. Prep. Proced. Int.* **17**, 317–384 (1975).
170. Turnbull, M.D., Hatter, G., and Ledgerwood, D.E. *Tetrahedron Lett.* **25**, 5449–5452 (1984).
171. Evans, D.A., Chapman, K.T., and Carreira, E.M. *J. Am. Chem. Soc.* **110**, 3560–3578 (1988).
172. Chen, K.-M., Hardtmann, G.E., Prasad, K., Repic, O., and Shapiro, M.J. *Tetrahedron Lett.* **28**, 155–158 (1987).
173. Mori, Y., Kuhara, M., Takeuchi, A., and Suzuki, M. *Tetrahedron Lett.* **29**, 5419–5422 (1988).
174. Evans, D.A., Gauchet-Prunet, J.A., Carreira, E.M., and Charette, A.B. *J. Org. Chem.* **56**, 741–750 (1991).
175. Still, W.C. and Barrish, J.C. *J. Am. Chem. Soc.* **105**, 2487–2489 (1983).
176. Männig, D. and Nöth, H. *Angew. Chem., Int. Ed. Engl.* **24**, 878–879 (1985).
177. Evans, D.A., Fu, G.C., and Hoveyda, A.H. *J. Am. Chem. Soc.* **110**, 6917–6918 (1988).
178. Burgess, K. and Ohlmeyer, M.J. *J. Org. Chem.* **53**, 5178–5179 (1988).
179. Burgess, K., van der Donk, W.A., and Ohlmeyer, M.J. *Tetrahedron:Asymmetry* **2**, 613–621 (1991).
180. Evans, D.A. and Fu, G.C. *J. Org. Chem.* **55**, 2280–2282 (1990).
181. Burgess, K., van der Donk, W.A., and Kook, A.M. *J. Org. Chem.* **56**, 2949–2951 (1991).
182. Evans, D.A., Fu, G.C., and Hoveyda, A.H. *J. Am. Chem. Soc.* **114**, 6671–6679 (1992).
183. Evans, D.A., Fu, G.C., and Anderson, B.A. *J. Am. Chem. Soc.* **114**, 6679–6685 (1992).
184. Schreiber, S.L. *Chem. Scr.* **27**, 563–566 (1987).
185. Ikemoto, N. and Schreiber, S.L. *J. Am. Chem. Soc.* **114**, 2524–2536 (1992).
186. Narasaka, K. *Synthesis* 1–11 (1991).
187. Tomioka, K. *Synthesis* 541–549 (1990).
188. Guseinov, M.M., Akhmedov, I.M., and Mamedov, E.G. *Azerb. Khim. Zh.* 46 1976; *Chem. Abstr.* **85**, 176 925 1976.
189. Oppolzer, W., Rodriguez, I., Blagg, J., and Bernardinelli, G. *Helv. Chim. Acta* **72**, 123–130 (1989).
190. Bednarski, M. and Danishefsky, S. *J. Am. Chem. Soc.* **105**, 6968–6969 (1983).
191. Chapuis, C. and Jurczalk, J. *Helv. Chim. Acta* **70**, 436–440 (1987).
192. Narasaka, K., Iwasawa, N., Inoue, M., Yamada, T., Nakashima, M., and Sugimori, J. *J. Am. Chem. Soc.* **111**, 5340–5345 (1989).
193. Corey, E.J., Imai, N., and Zhang, H.-Y. *J. Am. Chem. Soc.* **113**, 728–729 (1991).
194. Gao, Q., Maruyama, T., Mouri, M., and Yamamoto, H. *J. Org. Chem.* **57**, 1951–1952 (1992).

4 Carbohydrates

In this chapter, we shall inspect the occurrence of the various forms of carbohydrates in Nature. The utilization of sugars as sources of and storage vehicles for energy would also belong to the subjects covered in this chapter, but since they form the basis of biochemistry (and, to some extent, organic chemistry) they will not be covered in detail. However, it is reasonable to remind ourselves of some facts concerning the structures and nomenclature of carbohydrates.

Carbohydrates are formed as a result of the photosynthetic function of plants, algae and bacteria. These organisms can utilize atmospheric carbon dioxide, which by the action of photosynthetic enzymes is converted to a chemically useful form, i.e. carbohydrates. The green leafed plants and the blue-green algae of the oceans are typical examples of efficient light harvesting systems.

In the organisms, glucose is broken down through the action of enzymes. This phenomenon is called glycolysis. The initial product from glucose is pyruvic acid, which can be further broken down into acetic acid (acetyl coenzyme A, see Section 7.1). Simultaneously, a large amount of energy (62 kJmol^{-1}) is liberated, and this energy is tied in chemical form to adenosine triphosphate (ATP). The breakdown of carbohydrates is further continued in the citric acid (Krebs) cycle, where glucose is oxidized to carbon dioxide. The combined action of glycolysis and the citric acid cycle can recover 1160 kJmol^{-1} of the energy bound in glucose, which corresponds to *ca* 40 % efficiency (the heat of formation of glucose is 2870 kJmol^{-1}).

4.1 Monosaccharides

The names and structures of the common monosaccharides of three (trioses), four (tetroses), five (pentoses) and six (hexoses) carbon atoms are shown in their Fischer projection forms in Figure 4.1. The natural carbohydrates all belong to the D series, i.e. they are derived from D-glyceraldehyde.

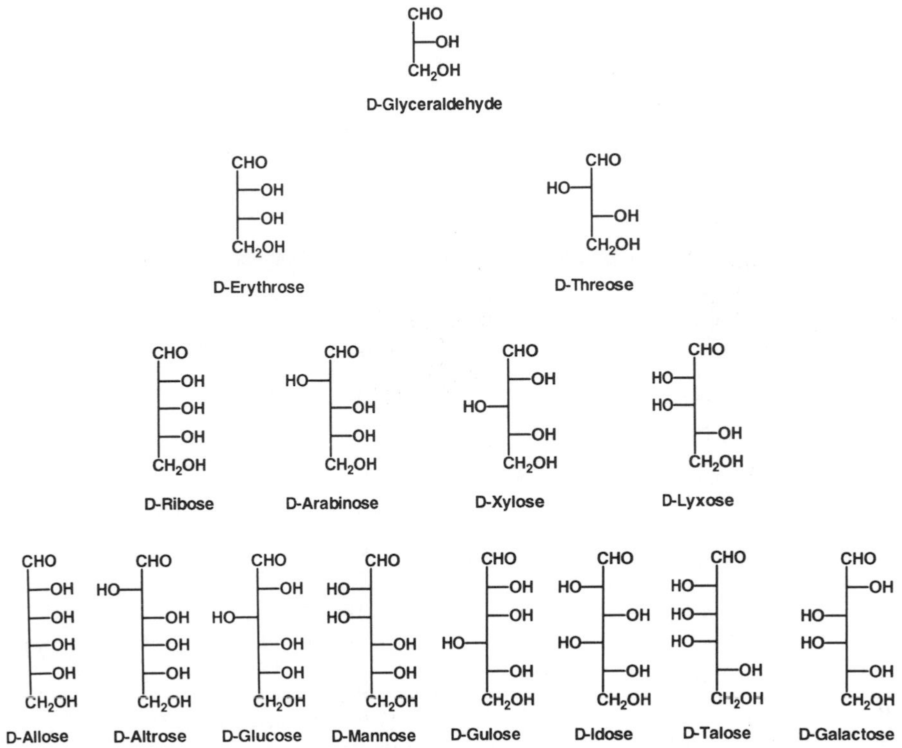

Figure 4.1

Carbohydrates with five or more carbon atoms seldom occur in the open chain form, since they cyclize to the cyclic hemiacetal. The five membered cyclic forms are called furanoses and the six membered ones pyranoses. The cyclic forms of ribose (a furanose) and glucose (a pyranose) are shown in Figure 4.2.

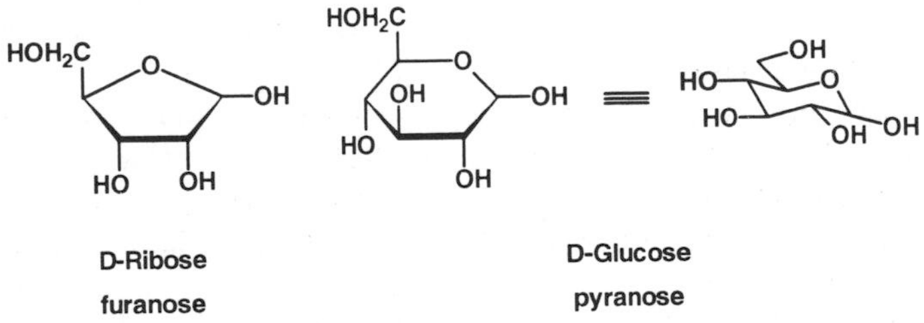

Figure 4.2

The structural chemistry of carbohydrates is very varied. Glucose itself can be dimerized in 25 different ways; similarly, a glucose trimer can be formed in 176 different ways (not including anomeric isomers). The more complex carbohydrates are usually represented in a shorthand form; for instance, fructofuranose is Fruf, glycopyranose is Glcp and galactopyranose is Galp. The common carbohydrates are listed in Table 4.1.

Table 4.1 The common carbohydrates

Name	Structure
Monosaccharides	
D-Glyceraldehyde	D-Glycero-triose
D-Erythrose	D-Erythro-tetrose
D-Threose	D-Threo-tetrose
D-Arabinose	D-Arabino-pentose
D-Lyxose	D-Lyxo-pentose
D-Ribose	D-Ribo-pentose
D-Xylose	D-Xylo-pentose
D-Allose	D-Allo-hexose
D-Altrose	D-Altro-hexose
D-Galactose	D-Galacto-hexose
D-Glucose	D-Gluco-hexose
D-Gulose	D-Gulo-hexose
D-Idose	D-Ido-hexose
D-Mannose	D-Manno-hexose
D-Talose	D-Talo-hexose
Disaccharides	
Cellobiose	β-D-Glcp-(1-4)-D-Glc
Lactose	β-D-Galp-(1-4)-D-Glc
Maltose	α-D-Glcp-(1-4)-D-Glc
Saccharose	α-D-Glcp-(1-2)-β-D-Fruf
Trisaccharides	
Kestose	β-D-Fruf-(2-6)-β-D-Fruf-(2-1)-α-D-Glcp
Raffinose	α-D-Galp-(1-6)-α-D-Glcp-(1-2)-β-D-Fruf

The hexopyranoses are shown in the conventional chair representation in Figure 4.3. In all cases, the anomeric hydroxyl group at C-1 is drawn out with a wavy line to indicate that both α- and β-anomers are possible. Of the remaining substituents in the tetrahydropyran ring, glucose has all its substituents equatorial, and is thus thermodynamically the most stable one. Mannose, allose and galactose each have one axial substituent, and altrose, talose and gulose have two axial substituents. Idose has all three remaining hydroxyl substituents in axial positions. Note that epimerization of the C-5 hydroxymethyl appendage leads to the enantiomeric series; for example, isomerization of glucose leads to the antipode of idose.

All the sugars we have discussed so far contain an aldehyde function. Such carbohydrates are known as aldoses, as distinct from ketoses which contain a ketonic carbonyl group. The aldoses react with ammoniacal silver solutions to form a silver mirror and the corresponding carboxylic acid, or aldonic acid. Because of this reactivity, the terminal aldoses are also called reducing sugars.

In carbohydrates, the formation of the six membered ring in the aldohexose series leads to the generation of a new stereogenic centre. The two stereoisomeric forms are

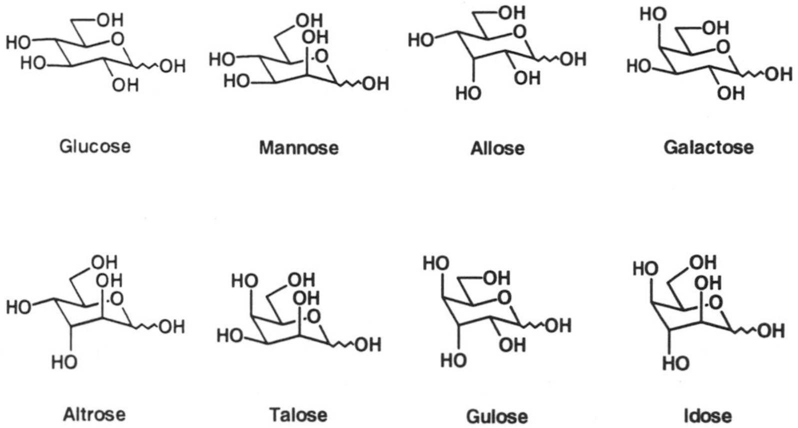

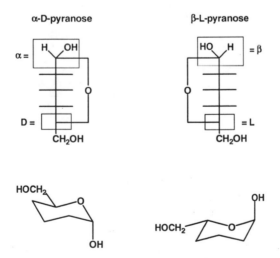

Figure 4.3

known as anomers, and they are conventionally distinguished as the α- and β-anomers. The early definition of the nomenclature was based on optical rotation: the α-anomer was the one (in the D series) which showed higher optical rotation. This was based initially on the optical rotatory properties of the crystalline anomers of glucose. A more rigorous definition is based on structures: in the D series, the β-form is the one having the hemiacetal hydroxyl group on the left in the Fischer projection. In modern terminology, we designate the α-anomer as that having an axial hydroxyl group and the β-anomer as that having an equatorial hydroxyl group (Figure 4.4).

Figure 4.4

The anomeric effect is a phenomenon typical to carbohydrate chemistry (Figure 4.5). The anomeric effect relates to the propensity of the C-1 substituent in a pyranose ring to adopt an axial position, even though the steric effects are not favourable. The phenomenon is not restricted to pyranoses and is common to all cyclic acetals.

Anomeric effect

Figure 4.5

The anomeric effect can be viewed to have its origins either in the stabilization or destabilization of orbital interactions. The axial C—OR bond lies in the same plane as the axial lone pair on the ring ether oxygen. Delocalization of electron density from the p-orbital is possible, stabilizing this arrangement. In the corresponding equatorial C—OR anomer such a hyperconjugative effect is not possible. The same phenomenon can also be explained by dipolar interactions: in a β-glycoside (equatorial C—OR) the dipoles point in the same direction (not favoured), whereas in the α-glycoside (axial C—OR) the dipoles tend to cancel each other (favoured).

In terms of frontier molecular orbital theory, the anomeric effect is explained by the stabilization provided by overlap of a lone pair on the ring oxygen atom with the antibonding σ*-orbital of the exocyclic C—O bond.[1] When the C—O bond is axial, the σ*-orbital is favourably oriented to overlap with the axial p-type lone pair on oxygen. Since the antibonding orbital for a C—O bond is lower in energy than the corresponding antibonding orbital for a C—H bond, the overlap with the lone pair is more effective for the more electronegative substituent. The (n_O–σ^*_{COR}) interaction is bonding between carbon and oxygen, which strengthens and shortens the ring C—O bond. Charge transfer to the σ^*_{COR}-orbital also weakens and lengthens the axial C—OR bond (Figure 4.6).

σ* orbital for C-OR bond:

Figure 4.6

Another term, *exo*-anomeric effect, refers to the favoured conformation of the alkoxy substituent O—R. The *exo*-anomeric effect has a similar origin to the anomeric effect; however, the *exo*-anomeric relates to the preferred orientation of the (axial, or α) O—R bond. Three possible staggered conformations can be considered, **A**–**C** in Figure 4.7.

In conformations **A** and **C**, one of the lone pairs of the exocyclic oxygen is disposed in an antiperiplanar fashion to the ring C—O bond, thus making possible a similar interaction as discussed above for the anomeric effect. Conformation **B** lacks such a possibility, and is thus disfavoured against the other two alternatives. Distinction between conformations **A** and **C** can be seen to arise from unfavourable interactions of the R group with the axial hydrogens (axial 1,3-strain) in the case of **C**. Conformation **A** is thus the most favoured. However, the energy differences between the various conformations are often rather small, and equilibration of the anomers is thus a facile process.

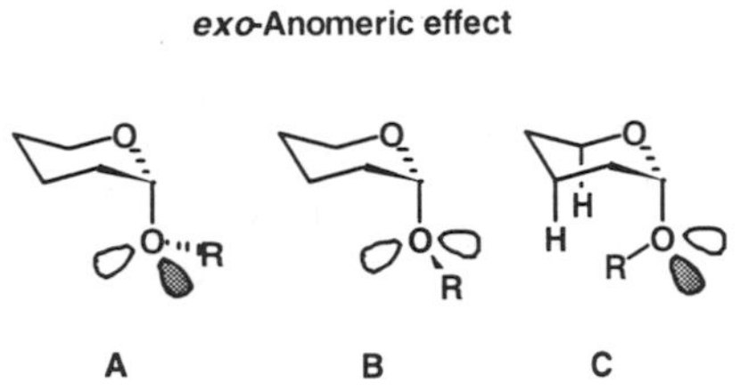

exo-**Anomeric effect**

A B C

Figure 4.7

4.1.1 Acidic Sugars

Acidic sugars fall into three main types, depending on which carbon atom has been oxidized: in aldonic acids, the aldehydic carbonyl group is oxidized to carboxylic acid; in uronic acids the terminal hydroxymethyl is oxidized to the carboxylic acid; and in saccharic acids both groups are oxidized to the carboxylic acid.

By far the most important group of acidic sugars is the uronic acids, which commonly occur as the hexuronic acids. These are intermediates in the biosynthesis of pentoses from hexoses. Many of these occur in the gums of plants, are used as building blocks for the formation of bacterial cell walls, and are incorporated into proteoglycans (copolymers of proteins and carbohydrates). Vitamin C (ascorbic acid) is a member of this structural class (Figure 4.8).

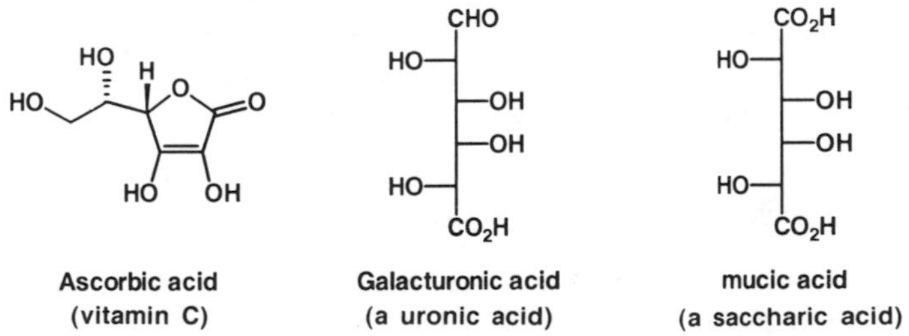

Ascorbic acid   Galacturonic acid   mucic acid
(vitamin C)   (a uronic acid)   (a saccharic acid)

Figure 4.8

4.1.2 Sugar Alcohols

There are two main types of sugar alcohols. Alditols are formed through the reduction of the aldehyde function of a carbohydrate. The sweetener xylitol (wood sugar alcohol) is perhaps the best known example of these (Figure 4.9). D-Mannitol is present in seaweed, and D-glycerol and D-ribitol occur in the polysaccharides of microbes. The other group of sugar alcohols is that of the cyclic derivatives, the cyclitols (Figure 4.10). Of the deoxy sugars, deoxyribose is a constituent of deoxyribonucleic acid (DNA). L-Fucose (6-deoxy-L-galactose) and L-rhamnose (6-deoxy-L-mannose) are important components of some bacterial cell walls. Recent information on protein glycosylation also indicates an important role for fucose in the signal transduction processes in mammalian cells. The common deoxy sugars also include carbacyclic cyclohexane derivatives, whose

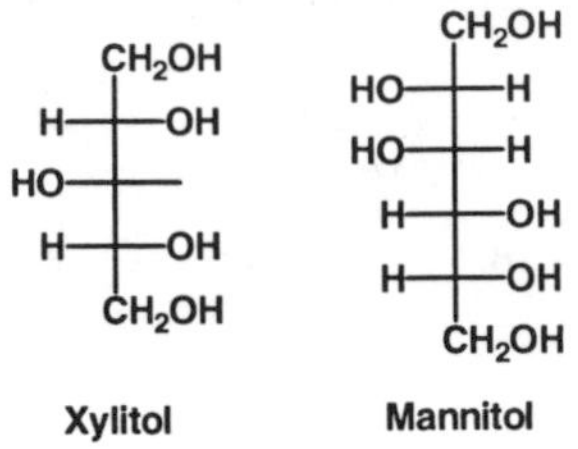

Figure 4.9

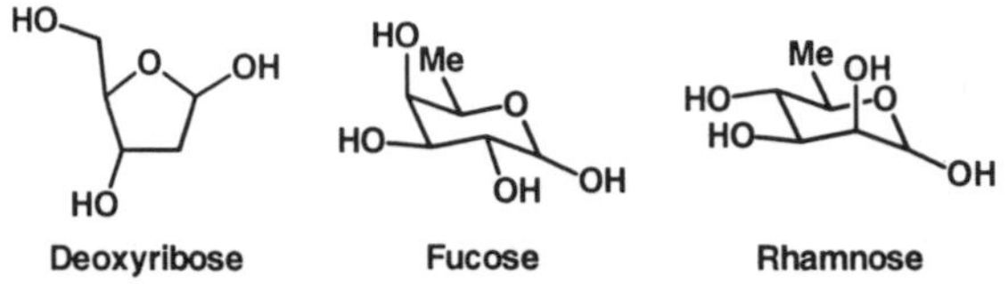

Figure 4.10

1,2,3,4,5,6-cyclohexanehexols are known as inositols (From Greek *inos* = fibre, muscle). There are altogether nine stereoisomers for inositol, of which only two are optically active. *myo*-Inositol is the commonest inositol, and it is present in practically all animal and plant species. In animals and micro-organisms, *myo*-inositol is usually a part of the phospholipids, whereas in plants it usually occurs as its hexaphosphate, phytic acid. Of the other inositols, only D-*chiro*- (D-pinitol), L-*chiro*-(L-quebrachitol) and *scyllo*-inositol occur naturally. We shall return to the cyclitols in Section 4.6.

4.1.3 Amino Sugars

Nearly a century ago, several antibiotics contained a modified carbohydrate unit which contained an amino group. Glucosamine (2-amino-2-deoxy-D-glucose) was known to be a constituent of lobster shells, and galactosamine (2-amino-2-deoxy-D-galactose) a component of cartilage (Figure 4.11). Extensive studies in the antibiotic field started to reveal the multitude of amino sugars present in nature. The most common amino sugar, glucosamine, occurs in its *N*-acetylated form in polysaccharides, glycoproteins and proteoglycans as well as chitin. Only a few amino sugars occur in their free forms, as they are usually components of complex antibiotics and oligo- and polysaccharides.

Many rare amino sugars, such as 3-amino, 4-amino and diamino sugars, are present in a number of antibiotics. Examples (Figure 4.12) are kanosamine (3-amino-3-deoxyglucose, part of kanamycins), sibirosamine (from the potent antitumor antibiotic sibiromycin), and perosamine (from the antifungal heptaenic antibiotic perimycin).

Some of the higher monosaccharides occur as their amino derivatives. 2-Acetamido-3-*O*-(1-carboxyethyl)-2-deoxy-D-glucose (2-acetamido-2-deoxymuraminic acid) is a component of bacterial cell walls; 5-acetamido-3,5-dideoxy-D-glycero-D-galacto-2-nonulopyranic acid (*N*-acetylneuraminic acid) and the corresponding *N*-glycolylneuraminic acid are found in many glycoproteins.

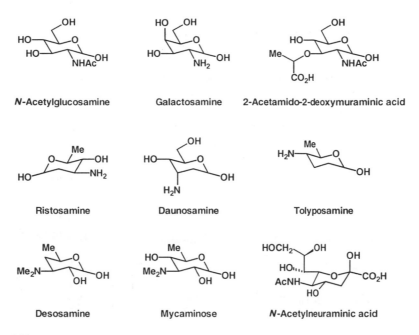

Figure 4.11

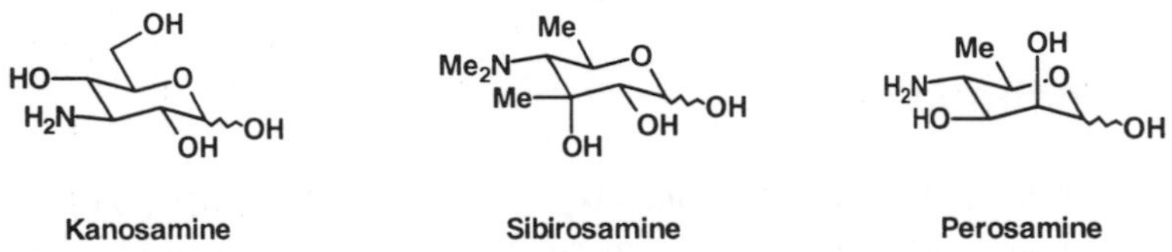

Figure 4.12

2-Amino-2-deoxy sugar derivatives can be synthesized from glycals through [4+2] cycloaddition with azodicarboxylate.[2] Photocatalysed cycloaddition of azodicarboxylate with the glycal gives the cycloadduct which can be decomposed to the 2-amino-2-deoxy sugar by treatment with acidic methanol followed by reduction of the hydrazine (Raney nickel) and acetylation (Scheme 4.1). Silicon protecting groups are essential, as acetylated glycals give poor yields and greatly eroded diastereoselectivity in the cycloaddition step.

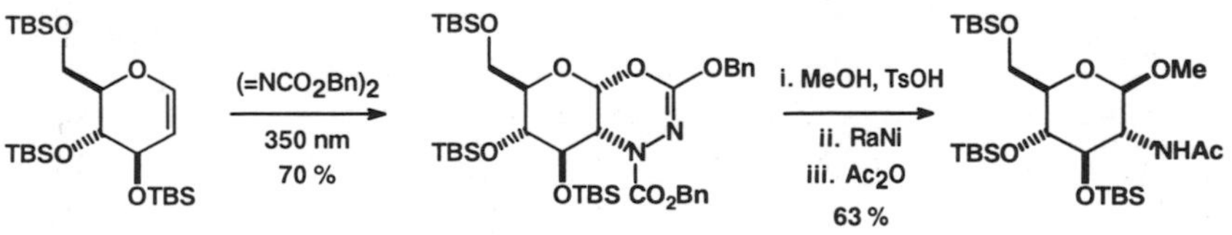

Scheme 4.1

Glycosylated 2-amino-2-deoxy sugars can also be synthesized directly from glucals using a modification of the oxidation technique developed by Danishefsky for oligosaccharide synthesis from glucals (see Section 4.2).[3,4] Treatment of the glucal with iodonium di-*sym*-collidine perchlorate and benzenesulfonamide gives the *trans*–diaxial iodosulfonamide which, upon treatment with a carbohydrate and a base, gives the 2-sulfonamido-β-glycoside via the aziridine (Scheme 4.2).[5]

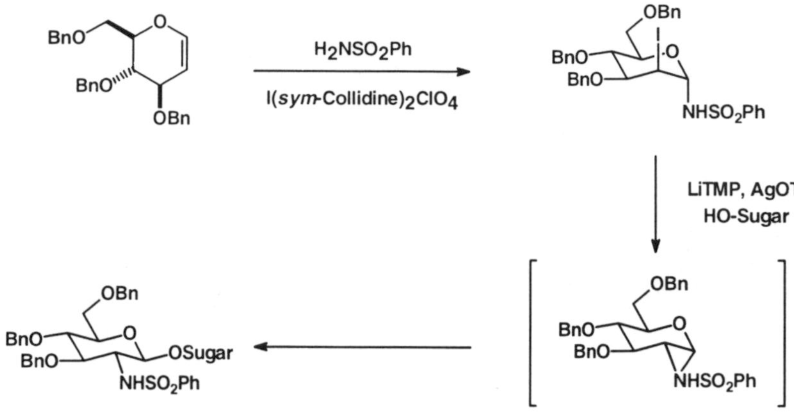

Scheme 4.2

4.2 Polysaccharides

Carbohydrates provide fascinatingly complex starting materials for biopolymers. If one considers the possibilities for structural isomers, the existence of several sites for coupling can give rise to a much larger number of polysaccharide isomers than is available for either polypeptides or polynucleic acids, as shown in Table 4.2.[6]

Table 4.2 Isomers of biopolymers

Product	Structure	Number of isomers	
		Peptides, nucleic acids	Saccharides
Monomer	Z	1	1
Dimer	Z_2	1	11
Trimer	Z_3	1	120
Tetramer	Z_4	1	1424
Pentamer	Z_5	1	17 872
Monomer	Z	1	1
Dimer	YZ	2	20
Trimer	XYZ	6	720
Tetramer	WXYZ	24	34 560
Pentamer	VWXYZ	120	2 144 640

The structural versatility is an added bonus if one considers the possibility for information storage, and carbohydrate units connected to proteins definitely play a central role in cellular recognition processes. Simultaneously, the biosynthesis of polysaccharides is much more complicated than that of polypeptides or polynucleotides—typically the enzymatic processes used for connecting two carbohydrate units require a highly specific enzyme, and therefore polysaccharides with a defined cellular function are still unknown. However, we shall see in the next section that glycosylation of proteins is much better understood.

Starch, cellulose and pectins are the commonest polysaccharides to be found in plants. Teichoic acid (sugar phosphates) and mureins (branched copolymers of amino sugars and peptides) occur widely in the cell walls of Gram-positive bacteria.

The cell walls of Gram-negative bacteria are structurally very complex. The cell wall peptidoglycan is surrounded by a cytoplasmic membrane, which is further surrounded by an outer membrane-like structure. This is composed of proteins, phospholipids, lipoproteins and lipopolysaccharides (LPS) which are not covalently bound to the peptidoglycan. These LPS are the basis for the species specific immune reactions. The part of LPS which is not responsible for the antigenic activity is called the core polysaccharide. This is bound covalently to the lipid material. The core polysaccharide is structurally much simpler than the antigenic regions, and it typically contains lipid A (KDO is 3-deoxy-D-mannooctopyranulosonic acid; Figure 4.13).

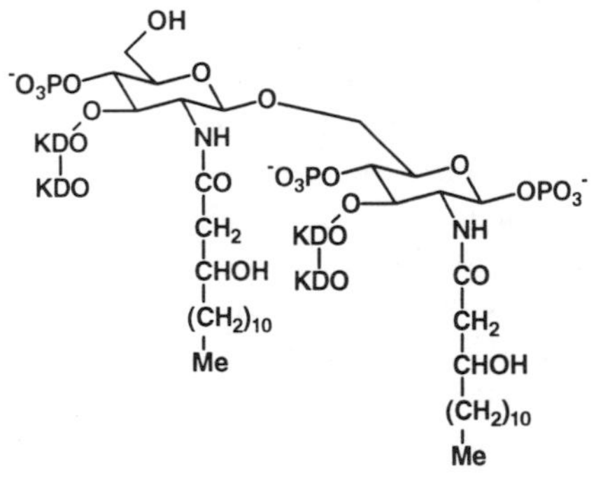

Partial structure of Lipid A

Figure 4.13

In the synthesis of oligo- and polysaccharides, formation of the glycosidic bond is traditionally conducted with a free alcohol (glycosyl donor) by converting the hemiacetal hydroxyl group to a good leaving group with any of a number of activating reagents,[6] and allowing this to react with a carbohydrate component with a free hydroxyl group (glycosyl acceptor) (Scheme 4.3).

Acid activation of the glycosyl donor gives rise to the protonated intermediate which can be converted to the glycosyl halide (the Koenigs–Knorr procedure, route A). This can then be coupled with the alcohol ROH in the presence of, for example, silver salts to give the disaccharide. Alternatively, the protonated intermediate can also be reacted

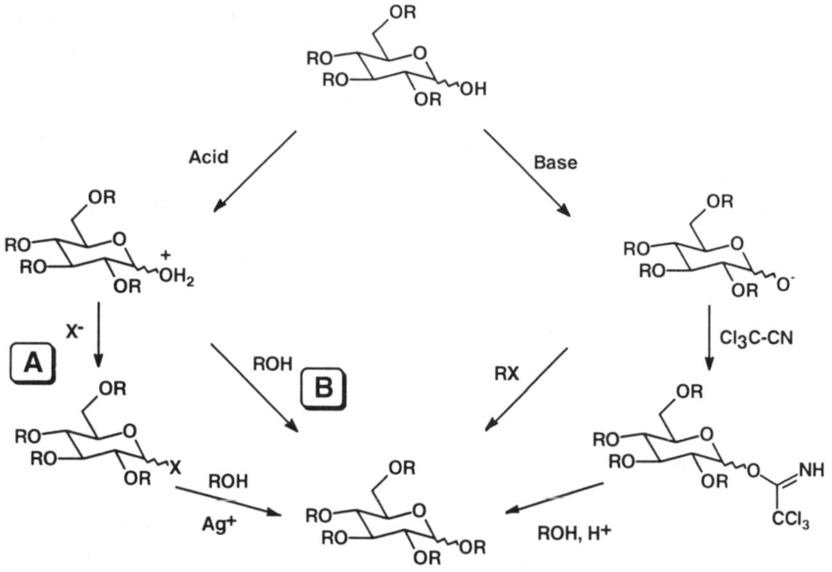

Scheme 4.3

directly with the alcohol ROH (Fischer–Helferich coupling, route B). The latter route is hampered by its reversibility, making it less attractive for polysaccharide synthesis.

Base activation of the glycosyl donor gives an alkoxide, which can participate in ring–chain tautomerism. However, this intermediate can be trapped with trichloroacetonitrile to give the imidate which reacts under acid catalysis with alcohols to give the disaccharide.

The Koenigs–Knorr coupling suffers from several shortcomings: the formation of the halogenose requires rather drastic reaction conditions, the halogenose is thermally unstable and prone to hydrolysis, and the need for heavy metal salts is an obvious problem, especially in large scale work. Several new alternatives have been sought to overcome these problems, including the use of thiol or fluorine activation. A recent efficient method for glycosidation is based on the use of glycosyl fluorides.[7] The fluoro sugars can be conveniently synthesized from the corresponding phenylthio sugars by treatment with NBS and diethylaminosulfur trifluoride (DAST), or directly from the free alcohol by treatment with DAST (Scheme 4.4).[8] Coupling is effected with tin(II) chloride–silver perchlorate activation. The method is mild enough so that most protecting groups and glycosidic linkages survive intact.

Direct oxidative coupling of glycals is also possible, as shown by Danishefsky utilizing one of two powerful alternative. Oxidative coupling can be effected with iodonium perchlorate (Scheme 4.5).[3] Careful choice of the protecting groups directs the bond formation: the acyl protected glycal will donate its free hydroxyl groups for bond formation with the ether protected glycal, but will not react with itself. This coupling strategy produces the diaxial α-linked 2-iodoglycosides. The formation of β-linked glycosides requires a different strategy.

Epoxidation of glycals incorporating non-participating protecting groups with dimethyloxirane leads to highly stereoselective epoxide formation ($\alpha:\beta$ ratio 20:1 with

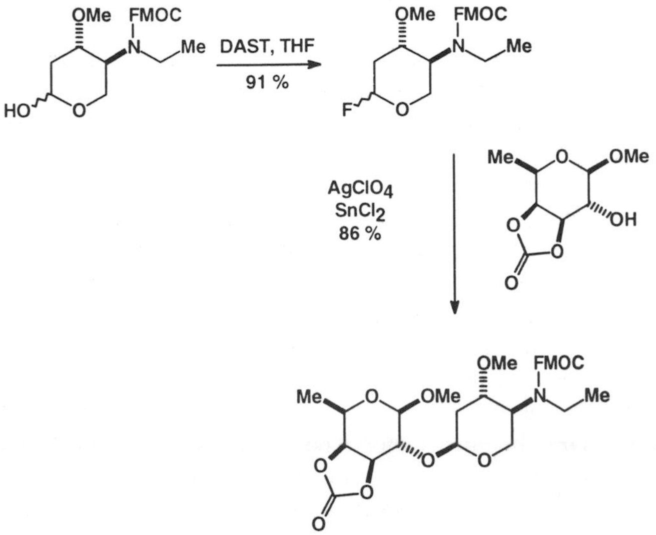

Scheme 4.4

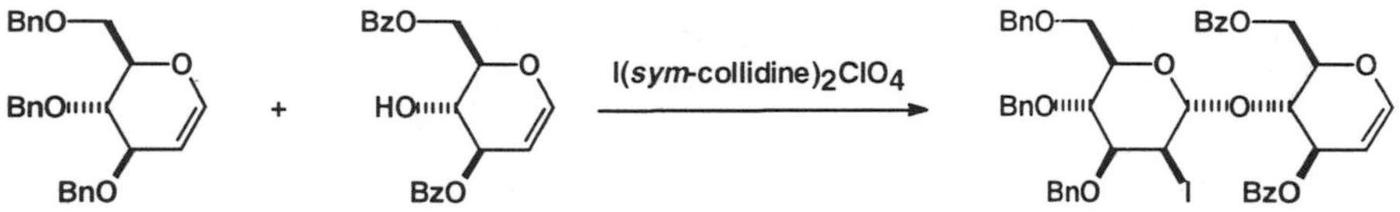

Scheme 4.5

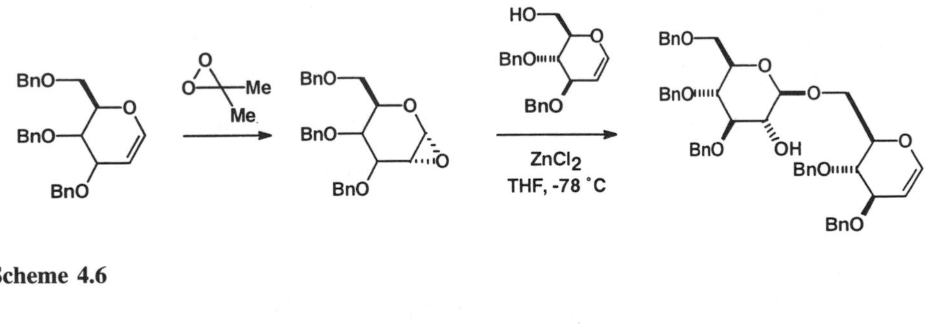

Scheme 4.6

benzyl protection; with *tert*-butyldimethylsilyl (TBS) protection, only α-epoxide was observed). The epoxide then can be coupled with the alcohol in the presence of anhydrous zinc chloride to give the β-glycoside as the sole product (Scheme 4.6). The yields are low (50–58 %) owing to the instability of the epoxide towards the Lewis acid.

4.3 Glycoproteins and Proteoglycans

The term glycoprotein is often used to refer to all the macromolecular complexes of proteins and carbohydrates. Such a broad definition, however, mixes the true glycoproteins with proteoglycans and carbohydrate–protein complexes. In true glycoproteins, the protein chain is connected to a branched polysaccharide; in proteoglycans the oligosaccharide is unbranched.

Proteins perform their physiological function usually only in specific areas of the cell. The newly synthesized protein is transported from the rough endoplasmic reticulum to its site of function, guided by highly selective recognition processes, which at least in eukaryotic cells is dependent on glycoproteins. The carbohydrate part of the glycoprotein plays a key role in this process. On the one hand the carbohydrate moiety can protect the protein from degradation: the copper-transport protein ceruloplasmin has a biological half-life of 54 h, whereas its deglycosylated analogue asialoceruloplasmin has a half-life of less than 5 min. The other major function of the carbohydrate portion is within recognition and control processes such as cell growth and differentiation. Tumour cell membranes have an altered glycoprotein structure, and these conjugates are, in part, tumour associated antigens. Also the blood group antigens are glycoproteins.[9] Some glycoproteins are classified according to their functions in Table 4.3.

Table 4.3 Some glycoproteins with their functions

Suggested function	Glycoprotein
Enzyme	Cholinesterase
	Bromelain
	Ficin
	Ribonuclease
	Yeast invertase
Nutrient storage	Casein
	Ovalbumine
Hormone	Erytropoietine
	FSH (follicle stimulating hormone)
	LH (luteinizing hormone)
	Thyroglobulin
Plasma and serum	α-, β- and γ-glycoproteins
Protective mechanism	Fibrinogen
	Immunoglobulins
	Interferon
Structural proteins	Bacterial cell walls
	Collagen
	Extensin (plant cell wall)
Toxin	Fungal mycotoxins
	Ricin
Transport	Ceruloplasmin
	Haptoglobine
	Transferrin
Unknown	Avidine (egg white)
	Blood group antigens

Proteoglycans are building blocks for connective tissue. Disturbances of proteoglycan metabolism lead to a broad range of diseases known as hyperglycosaminoglycanuria. Typical to these diseases is increased secretion of proteoglycans. Clinical symptoms vary from loss of sight owing to opacity of the cornea to child death.

The glycosidic bonds are typically formed through either asparagine (amide bond formation with glucosamine), serine (β-glucosyl ether) or the serine/threonine α-glycosidic bond with N-acetylgalactosamine. Typical structural types are shown in Figure 4.14. The syntheses of N- and O-glycosyl amino acids have been reviewed.[10]

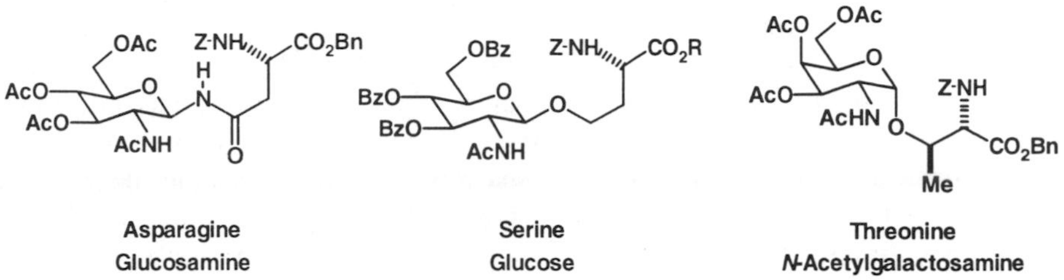

Asparagine
Glucosamine

Serine
Glucose

Threonine
N-Acetylgalactosamine

Figure 4.14

Glycosylation of proteins is a post-translational modification for the already constructed protein chain.[11] The core carbohydrate portion is first assembled on a polyprenol, dolichol (Figure 4.15), and then transferred to the nascent polypeptide chain in the rough endoplasmic reticulum. Final maturation occurs primarily in the Golgi apparatus of the cell by highly specific glycosyltransferases which selectively transfer the peripheral sugars (N-acetylglucosamine, galactose, fucose and sialic acid = N-acetylneuraminic acid) to the pentaglycosyl core structure.

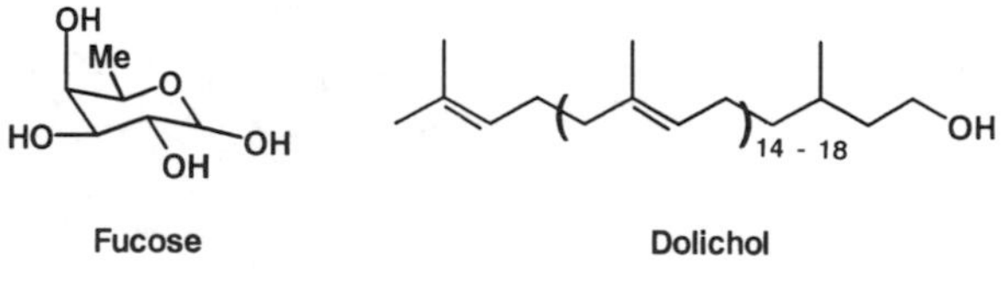

Fucose

Dolichol

Figure 4.15

4.4 Glycolipids

Glycolipids occur widely in nature, but they still represent only a small fraction of the total lipids. Glycolipids form parts of membranes and their actual role is not fully understood. They are known to participate in the biosynthesis of glycoproteins and complex polysaccharides: the polysaccharide is built on a glycolipid, from where it is transferred to a protein or carbohydrate at the end of biosynthesis. It has also been observed that glycolipids have a regulatory role in the synthesis of proteoglycans, and they also inhibit the effects of toxic and antiviral agents.

Other toxins, such as tetanus and cholera toxins, bind the carbohydrate part of glycolipids. Gangliosides inhibit the action of these toxins, which facilitates the protection against these toxins. Interferon interacts with gangliosides, which increases the antiviral effectivity of interferon.

The simplest glycolipids to occur in mammalian tissues are monoglycosyl ceramides, or cerebrosides. In the cerebrosides of the brain, sphingosine (or sphinganine) is usually glycosylated with D-galactose, in serum with D-glucose. The lipids are typically long chain saturated (behenic acid, $C_{21}H_{43}COOH$, lignoceric acid, $C_{23}H_{47}COOH$), unsaturated (nervonic acid, Δ^{15}-$C_{24:1}$) or hydroxy acids (cerebronic acid, $C_{22}H_{45}CHOHCOOH$).

Gangliosides (e.g. Figure 4.16) are sialic acid containing glycosphingolipids. They are composed of a hydrophilic oligosaccharide attached to a double lipophilic tail named ceramide, and to one or more 5-acetamido-3,5-dideoxy-D-glycero-D-galacto-2-nonulopyranoic (sialic) acid residues. The number of the sialic acid residues determines the polarity of the ganglioside. About 50 types of ganglioside are known, and they are typically constituents of all mammalian somatic cell membranes. Gangliosides also seem to play an important role in nerve growth, nerve regeneration and in brain functions: the mammalian brain cortex has the highest relative amount of gangliosides, *ca* 10 times that found in extraneural organs. Gangliosides also occur in the pancreas, spleen, liver, kidneys and most prevalently in the grey matter of the brain. Accumulation of the lipids, lipidosis, is a group of serious genetic diseases of which a dozen or more distinct disorders are known. Tay–Sachs disease (affects the brain), and Gaucher's disease (affects the spleen and liver) are two examples for which the enzymatic malfunctions are known. For the former the defective enzyme is hexosaminidase A, an enzyme that normally hydrolyses galactose residues from ganglioside GM_2. In the case of Gaucher's disease, glucocerebrosidase activity is deficient.

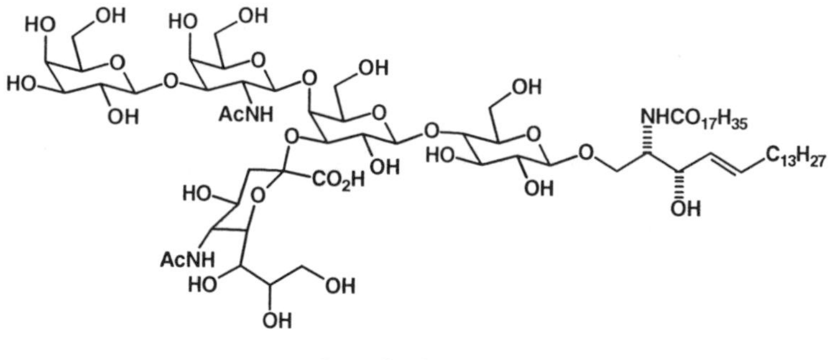

Ganglioside GM$_1$

Figure 4.16

4.5 Sugar Antibiotics

Many antibiotics contain rare carbohydrate units. Nucleoside antibiotics usually interfere with the DNA/RNA synthesis. Owing to this activity, they are also very toxic, which is reflected in their restricted clinical applicability. Puromycin is a purine analogue, blastmycin is a pyrimidine analogue, and showdomycin, coformycin and pentostatin are modified nucleoside antibiotics (Figure 4.17).

Sugars can also be connected to several aromatic aglycons which exhibit antitumoral activity. Daunomycin (also known as daunorubicin) and adriamycin (doxorubicin) are members of the antitumoural anthracycline antibiotics (Figure 4.18).

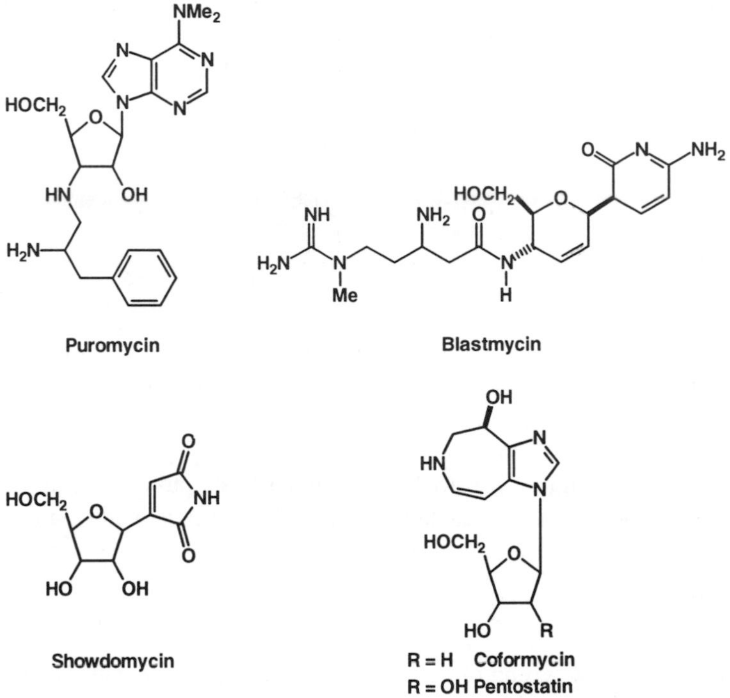

Puromycin

Blastmycin

Showdomycin

R = H   Coformycin
R = OH Pentostatin

Figure 4.17

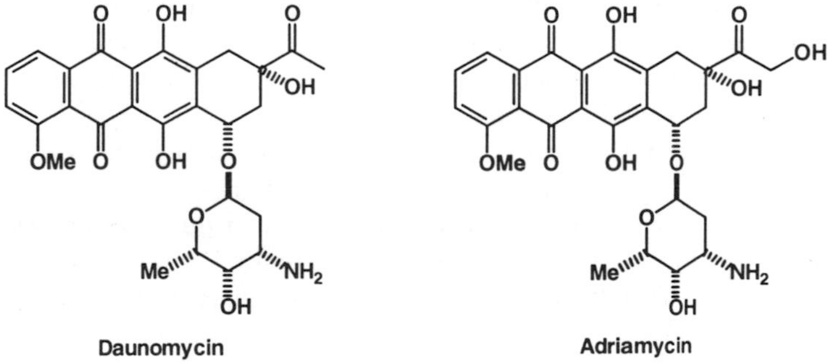

Daunomycin

Adriamycin

Figure 4.18

In macrolide antibiotics, a macrocyclic lactone, formed via the polyketide pathway, is usually glycosidated and connected to a carbohydrate unit, typically with an amino sugar. These are effective against Gram-positive bacteria, but their effects on Gram-negative bacteria are usually weak. Erythromycin, the leucomycins (see Chapter 7 for structures) and avermectin B_{1a} (Figure 4.19) are typical examples of such antibiotics.

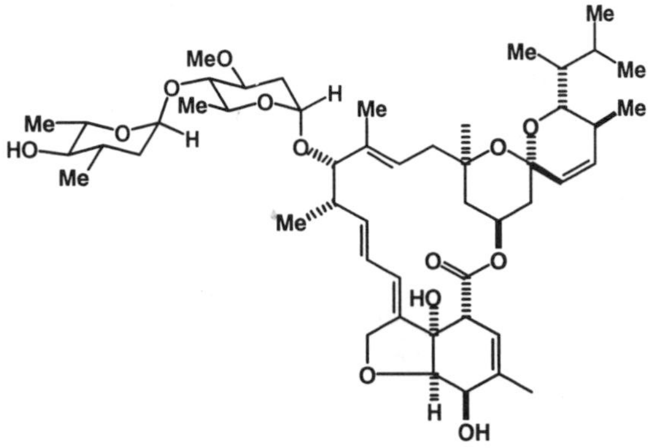

Avermectin B$_{1a}$

Figure 4.19

Other antibiotics containing a carbohydrate unit are streptomycin and vancomycin. Lincomycin and clindamycin also contain a rare 4-propylproline unit (Figure 4.20).

4.6 Cyclitols

The cyclitols are a diverse group of cyclic, polyhydroxylated compounds which usually have a cyclohexane skeleton. Their biological activities are as varied as their structures, and many of them occur in phosphorylated form. Phosphorylated inositols have been shown to act as second messengers in many intracellular signal transduction processes by mediating the release of calcium from non-mitochondrial stores. Aminodeoxyinositols and -conduritols occur in the aminocyclitol antibiotics, and a number of conduritol (Figure 4.21) derivatives have important physiological actions, such as glycosidase inhibition, antifeedant, antibiotic, tumourstatic and growth-regulating activities. The synthetic activity is further boosted by the fact that various hydroxylated cyclohexene derivatives act as glycosidase inhibitors.[12,13]

Synthesis of the cyclitols has seen a renaissance during the past few years, mainly thanks to the rapid development of novel strategies for the construction of polyhydroxylated systems in a stereochemically homogeneous manner. An interesting development is the utilization of the microbial oxidation of benzene and its derivatives. Oxidation of benzene with *Pseudomonas putida* 39-D produces the corresponding *cis*-cyclohexanedienediol.[14,15]

Ley has applied this microbial oxidation to the synthesis of pinitol (Scheme 4.7), a feeding stimulant for the larvae of the yellow butterfly *Eurema hecabe mandarina* and

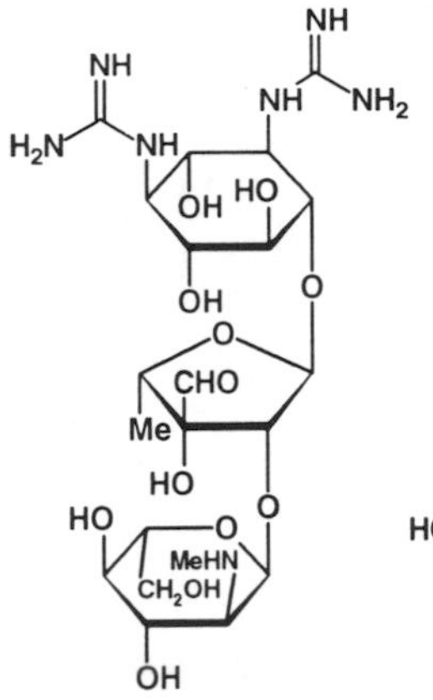

Streptomycin

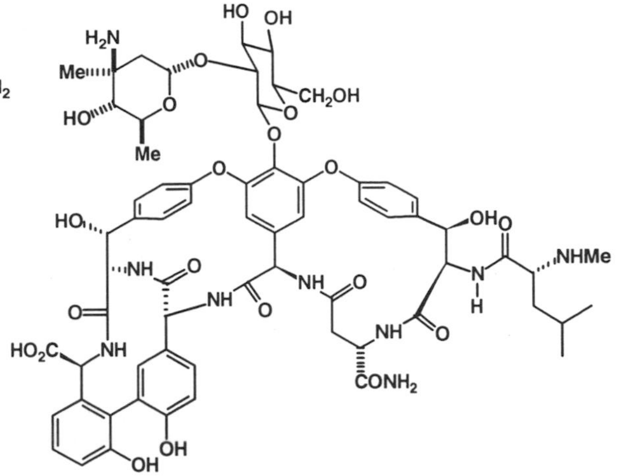

Vancomycin

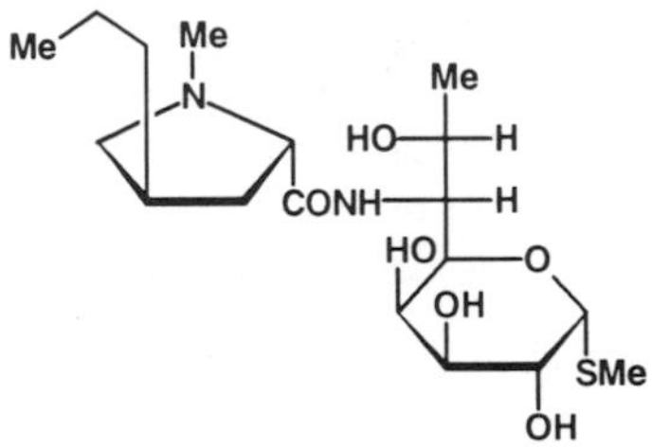

Lincomycin

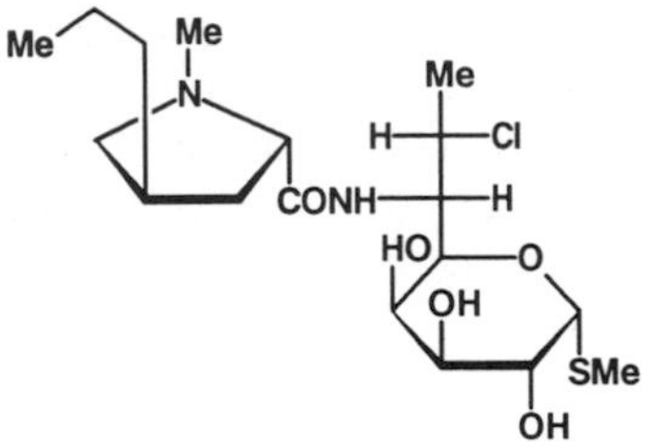

Clindamycin

Figure 4.20

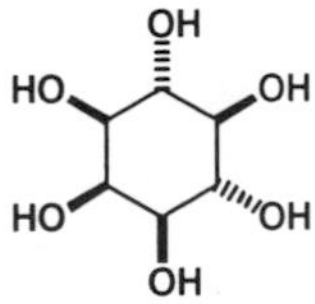

myo-Inositol

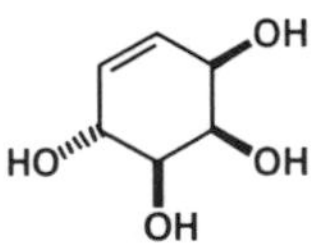

(-)-Conduritol

Figure 4.21

an inhibitor of larval growth of *Heliothis zea* on soybeans.[16] The oxidation product was protected as the dibenzoate and epoxidized with mCPBA to give a mixture of the desired *trans* epoxide (73 %) and the undesired *cis* epoxide (17 %), the latter presumably resulting from the attack of the peracid being directed by the benzoate group. The major product was then subjected to epoxide opening to give the expected product with methoxide attack at the distal end of the epoxide. Final *cis* hydroxylation of the double bond (OsO_4, NMMO) gave the triols in a 5:1 ratio with the desired isomer predominating (through attack of the oxidizing agent from the less hindered α-face).

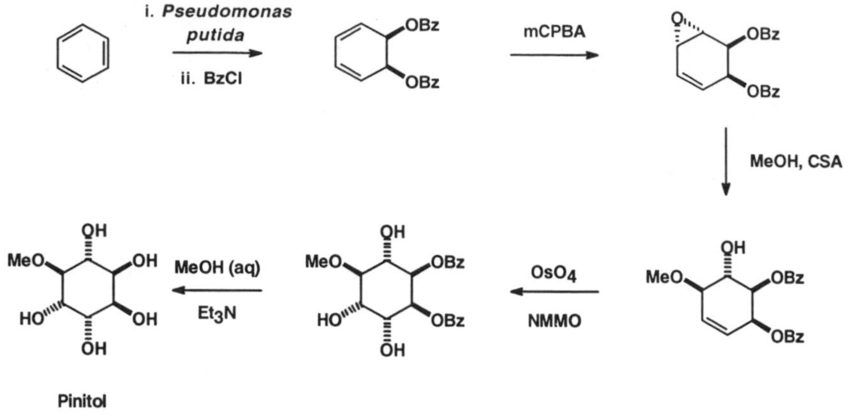

Scheme 4.7

The microbial oxidation is both stereospecific and enantioselective. If the benzene ring carries a substituent (typically a halogen), the product will be a single enantiomer (such as the bromine containing compound in Scheme 4.8). These diol derivatives have been derivatized using the standard methodology (protection of the diols followed by sequential epoxidation/epoxide opening and *cis*-hydroxylation) to give (+)-pinitol. Reversal of the order of the oxidation processes gives access to the enantiomeric, non-natural (−)-pinitol.[17]

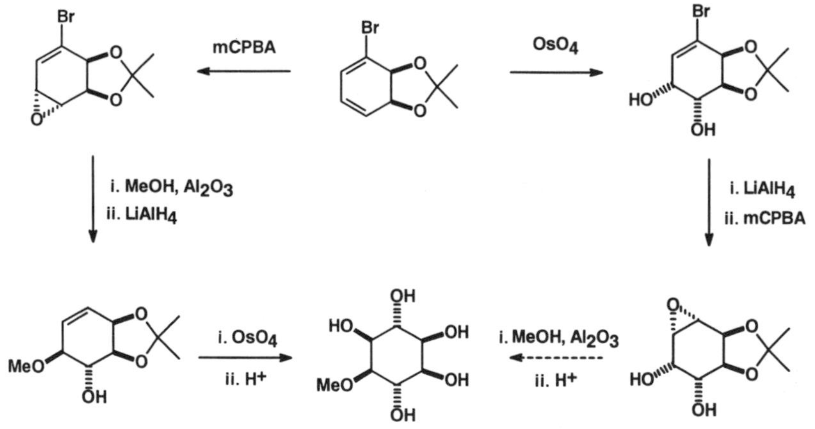

Scheme 4.8

A similar strategy has also been used in the synthesis of D- and L-myo-inositol 1,4,5-triphosphates, as well as the corresponding deoxy, fluoro and methyl derivatives (Scheme 4.9).[18] The addition of the nucleophile is completely stereoselective (from the less hindered α-face), but the regioselectivity is less impressive for the small nucleophiles (hydride and fluoride, *ca* 4:1 to 6:1 ratio).

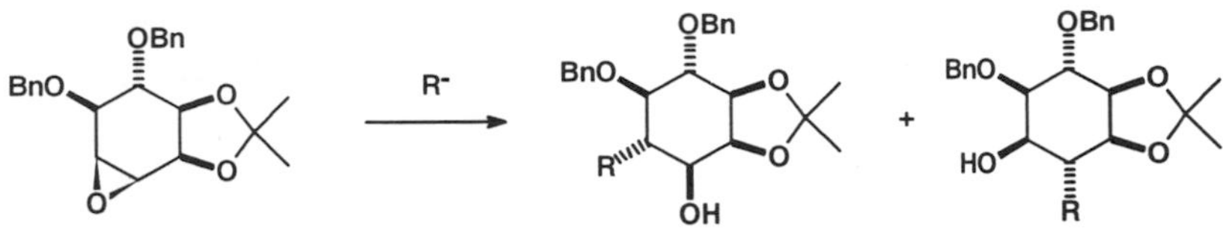

Scheme 4.9

R^-		
H^- (LiAlH$_4$)	76 %	12 %
F^- (TASF)	59 %	15 %
Me^- (Me$_2$Cu(CN)Li$_2$)	73 %	3 %

Starting from chlorobenzene, one can devise a rapid route to conduritols (Scheme 4.10).[19] Oxidation gives the *cis*-diol[20] which can be regio- and stereoselectively epoxidized with mCPBA (Henbest oxidation, the allylic alcohol directs the oxidation through coordination). Finally, epoxide opening can be achieved, again with high regioselectivity owing to polarity control.[21]

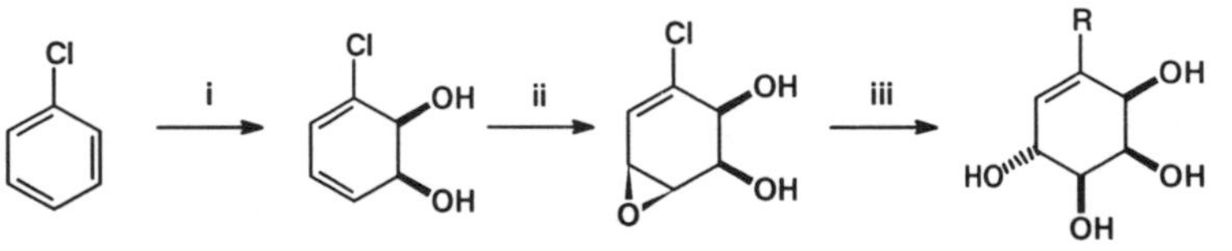

Reagents: i, *Pseudomonas putida*; ii, mCPBA, acetone, 61 %; iii, H$_2$O, TFA, 90 %; iv, Na/NH$_3$, 70 %.

Scheme 4.10

Employment of the Ferrier rearrangement gives access to the cyclitols from carbohydrates.[22] The enol acetate was derived from glucose (Scheme 4.11), and mercury(II) trifluoroacetate induced the Ferrier rearrangement to give the ketone. Stereoselective reduction of the inosose was achieved using intramolecular hydride delivery[23-25] with sodium triacetoxyborohydride to give the desired, differentially protected hexol, which was converted to the final product.

The stereocontrolled reduction occurs through the complex formed by the triacetoxyborohydride and the (necessary) free alcohol β to the ketone. The complex (Figure 4.22) delivers the hydride intramolecularly from the same face as the axial alcohol to give the desired equatorial alcohol contaminated with less than 10 % of the axial alcohol isomer.[26]

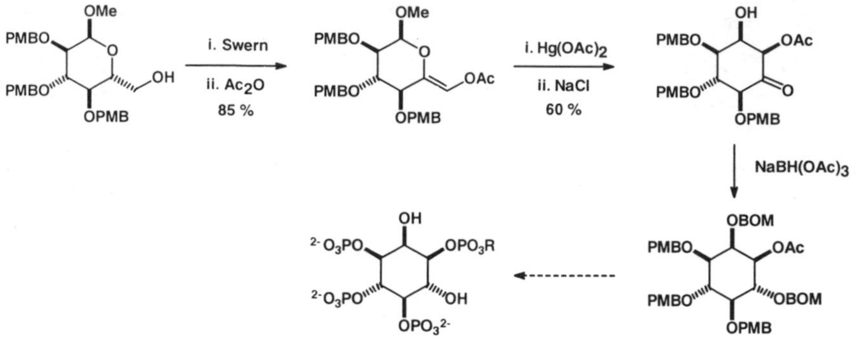

Scheme 4.11

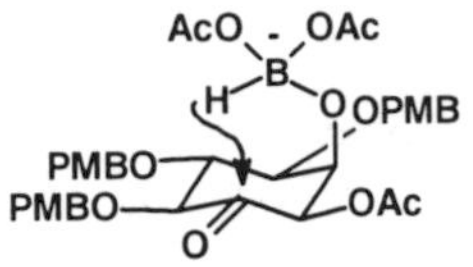

Figure 4.22

References

1. David, S., Eisenstein, O., Hehre, W.J., Salem, L., and Hoffmann, R. *J. Am. Chem. Soc.* **95**, 3806–3807 (1973).
2. Leblanc, Y., Fitzsimmons, B.J., Springer, J.P., and Rokach, J. *J. Am. Chem. Soc.* **111**, 2995–3000 (1989).
3. Friesen, R.W. and Danishefsky, S.J. *J. Am. Chem. Soc.* **111**, 6656–6660 (1989).
4. Halcomb, R.L. and Danishefsky S.J. *J. Am. Chem. Soc.* **111**, 6661–6666 (1989).
5. Griffith, D.A. and Danishefsky, S.J. *J. Am. Chem. Soc.* **112**, 5811–5819 (1990).
6. Schmidt, R.R. *Angew. Chem., Int. Ed. Engl.* **25**, 212–235 (1986).
7. Nicolaou, K.C., Dolle, R.E., Papahatjis, D.P., and Randall, J.L. *J. Am. Chem. Soc.* **106**, 4189–4192 (1984).
8. Nicolaou, K.C., Groneberg, R.D., Miyazaki, T., Stylianides, N.A., Schulze, T.J., and Stahl, W. *J. Am. Chem. Soc.* **112**, 8193–8195 (1990).
9. Kunz, H. *Angew. Chem., Int. Ed. Engl.* **26**, 294–308 (1987).
10. Garg, H.G. and Jeanloz, R.W. *Adv. Carbohydr. Chem. Biochem.* **43**, 135–201 (1985).
11. Presper, K.A. and Heath, E.C. In *The Enzymology of Post-translational Modification of Proteins* (Freedman, R.B. and Hawkins, H.C., Eds.) Academic Press: New York, 1985, pp. 53–93.
12. Barton, D.H.R., Dalko, P., and Gero, S.D. *Tetrahedron Lett.* **32**, 2471–2474 (1991).
13. McIntosh, M.C. and Weinreb, S.M. *J. Org. Chem.* **56**, 5010–5012 (1991).
14. Gibson, D.T., Hensley, M., Yoshika, H., and Mabry, R.J. *Biochemistry* **9**, 1626–1630 (1970).
15. Gibson, D.T., Mahaderan, V., and Davey, J.F. *J. Bacteriol.* **119**, 930–936 (1974).
16. Ley, S.V. and Sternfeld, F. *Tetrahedron* **45** 3463–3476 (1989).
17. Hudlicky, T., Price, J.D., Rulin, F., and Tsunoda, T. *J. Am. Chem. Soc.* **112**, 9439–9440 (1990).
18. Ley, S.V. Parra, M, Redgrave, A.J., Sternfeld, F., and Vidal, A. *Tetrahedron Lett.* **30**, 3557–3560 (1989).

19. Carless, H.A. *J. Chem. Soc., Chem. Commun.* 234–235 (1992).
20. Boyd, D.R., Dorrity, M.R.J., Hand, M.V., Malone, J.F., Sharma, N.D., Dalton, H., Gray, D.J., and Sheldrake, G.N. *J. Am. Chem. Soc.* **113**, 666–667 (1991).
21. Ho, T.-L. *Polarity Control for Synthesis* John Wiley & Sons: New York, 1991.
22. Estevez, V.A. and Prestwich, G.D. *J. Am. Chem. Soc.* **113**, 9885–9887 (1991).
23. Saksena, A.K. and Mangiaracina, P. *Tetrahedron Lett.* **24**, 273–276 (1983).
24. Turnbull, M.D., Hatter, G., and Ledgerwood, D.E. *Tetrahedron Lett.* **25**, 5449–5452 (1984).
25. Evans, D.A., Chapman, K.T., and Carreira, E.M. *J. Am. Chem. Soc.* **110**, 3560–3578 (1988).
26. Bender, S.L. and Budhu, R.J. *J. Am. Chem. Soc.* **113**, 9883–9885 (1991).

5 Amino Acids, Peptides and Proteins

Peptides and proteins play a central role in the function of cells and organs. Large proteins can support the structure of the cell. Proteins catalysing chemical transformations are called enzymes. They can participate in the transfer of information and signals between cells, whence they are called receptors. Enzymes are extremely sensitive in recognizing the reacting molecules, and by way of their intrinsic asymmetric structure they exhibit high discrimination towards optical antipodes. Smaller peptides can be vital signal transduction components, e.g. in the central nervous system (neurotransmitters) and hormonal activity (peptide hormones).

Whether one considers a large enzyme complex composed of several peptide chains, which catalyses chemical reactions, or a small neurotransmitter peptide, they share the common feature of being constructed of amino acids.

In this chapter we will take a look at the amino acids and polypeptides, especially in terms of structure and synthesis. We shall finally briefly review enzymes, their properties and function, and the factors controlling them.

5.1 Amino Acids

Natural peptides and proteins are built from 20 so-called natural amino acids (Figure 5.1). The amino acids are, with the exception of glycine, optically active; in other words,

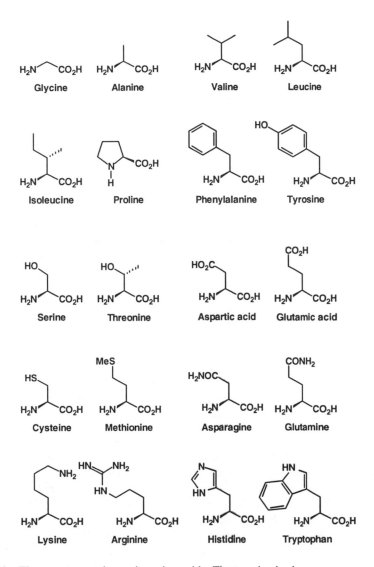

Figure 5.1 The twenty proteinogenic amino acids. The L series is shown

their α-carbon is asymmetrically substituted. With the exception of certain microbial products, all natural amino acids belong to the same stereochemical series (L), where the absolute stereochemistry of the chiral carbon is S. Cysteine makes an exception, where the stereochemical designation is R—this, however, is owing to the nature of the systematic nomenclature. Two amino acids, threonine and isoleucine, have a second chiral centre. Again, one diastereomer of each of these is predominant in Nature.

An important feature for the chemistry of the amino acids is that they contain both a basic (NH$_2$) and an acidic (COOH) group. Because of the difference in the pK_a values of the carboxylic acid and amino groups, at neutral pH they both tend to be in their ionized forms. As a result, each amino acid has a characteristic pH where the amino acid occurs as a zwitterion. This pH, the isoelectric point, is the arithmetic mean of the two

Table 5.1 Natural proteinogenic amino acids.

Amino acid	Three-letter	One-letter	MW [a]	*E. coli*	pI [b] [c]	Essential
Acidic:						
Aspartic acid	Asp	D	114	9.9	2.8	
Glutamic acid	Glu	E	128	10.8	3.2	
Neutral:						
Cysteine	Cys	C	103	1.8	5.1	
Asparagine	Asn	N	114		5.4	
Phenylalanine	Phe	F	147	3.3	5.5	+
Threonine	Thr	T	101	4.6	5.6	+
Serine	Ser	S	87	6.0	5.7	
Glutamine	Gln	Q	128		5.7	
Methionine	Met	M	131	3.8	5.7	+
Tyrosine	Tyr	Y	163	2.2	5.7	
Tryptophan	Trp	W	186	1.0	5.9	+
Glycine	Gly	G	57	5.7	6.0	
Alanine	Ala	A	71	13.0	6.0	
Valine	Val	V	99	6.0	6.0	+
Leucine	Leu	L	113	7.8	6.0	+
Isoleucine	Ile	I	113	4.4	6.0	+
Proline	Pro	P	97	4.6	6.3	
Basic:						
Histidine	His	H	137	0.7	7.5	+
Lysine	Lys	K	129	7.0	9.6	+
Arginine	Arg	R	157	5.3	11.2	+

[a] MW = molecular weight. [b] Occurrence in *Escherichia coli* as a percentage. [c] Isoelectric point.

pK_a values, and it is also the pH where the solubility of the amino acid is lowest. In Table 5.1, the amino acids are classified according to their chemical nature as acidic, neutral and basic amino acids. As one can observe, the isoelectric point of the neutral amino acids is slightly on the acidic side. The *essential amino acids*, which cannot be synthesized by the human body but have to be obtained with nutrients in order to prevent a negative nitrogen balance, are marked with an asterisk. Also included in the table are the three- and one-letter codes commonly used for the amino acids.

As stated previously, the natural amino acids usually occur in their L form. Strictly speaking this is not true, as several microbes also contain D-amino acids. These are usually constituents of the bacterial cell walls, especially in the rigid peptidoglycan net surrounding the cytoplasmic membrane, which is responsible for the shape and strength of the bacterial cell.

Besides the D enantiomers of the 20 amino acids, a large number of amino acids of various structures are also known. Several hundred of them occur naturally, but their functions are widely unknown.[1] Structurally and biogenetically the simplest ones are just modifications of the 20: 4-hydroxyproline and 4-hydroxylysine, constituents of collagen tissue, are produced post-translationally by oxygenation of the intact peptide. The newly introduced hydroxyl groups participate in the formation of the cross-links between the peptide chains, giving collagen its characteristic strength. *N*-Methylated amino acids are also found in many peptides and proteins. At least in some cases, the methylation is used to impart upon the peptide certain structural characteristics. More complicated and rare amino acids include the cyclopropane containing amino acids. 1-Aminocyclopropanecarboxylic acid (Figure 5.2) is the biogenetic precursor of the plant growth factor ethylene. α-(Methylenecyclopropyl)glycine and hypoglycine occur in the unripe fruit of *Blighia sapida* and litchi seeds. These compounds are such powerful hypoglycaemic agents that ingestion of the unripe fruit can be lethal.

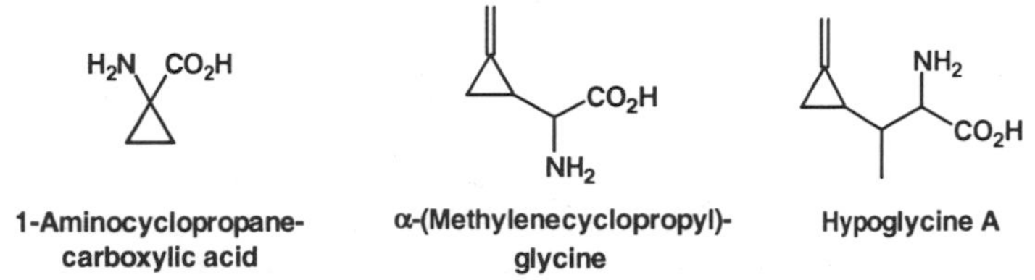

1-Aminocyclopropane-
carboxylic acid

α-(Methylenecyclopropyl)-
glycine

Hypoglycine A

Figure 5.2

Several pyrrolidine containing amino acids of the kainic acid type are also known. Kainic acid (Figure 5.3) has been isolated from the alga *Digenea simplex*, and it is being used extensively as a research tool in neurobiology to impart selective lesions in the brain and to explore the physiological pharmacology of excitatory transmission. The isoxazole containing amino acid ibotenic acid (Figure 5.3) is, like kainic acid, an agonist of the glutamate receptor.

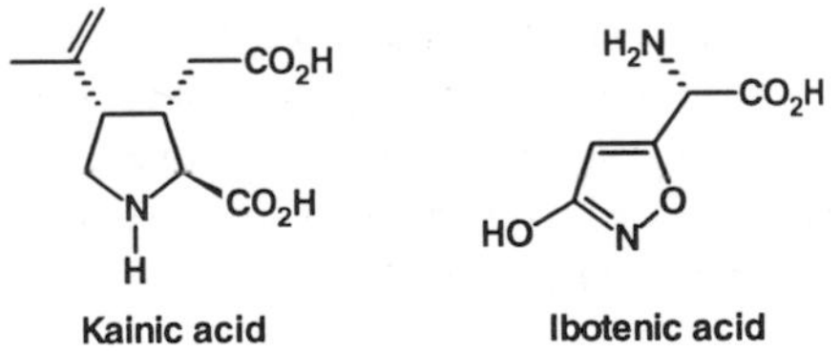

Kainic acid

Ibotenic acid

Figure 5.3

Unsaturated amino acids, like MeBmt, a constituent of the immunosuppressant antibiotic cyclosporin A, and ADDA, $(2S,3S,8S,9S)$-3-amino-9-methoxy-2,6,8-trimethyl-10-phenyldeca -(4E,6E)-dienoic acid, a part of the cyanobacterial hepatotoxins microcystin-LR and nodularin, are examples of more complex amino acids (Figure 5.4).

Condensation of an amine with a carboxylic acid leads to the formation of an amide bond. When this bond exists in a peptide or a protein, it is called a peptide bond. In the following, we shall briefly discuss peptides and proteins.

5.2 Peptides and Proteins

As an example of peptides, we shall inspect the strongly blood pressure raising CCK-7 (Figure 5.5), the *C*-terminal heptapeptide of cholecystokinin. Cholecystokinin diminishes

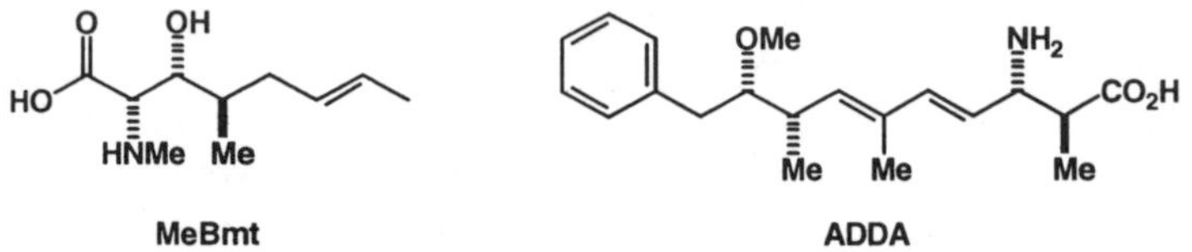

MeBmt　　　　　　　　　　　　ADDA

Figure 5.4

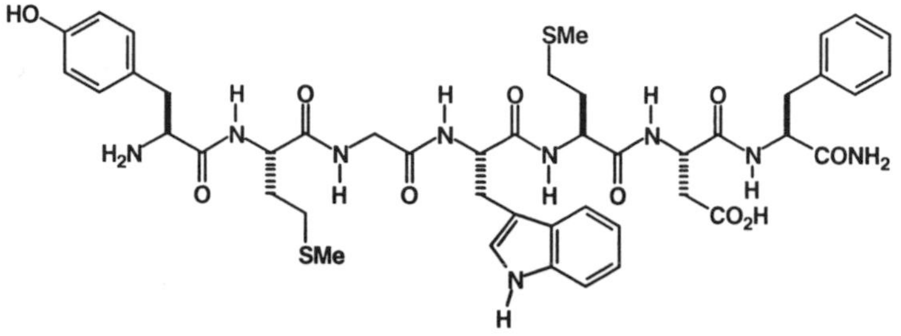

Tyr-Met-Gly-Trp-Met-Asp-Phe-NH₂
CCK-7

Figure 5.5

appetite, and this property is being utilized for the development of treatments for both anorexia and obesity.

The structure of a peptide is usually represented by drawing the *N*-terminus to the left, and the *C*-terminus to the right. In naming, one follows the same order. For larger proteins, it is difficult to use the systematic name, and one usually uses either the three- or one-letter codes for the amino acid residues. The larger proteins, and especially the functional proteins and receptors, are often formed from several hundred amino acid residues.

Each amino acid has its characteristic chemical and physical properties. When the amino acids combine to form a polypeptide, this also has its own characteristic properties which are dependent on the amino acid sequence. This sequence order of amino acids is called the *primary structure*, and it describes the order in which the amino acids are joined together. However, the primary structure does not give any information on the three-dimensional structure of the peptide.

The sequence can be determined in a straightforward manner by splitting one amino acid residue at a time, and analysing the amino acid thus obtained. The Edman degradation is a common method for the determination of the sequence starting from the *N*-terminal end. Phenylisocyanate, or more conveniently phenylisothiocyanate, is reacted with the free *N*-terminal amino group to form a thiourea. Under acidic conditions, the sulfur atom displaces the protonated amino group to give a thiazolidinone, and a peptide with one less amino acid residue. Acid catalysed opening of the thiazolidinone followed by re-closure to reveal the phenylthiohydantoin concludes the sequence (Scheme 5.1). The degradation methods are very highly developed, based on classical organic analytical reactions, and the sequence analysis is usually performed with a fully automated instrument.[2]

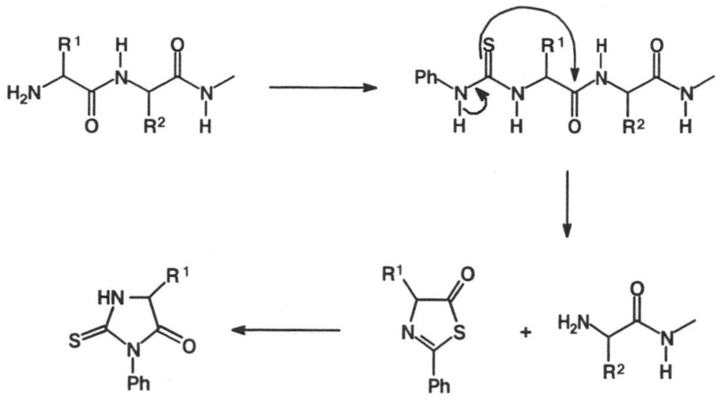

Scheme 5.1

The primary structure of peptides is not the only, or even the most important, feature for the function of the peptide. The peptide chain folds into various possible conformations, whose relative stabilities are highly dependent on local and global energetic components, such as hydrogen bonding and van der Waals forces. Although these interactions are typically small, their combined action can lead to remarkably high energetic contributions. If one conformation is more favourable than the others, one talks about a local or a global energy minimum. This simply means that the conformation is energetically more favourable than the other ones close to it on the reaction path (local minimum) or it is the absolute energy minimum (global minimum). The global minimum conformation is the one that can be expected to be obtained on isolation of the protein. This folded form is called the *tertiary structure* (if the protein is formed from more than one peptide chain, one talks about a quaternary structure for the intact protein complex, and the tertiary structure refers to the three-dimensional structures of the individual peptide chains). Tertiary structure is thus the structure that best describes the protein in its natural environment. This structure is often so stable that it can be isolated, and even crystallized in its original form.

Between the primary and the tertiary structures there is another level of structural composition, the *secondary structure*, which refers to the local structures of spans of several amino acids long sequences. These include structurally similar, often repeating, 'units', helices, sheets and turns.

At the secondary structure level, one can distinguish several structural motifs which are generally found in the protein structures. α-Helix and β-sheet are the oldest ones of these.[3] The regions joining these were not so long ago described simply as random loops. However, they are neither random nor loops. Recently, protein structures have been more thoroughly studied and found to contain certain often repeated patterns, including β-turns.[4,5] Approximately 60 % of the β-turns are located on the surface. A crystallographic study at high resolution established that the active site of an aspartic proteinase from *Rhizopus chinensis* contains two β-turns: both of the catalytic asparagines are involved in turns. However, the pronounced occurrence of β-turns on the protein surface makes them viable candidates for molecular recognition, as exemplified by antigenic recognition, cell–cell recognition and protein–DNA recognition.[6]

Peptides provide smaller molecular assemblies amenable to detailed structural characterization by direct physical methods. The α-helices[3] are formed through an intramolecular hydrogen bond connecting the amino group of the residue $i + 4$ to the carbonyl group of the residue i through a hydrogen bond. A 13 membered, hydrogen bonded ring is formed. The structure of a β-turn is shown (in stereo) in Figure 5.6. In the apparently random regions of the peptide we can distinguish regions where the protein chain folds back on itself and forms a turn. Two major categories of turns are defined: β-turn (reverse turn, β-bend, inverse turn, U-turn) and γ-turn. The former is distinguished by a hydrogen bond between residue $i(C===O)$ and $i + 3(N—H)$ (forming a 10 membered ring), and the latter similarly between $i(C===O)$ and $i + 2(N—H)$ (forming a seven membered ring).[7]

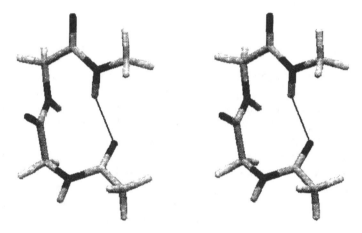

Figure 5.6

Each amino acid sequence has a different propensity to favour the different structural motifs (α-helix, β-sheet, various forms of turns). The folding of a long peptide chain is not, however, governed only by the folding of the shorter peptide sequences, but often long range effects (hydrogen bonding, salt bridges, van der Waals forces, etc.) can cause changes in the overall shape.

Thus two α-helices can be joined to each other by a sequence which on its own would assume a more or less random form. The associative forces between the two helices can help this sequence to assume a particular type of turn. The turns, and especially the combinations of turns, sheets and helices, have been shown to be important for many recognition processes in life. As one continues the combination of secondary structural motifs, the final result is the tertiary structure of the protein.

The final stage for the structure of a multiprotein complex is the *quaternary structure*. This describes the overall structure of a complex formed from two or more peptide chains which are joined together by non-covalent bonds to form the protein. Haemoglobin is such a multi-chain protein complex. It is formed from four chains, each in its own tertiary structural form, which are then joined to form a complex which in turn binds four molecules of haemin. It is only this very large complex of eight individual components that can assume the functional form to be able to transport oxygen in the blood, and even then only provided that it is correctly assembled. Even very small changes in the structure of the peptide chain can dramatically alter the function of a complete protein complex. Haemoglobin S is a mutant of haemoglobin, which has suffered a single point mutation at its sixth amino acid residue (Glu to Val). This causes changes in the three-dimensional structure of the haemin complex, with the end result that at low oxygen pressures the red blood cells change their shape from a biconcave disc into an elongated form. These sickle-shaped cells can block the capillaries and thus restrict blood circulation, causing serious tissue damage. Sickle cell anaemia, as the disease is known, is a serious hereditary disease, especially in central Africa.

5.3 Enzymes and Receptors

Enzymes are naturally occurring polypeptides which catalyse reactions. The most important property of the enzymes for their function is their outstanding substrate specificity. Out of a set of structurally closely related substrates a typical enzyme can select only one whose reaction it will catalyse. This is a key feature for the function of the enzymes, and also for the direction evolution has taken from primordial times.

In order for us to understand why enzymes are so selective, we must inspect the underlying reasons. As we have learned, each peptide chain adopts an individual form and shape, a three-dimensional structure. Enzymes (and receptors) recognize their substrates in the active centre, which can be regarded as a glove or a pocket. The substrate can bind only into a pocket of the right size and shape. This is, in simplified form, the lock and key model, which was the prevailing theory for enzyme–substrate recognition until the 1960s. Currently we know that this is only partly true, as the key also forces the lock to change its spatial structure (Figure 5.7).

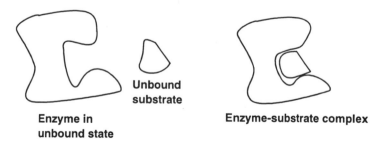

Enzyme in
unbound state

Unbound
substrate

Enzyme-substrate complex

Figure 5.7

As the substrate binds into the active site, there is a structural change in the overall complex and in each binding partner, as a result of which a suitable catalytic group of the enzyme is brought close to the reactive site in the substrate; a chemical reaction then takes place. After the chemical transformation has occurred, the modified substrate (product) dissociates from the active site since it cannot fit the site as perfectly as the substrate can. The active site is believed to be most closely complementary (binding tightest) to the transition state of the reaction, which is equivalent to lowering the free energy of the transition state. According to the Gibbs function, this has a rate accelerating effect, which in many enzyme catalysed reactions has been observed to be of the order of 10^9. The reaction chain can proceed over and over again, and one can speak of a truly catalytic reaction. Since the active site is very selective owing to its structural features, the possibilities for enzyme catalysis of organic reactions has been explored quite widely recently.[8]

We have already mentioned another type of proteins, the receptors. These are structurally akin to the enzymes. The most notable difference is that unlike enzymes, receptors do not participate in chemical reactions.

The binding of an active compound to the receptor site causes a structural change not unlike that observed in the case of an enzyme discussed above. The receptor proteins are usually located in the cell membrane, as exemplified in Figure 5.8 for the platelet derived growth factor (PDGF) receptor.[9] Activation of the receptor leads to a complex cascade of events leading ultimately to intracellular phosphorylation by mitogen-activated protein kinases (thick arrow). This phosphorylation is modulated by a number of other signals, including input from the phorbol ester receptor protein, kinase C (PKC), and the guanosine triphosphate activated protein (GAP).

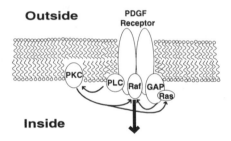

Figure 5.8

The function of the receptor is to receive and filter various signals coming from inside the cell or from its surroundings and then transmit the correct information further to another site, where usually an enzymatic reaction is affected. Only the correct agonist can trigger the transmission of the signal. Receptors, like enzymes, are abundant all over the organism. For instance, the function of the nervous system is based on a very delicate balance of a multitude of receptors.

The control function of receptors is important also in controlling the intracellular syntheses, whether of peptides or secondary metabolites. As a need for a particular chemical transformation is raised within the cell, this information is passed on through receptors and chemical messengers to the enzymes, which initiate (or turn off) the synthesis. This system can be used advantageously in medicine, since most maladies are caused by the malfunction of enzymes or receptors. There are also attempts to harness this property for the synthesis of natural products, employing bacterial cells, plant cell cultures and even isolated enzymes.

5.4 Chemical Modification of Peptides

After protein synthesis, many natural proteins have not yet achieved their final functional form—they must be subjected to post-translational modifications. A typical example is the existence of 4-hydroxyproline in collagen and plant cell membrane proteins. The hydroxy function is brought into proline, already incorporated in a protein, by an enzymatic reaction, and is a necessary condition for the cross-linking of the peptide chains to form a rigid matrix.

Especially in medicinal chemistry, the structural properties of peptide mimetics have found wide utility during the past decade or so. Many hormones and neurotransmitters are relatively small peptides. It is known that these signal transmitters bind to their respective receptors in a highly defined conformation. The utility of natural peptides, although they are usually readily available by isolation, solid phase synthesis or recombinant DNA techniques, is limited by several factors. They can easily assume a wide range of conformations, leading to multiple pharmacophore presentation and thereby lack of specificity. Natural-like peptides are also easily hydrolysed by thermal and proteolytic means. These factors lead to low oral activity, low membrane penetration, and often problems associated with potential antigenicity. Structural modifications on amino acid structures have been sought in order to cure at least some of these problems, and a wide variety of such 'conformationally constrained peptide mimics' are known. They often alleviate the problems, being less sensitive towards hydrolysis (by steric congestion at the α-centre, causing neo-pentyl type interactions at the tetrahedral intermediate of hydrolysis). The steric factors also impose strong conformational bias, and different kinds of secondary structural features can be dialled in by proper choice of substitution. The narrower conformational space available for the peptide also means that the problem of multiple pharmacophore presentation is lessened. Because of the stability and the possibility of adjusting the lipophilicity of the drug candidate, oral activity and membrane penetration can usually be improved. The major problem, of course, is that such peptidomimectics are available only through total synthesis. Typical types of conformationally constrained amino acid analogues include the bioisosteric replacements (Figure 5.9) and the α-carbon modifications (Figure 5.10).

An example of a conformationally constrained peptide mimic is compound L-364,718 (Figure 5.11), which has been designed using computer-aided molecular modelling and design methods. This molecule was designed to mimic the action of the active conformation of CCK-7.

Although practically all of the structure of the original active peptide has been abolished, L-364,718 binds to the same receptor site, thus blocking the action of CCK-7 (in other words it is an antagonist). The compound has an appetite increasing action, and its use for the treatment of anorexia is being studied.

5.5 Biosynthesis of Amino Acids

The main pathway for the biosynthesis of the aliphatic amino acids utilizes pyridoxamine as the source of ammonia. Scheme 5.2 shows the origin of the amino group from glutamic acid. Schiff base formation with pyridoxal phosphate (PALP) followed by proton shift

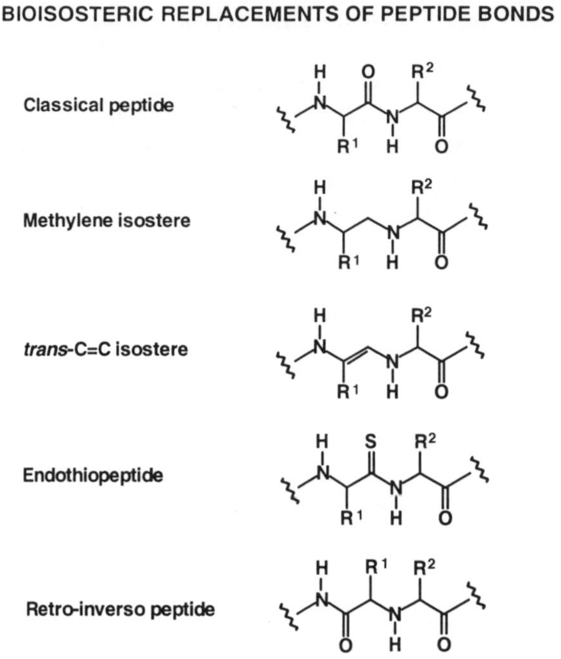

BIOISOSTERIC REPLACEMENTS OF PEPTIDE BONDS
Classical peptide
Methylene isostere
trans-C=C isostere
Endothiopeptide
Retro-inverso peptide

Figure 5.9

α-CARBON MODIFICATIONS
Classical peptide
α-Alkyl modification
Dehydropeptide
Cyclopropylogue
Aza/bora peptide
X = N, B

Figure 5.10

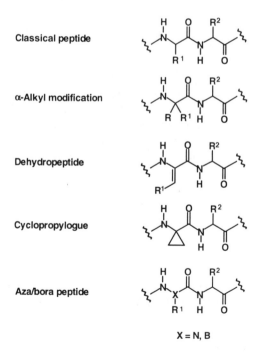

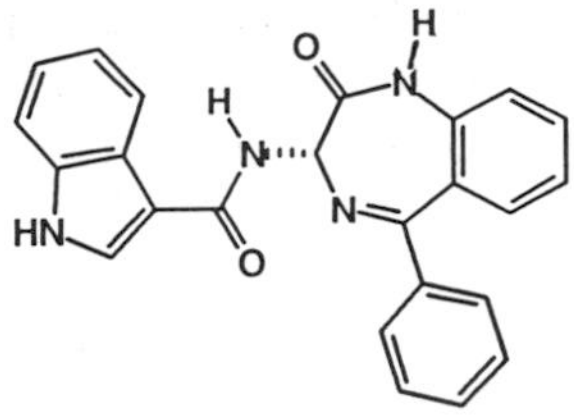

L-364,718

Figure 5.11

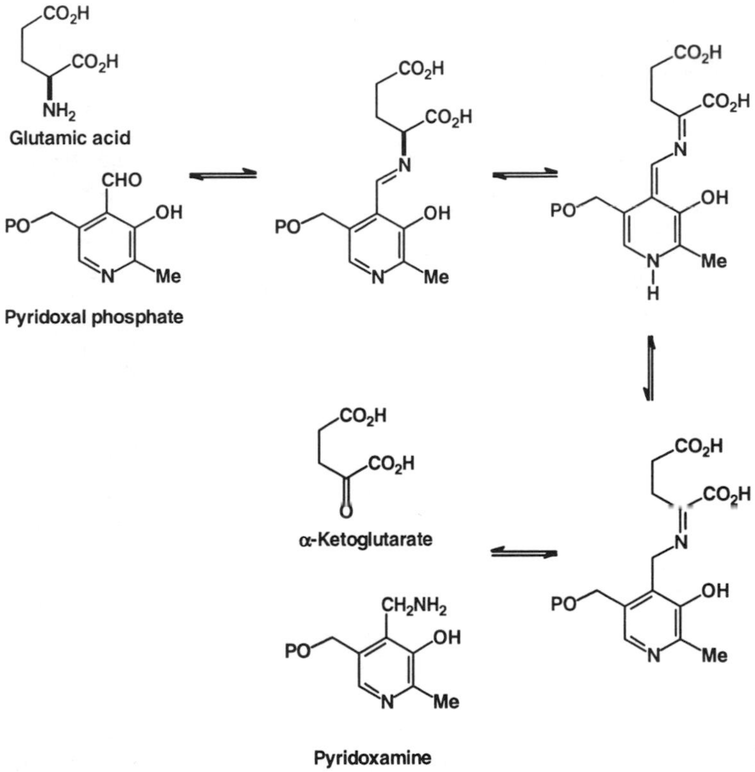

Scheme 5.2

gives the new α-keto carboxylate Schiff base, which is then hydrolysed to α-ketoglutarate and pyridoxamine. Reversal of the scheme and replacement of ketoglutarate with other α-keto acids gives access to the amino acids.

The aromatic amino acids are biosynthesized in a similar way from the corresponding α-keto acids. We shall now take a closer look at the ways the aromatic portions themselves are generated.

Aldol reaction of pyruvate (in the form of phosphoenol pyruvate, PEP; Scheme 5.3) and
D-erythrose-4-phosphate leads to the formation of a seven carbon sugar acid, 3-deoxy-
D-arabino-heptulosonic acid 7-phosphate (DAHP). This spontaneously loses phosphate,
and the enol form of the resulting diketo acid cyclizes to 3-dehydroquinic acid. NADH
mediated reduction gives quinic acid, whereas stereospecific dehydration (with the loss
of the hydrogen *syn* to the hydroxy group) gives 3-dehydroshikimic acid, and thereby
shikimic acid after reduction (NADH).

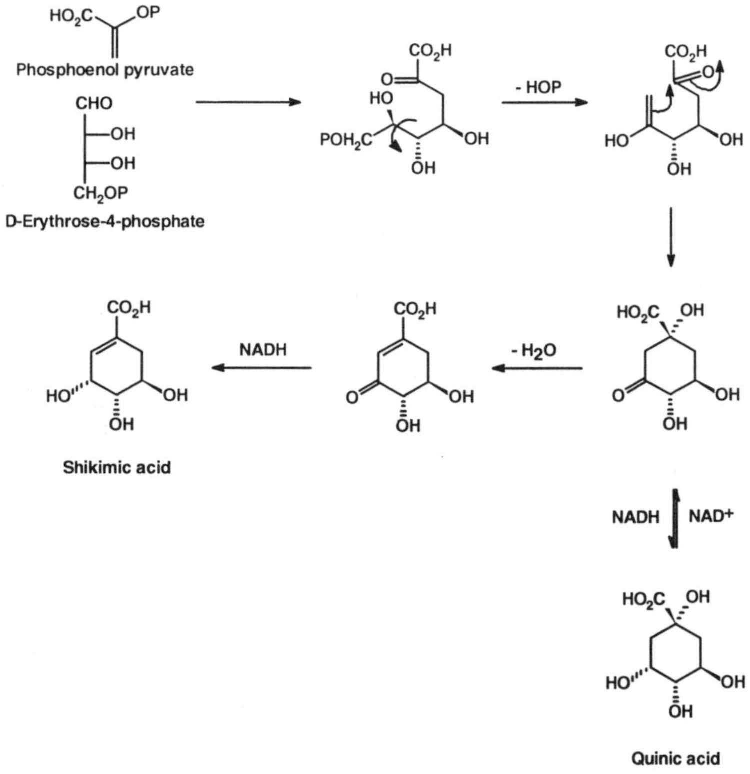

Scheme 5.3

The biosynthesis of the aromatic amino acids (phenylalanine and tyrosine) proceeds
via shikimic acid through chorismic acid and prephenic acid. Shikimic acid is
monophosphorylated at the pseudoaxial 3-hydroxyl group, and the 5-hydroxyl group
reacts with phosphoenol pyruvate to give chorismic acid after elimination (in a Grob
sense) of phosphoric acid (Scheme 5.4). This elimination is completely stereospecific
in that only the pro-*R* hydrogen at C-6 participates in the elimination. Chorismic acid
contains an allyl enol ether moiety required for the ensuing Claisen rearrangement which
in turn produces prephenic acid, the ubiquitous intermediate in the biosynthesis of the
aromatic amino acids phenylalanine and tyrosine.

Conversion of prephenic acid to the target amino acids requires only the adjustment
of the oxidation level and introduction of the amino function. Interestingly, this is done
via elimination of the chiral information in prephenic acid as follows. Simultaneous

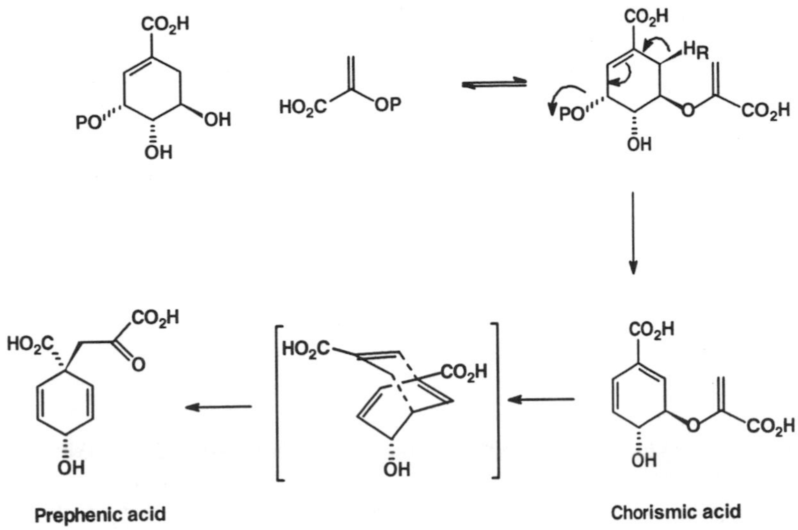

Scheme 5.4

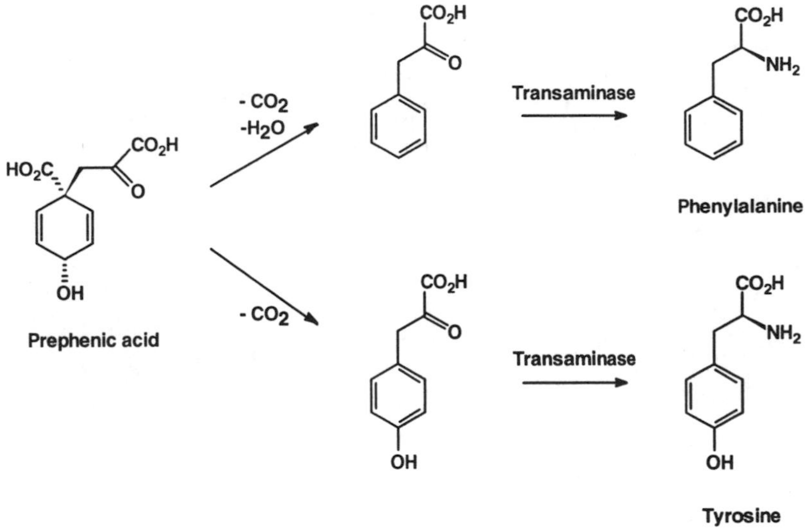

Scheme 5.5

loss of carbon dioxide and water from prephenic acid gives phenylpyruvic acid, which is transaminated enzymatically (pyridoxamine) to phenylalanine. Alternatively, the carboxyl group can be lost oxidatively, and the resulting *p*-hydroxyphenylpyruvic acid gives tyrosine upon transamination (Scheme 5.5).

The biosynthesis of tryptophan (and other indole derivatives) also starts with chorismate. Amination and loss of pyruvate leads to anthranilic acid (Scheme 5.6).

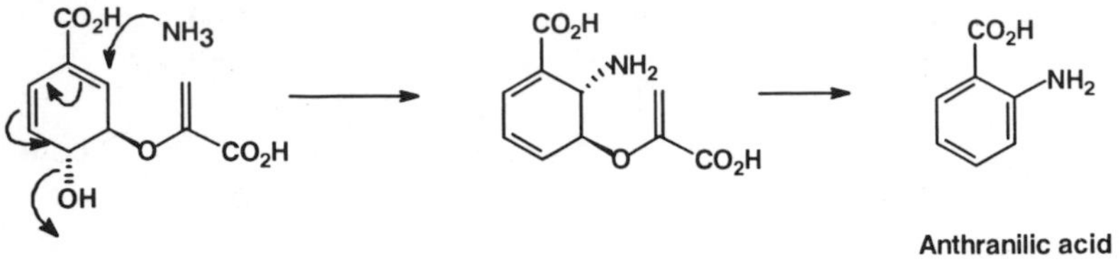

Scheme 5.6

Condensation of anthranilic acid with phosphoribosyl pyrophosphate gives an amino sugar intermediate. Opening of the cyclic hemiaminal followed by Amadori rearrangement, which transforms the α-hydroxyimine into an α-amino ketone, gives an intermediate suitable for the cyclization of the indole ring to give indole-3-glyceryl phosphate. Its reaction with the enamino acid derived from serine gives tryptophan after elimination of glyceraldehyde (Scheme 5.7). Remarkable in this synthesis is the fact that none of the original chiral centres of ribose or serine are conserved! The chiral centre of tryptophan (corresponding to that of serine) is formed by a stereoselective alkylation of the indole portion.

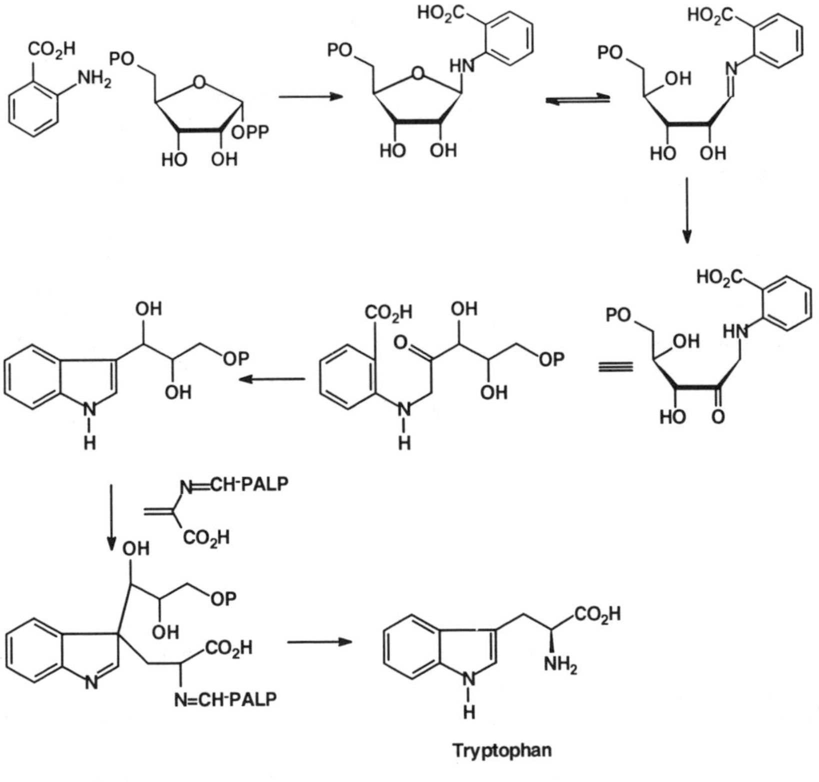

Scheme 5.7

5.6 Asymmetric Synthesis of Amino Acids

The asymmetric synthesis of amino acids has grown in importance over the past decade or so. Modified (non-natural) amino acids have been shown to be useful in various ways for the modification of existing peptide structures. The natural amino acid residues leave the peptide sequence vulnerable to enzymatic (and to some extent non-enzymatic) hydrolysis which can be overcome favourably through the use of α-alkylated amino acid analogues. Alkylation hinders the attack of a nucleophile considerably (the requisite tetrahedral transition state would suffer from a neo-pentyl type strain). Simultaneously, the chiral centre is also blocked against epimerization, which is often of considerable aid in the synthesis of such modified peptides. A further point to be taken into account is the fact that these substitutions usually have a pronounced effect on the conformational freedom of the two important peptide torsion angles, the Φ and Ψ angles, which define the secondary structure of the peptide. Roughly speaking, the (Φ,Ψ) angles for the various secondary structures are: right handed α-helix $(-60°, -60°)$, left-handed α-helix $(60°, 60°)$, parallel β-pleated sheet $(-120°, 120°)$, antiparallel β-pleated sheet $(-140°, 140°)$, extended chain $(180°, 180°)$, collagen helix $(-60°, 140°)$ and 3_{10}-helix $(-60°, -30°)$. The β-turns are characterized by the angles of two consecutive amino acid residues; for instance, for the type II β-turn the angles of the first are around $(-60°, 120°)$ and those of the second around $(80°, 0°)$. The angle ω is $0°$ for *cis* peptide bonds and $180°$ for *trans* peptide bonds. The direction of the side chain is described with the angle χ, which can vary to a much larger degree than the Φ, Ψ or ω angles (Figure 5.12).

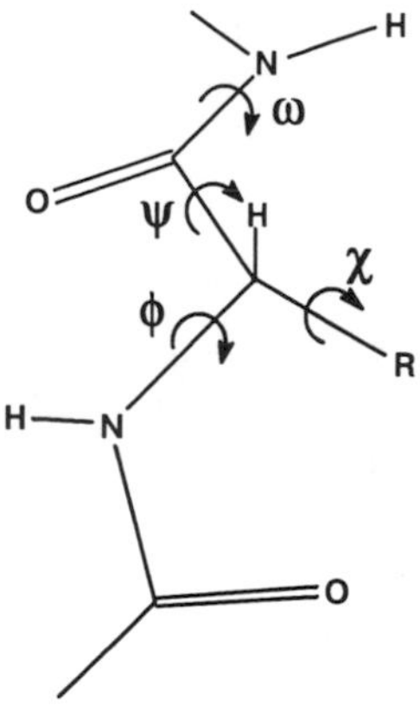

Figure 5.12

It is known that, for instance, α-methyl amino acids favour a β-turn-like structure, whereas α-ethylation often leads to a bias towards an extended, β-sheet-like substructure. The synthesis of α-alkylated amino acids in the chiral form is important, and numerous routes have been designed towards this end. We will briefly review some of these methods.

Perhaps the most general route is the one developed by Schöllkopf, which relies on the use of bislactim ethers.[10-12] The requisite reagents are simply derived from diketopiperazines through O-alkylation (e.g. with trimethyloxonium tetrafluoroborate). Treatment of the bislactim ether with a base yields the anion which is selectively alkylated on one side (the other face of the lactim ring is blocked effectively by the alkyl substituent on the remaining chiral centre). Hydrolysis of the product yields two amino acids of which the α-alkylated amino acid now formally has the original α-hydrogen replaced with the alkyl group (Scheme 5.8).

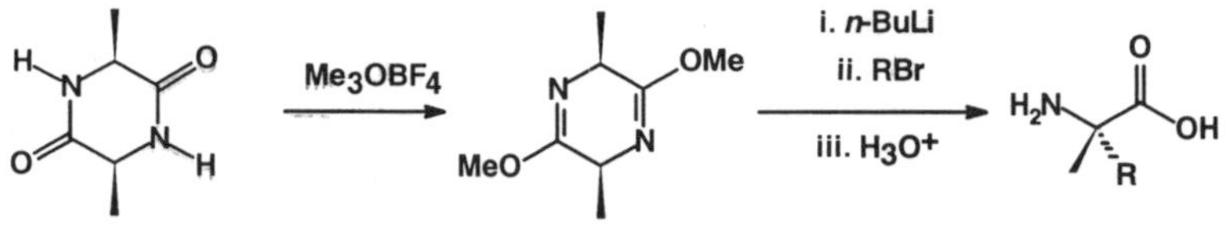

Scheme 5.8

The bislactim ethers also react with α,β-unsaturated carbonyl compounds in a Michael sense (Scheme 5.9) to give, after unravelling the amino acid moiety, β-substituted glutamic acid derivatives with high diastereoselectivity ($>150:1$).[13]

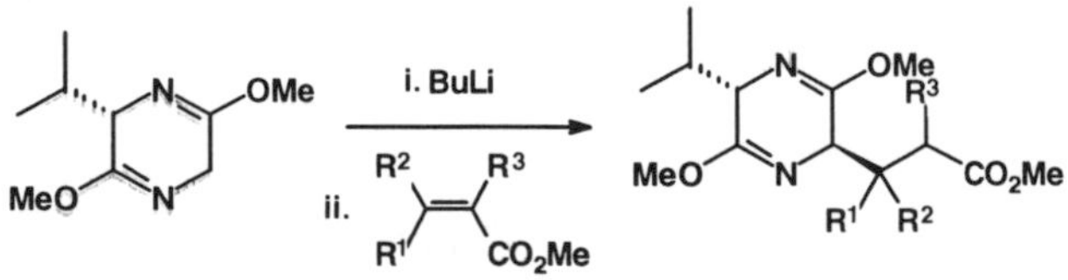

Scheme 5.9

Seebach has developed an ingenious method utilizing self-reproduction of chirality in an enolate alkylation in the context of α-alkyl amino acid synthesis.[14] In a typical application of this method, proline is first converted to an oxazolidinone with pivalaldehyde. A new chiral centre is created highly selectively in such a manner that the bulky tert-butyl group is placed exo (convex face) with respect to the oxazabicyclo[3.3.0]octane ring system (Scheme 5.10). Enolization of the lactone followed by enolate trapping with an electrophile leads to the alkylated product which can be liberated as the free amino acid by acidolysis. Enolization leads to the destruction of the original chiral centre through conversion to an sp^2 centre, and alkylation is directed solely on the convex face of the bicyclic system. Thus, the overall effect is the replacement of the α-hydrogen with an alkyl group with retention of configuration.

This method has been used also in a synthesis of more elaborate, conformationally constrained amino acid analogues for structural and antibody recognition studies.[15] The allylproline derivative was coupled with an amino acid, and processing of the olefin gave the spirocycle (Scheme 5.11). Lemieux–Johnson oxidation (OsO_4, $NaIO_4$) with reductive work-up gave the alcohol which was cyclized under Mitsunobu conditions.

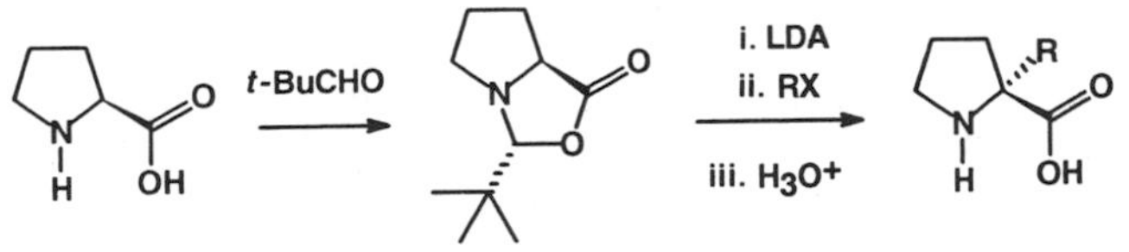

Scheme 5.10

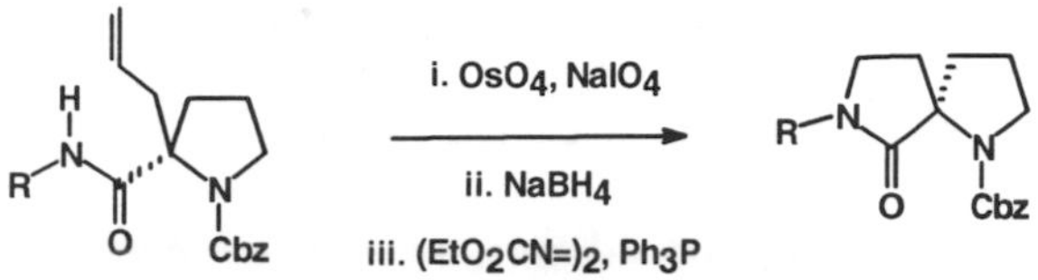

Scheme 5.11

The spirocyclic amino acid analogue functions as a rigid mimic of a β-turn, as shown by solution phase NMR studies as well as molecular modelling.

Williams has applied this alkylation in the enantioselective synthesis of the *Penicillium brevicompactum* toxin (−)-brevianamide (Scheme 5.12).[16] Allylation of the proline derivative was followed by the formation of the diketopiperazine moiety of the target. Ozonolytic cleavage of the allyl chain was followed by chain extension and eventual cyclization to furnish the target compound.

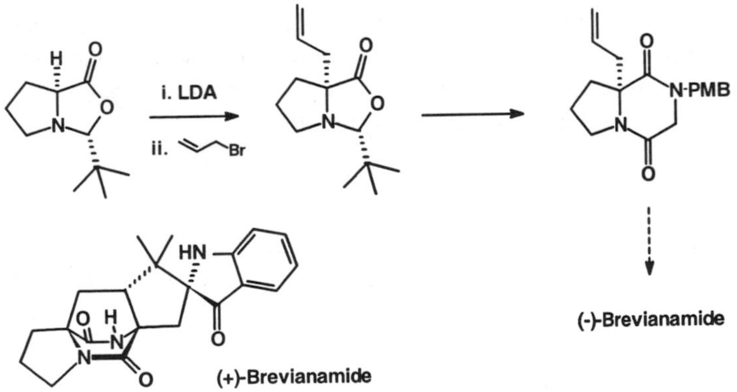

Scheme 5.12

The method has precedent in the synthesis of α-hydroxy acids, where the formation of a bicyclic intermediate is prohibited.[17] Alkylation of the dioxolanone (Scheme 5.13) furnishes the α-alkyl-α-hydroxy acid with retention, as in the case of amino acids, but this time the reasons for the stereoselectivity are reversed (the *tert*-butyl group directs the alkylation on the opposite face of the five membered ring).

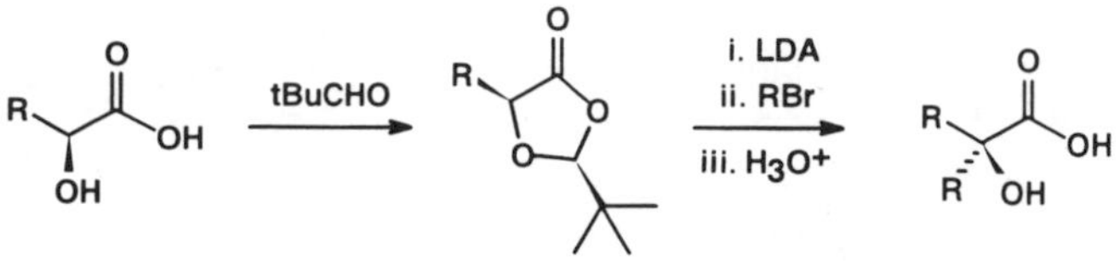

Scheme 5.13

An extension of the methodology allows the synthesis of both α- and β-amino acid derivatives starting with chiral glycine and β-alanine enolates.[18-20] Both the glycine and β-alanine enolates alkylate on the face opposite to the existing *tert*-butyl group (Scheme 5.14), in contrast to the bicyclic proline enolate case discussed previously.

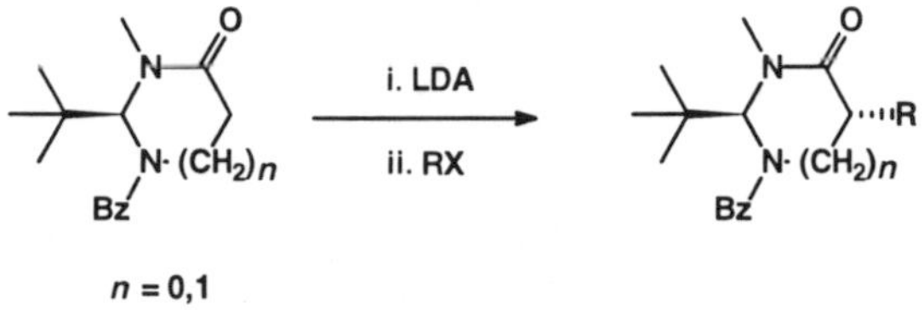

Scheme 5.14

Another method for introducing alkyl substituents at the α-position of an existing amino acid relies on the use of azetidinyl derivatives.[21] Deprotonation with LDA leads to the ester enolate which can be alkylated with high selectivity (typically >93 % de) with a variety of electrophiles. The β-lactam moiety (Scheme 5.15) is destroyed by hydrogenolysis followed by acidolysis to liberate the α-alkyl amino acid derivative. The overall applicability of this sequence is somewhat overshadowed by the fact that a rather elaborate route is needed to construct the starting material, and that after the requisite transformations the chiral auxiliary is completely lost.

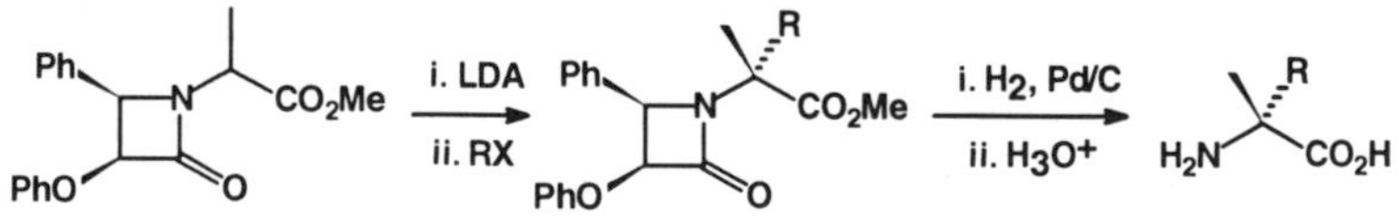

Scheme 5.15

For the synthesis of simple, non-natural amino acids bearing only a single alkyl substituent at the α-carbon, Williams has developed an interesting route relying on the use of 2,3-diphenylmorpholinones as starting materials (Scheme 5.16).[22] The chiral morpholinone is commercially available in both enantiomeric forms. Treatment with NBS gives the bromo compound with high stereoselectivity. This compound can be alkylated with retention of stereochemistry with suitable organometallic nucleophiles. Besides organozinc compounds (or zinc catalysis), allyl silanes and silyl enol ethers

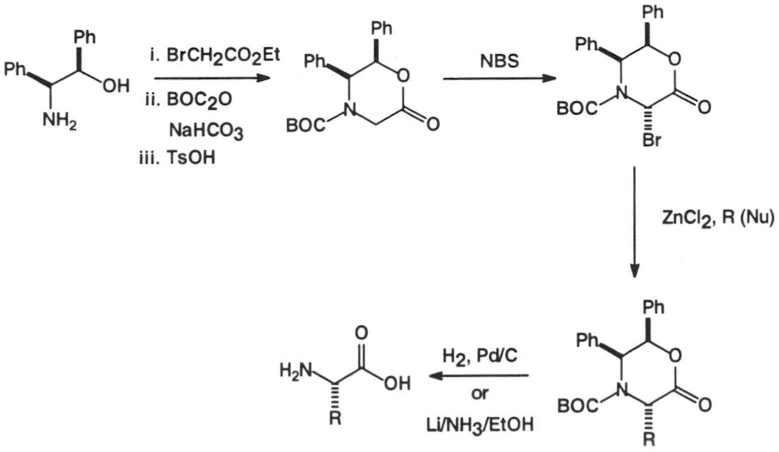

Scheme 5.16

also react. The chiral auxiliary is again cleaved destructively by hydrogenolysis of the benzylic bonds (catalytic hydrogenolysis or dissolving metal reduction) or oxidatively ($NaIO_4$) after deprotection of the nitrogen and hydrolysis of the lactone (TMS-I followed by aqueous acid). The overall enantioselectivities (diastereoselectivities at the stage of alkylation) are high but not exceptional. Allylsilane gave a modest 44 % de, and the TMS enol ether of acetophenone gave a 66 % de in the alkylation.

A more versatile route giving higher enantioselectivities is based on the alkylation of the oxazinone enolate with an electrophile.[23] Both mono- and disubstituted amino acids can be synthesized in high yields. In the first alkylation it is critical to use either lithium or sodium hexamethyldisilazane (HMDS), as stronger bases (including potassium HMDS) cause decomposition of the morpholinone (Scheme 5.17). The second alkylation correspondingly requires a stronger base. Activated electrophiles are also a necessary condition for the alkylations.

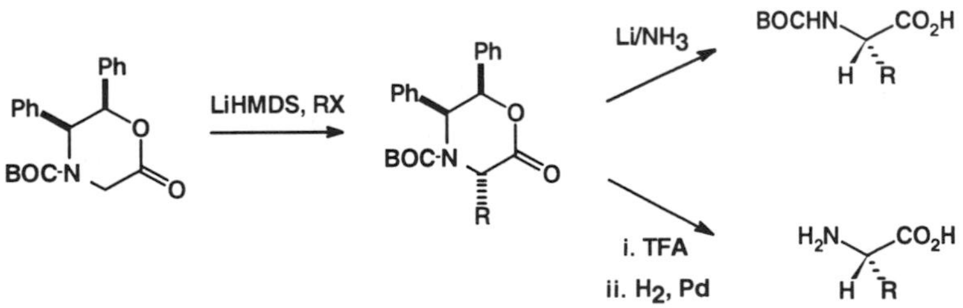

Scheme 5.17

The Evans chiral oxazolidinone auxiliaries are also well suited for use as chiral glycine enolate synthons. The utility of these chiral auxiliaries has been studied during the synthesis of MeBmt, the rare amino acid from cyclosporin A. As the chiral glycine equivalent the authors used the isothiocyanate derived from the chloroacetyloxazolidinone.[24] The enolate of the acyloxazolidinone must be formed with stannous triflate ($Sn(OTf)_2$) since the corresponding lithium and boron enolates gave disappointingly low diastereoselectivities. Tin(II) triflate secured high selectivities (91:9 to 99:1) for the formation of the *syn* aldol products. Epimerization is further suppressed by the formation of a cyclized adduct, preventing retro-aldol reaction as in the case of Seebach's MeBmt synthesis.[25] It is also of interest to note that existence of the chiral centre α to the aldehyde function did not alter the *syn/anti* selectivity of the reaction to any notable degree. The reaction can thus be regarded as being reagent controlled. When the reaction was performed using (2R)-2-methyl-5-hexenal, the intermediate aldol product could be transformed into MeBmt in three steps (Scheme 5.18).

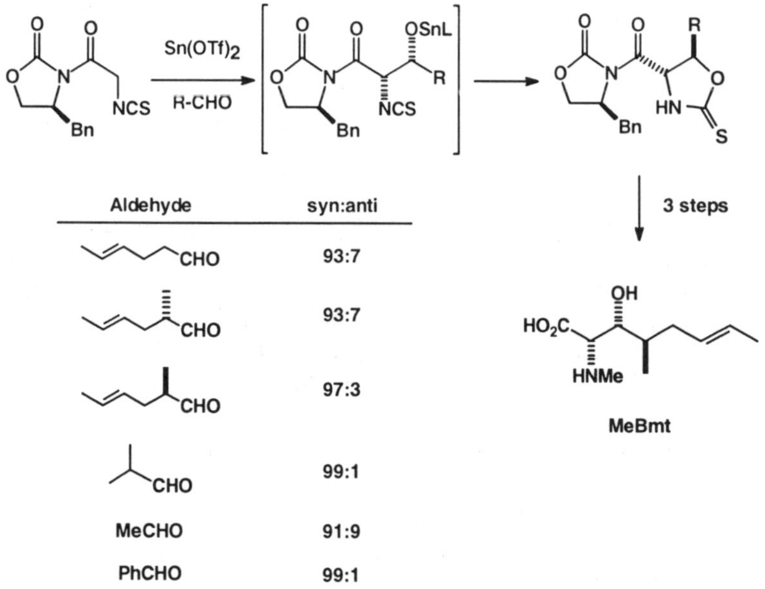

Aldehyde	syn:anti
⌇⌇CHO	93:7
⌇⌇CHO	93:7
⌇⌇CHO	97:3
CHO	99:1
MeCHO	91:9
PhCHO	99:1

Scheme 5.18

Direct amination of ester enolates can be used for the synthesis of α-amino acids. These methods rely on the use of a suitable chiral auxiliary on the acid equivalent, and the Evans chiral acyloxazolidinones can be used efficiently in this conversion. The oxazolidinyl enolates can be aminated directly with sulphonyl azide, or their corresponding α-bromo derivatives can be treated with azide, followed by reduction of the azido function to the amino group.

Chiral arylglycines are constituents of the glycopeptide antibiotics vancomycin and ristocetin. The arylglycine moiety is very prone to racemization owing to the enhanced acidity of the α-protons. Thus, their synthesis provides a challenge to any synthetic route designed for these antibiotics. The Evans oxazolidinones have been used with success in the synthesis of the parent antibiotics as well as some analogues.[26] Azidation (KHMDS, trisyl-N_3)[27] of the acyloxazolidinone yields a sulphonyltriazene intermediate which can be decomposed with potassium acetate to give the α-azido compound in high yield and

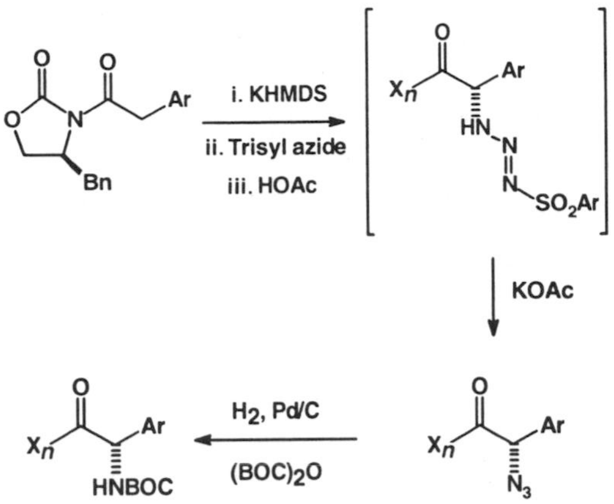

Scheme 5.19

high diastereoselectivity (88:12 to >95:5). Catalytic reduction of the azido group in the presence of di-*tert*-butyl pyrocarbonate gives the N-BOC amino derivative (Scheme 5.19).

The azidation has also been applied to the synthesis of diphthamide (Figure 5.13), the most complex post-translationally modified amino acid known to date.[28] Diphthamide is the target amino acid for the ADP ribosylation of protein synthesis elongation factor EF-2 triggered by the diphtheria toxin. The inhibition of protein synthesis is the explanation at the molecular level for the cytotoxicity of the diphtheria toxin.

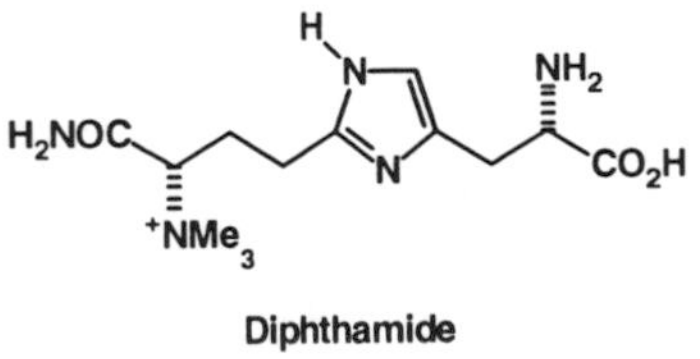

Diphthamide

Figure 5.13

Few practical sources of an electrophilic amino group exist. Chloronitrosoalkanes provide a feasible reagent for this conversion, and 1-chloro-1-nitrosocyclohexane has been used for the synthesis of α-aminocarbonyl compounds (Scheme 5.20).[29] The enolate derived from an acylated camphor sultam exhibited excellent facial selectivity (> 99 % de). The product hydroxylamine can be reduced to the amine with zinc.

The extremely high facial selectivity is explained to arise through a chelation controlled process where the $Z(O)$ enolate is held in the conformation shown in Figure 5.14. The formation of the $E(O)$ enolate is suppressed by unfavourable steric interactions with the camphor-3-methylene moiety. The lower Re face is now open for the approach of the electrophile, giving the observed high facial selectivity.

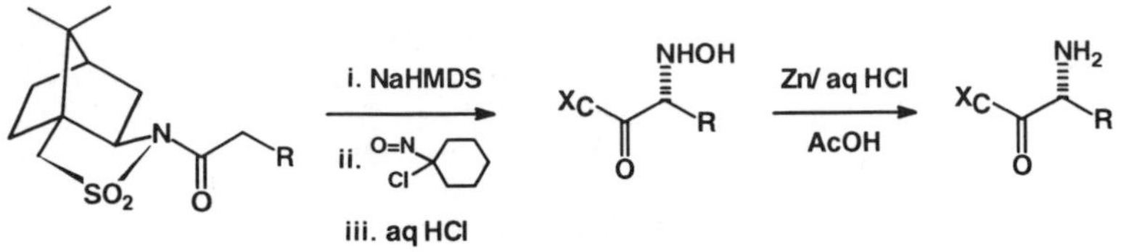

Scheme 5.20

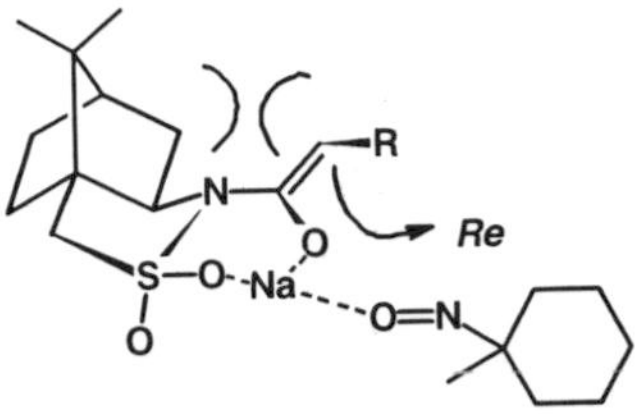

Figure 5.14

Utilizing a similar strategy, Oppolzer has also developed a chiral aminating reagent, based on the α-chloro-α-nitroso reagents, capable of aminating prochiral carbonyl compounds with high enantiofacial differentiation.[30] The product α-amino ketones can be reduced in high yields to the corresponding *anti*-β-aminols (Scheme 5.21), which are valuable synthetic intermediates.

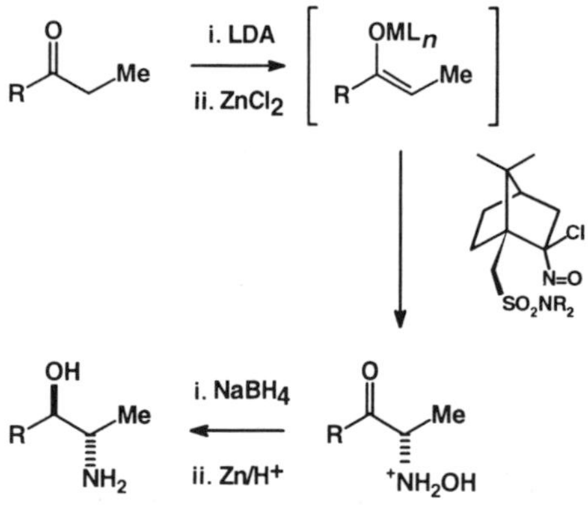

Scheme 5.21

The transmetallation of the enolate to the Zn enolate is necessary for the successful outcome of the reaction as the lithium enolates give eroded enantioselectivities. The stereochemical outcome has been rationalized as occurring though a chelated zinc transition state of the $Z(O)$ enolate (Figure 5.15). The corresponding $E(O)$ enolates derived from cyclic ketones or 2,6-dimethylphenyl propionate reacted sluggishly to give complex mixtures of products.[31,32]

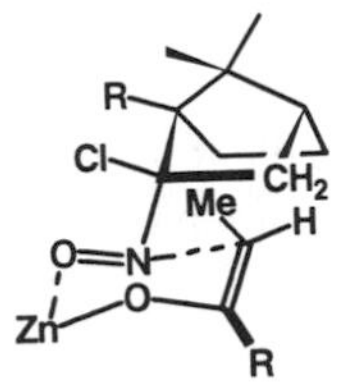

Figure 5.15

Catalytic asymmetric hydrogenation of α-aminoacrylic acid derivatives is an industrially important process. In this process a soluble Rh catalyst is used which is ligated with chiral diphosphine ligands. One of the earliest chiral diphosphines, DIOP (Figure 5.16), was developed by Kagan.[33] This catalyst system gave good levels of asymmetric induction, but further developments of the catalyst were needed before a commercial process could be achieved. The Monsanto process relies on the diphosphine DIPAMP which is chiral at phosphorus, and this is used in the synthesis of L-DOPA (Scheme 5.22).[34]

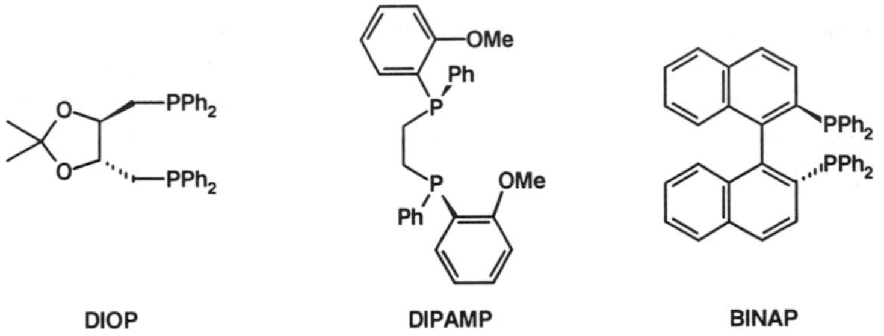

DIOP   DIPAMP   BINAP

Figure 5.16

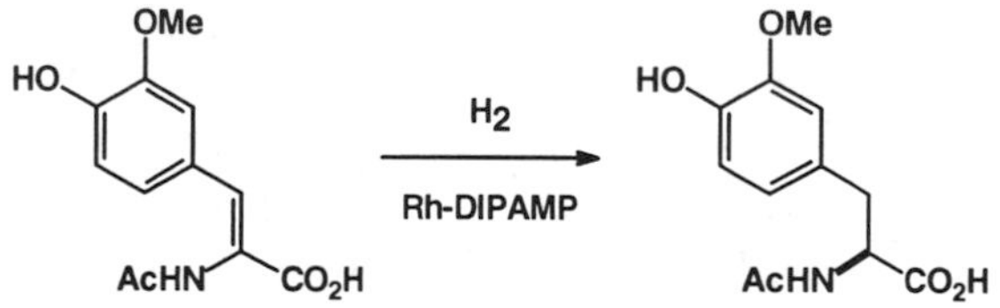

Scheme 5.22

Numerous other ligands have been synthesized with the aim of developing a more general catalytic system for the asymmetric reduction of double bonds. The mechanistic details of the reaction are sufficiently well understood so that this can help the development work. It is known that the enamide forms two complexes with the Rh–DIPAMP catalyst, of which the minor one reacts with hydrogen much faster than the major one.

Dynamic kinetic resolution has also been applied in the synthesis of amino acids. The α-centre of α-acetamido-β-ketobutyrate is susceptible to epimerization. The use of the RuN-BINAP catalyst (Scheme 5.23) in dichloromethane allows nearly complete enantio- and diastereoselection to give N-acetylthreonine in nearly quantitative yield (see Section 2.4).[35]

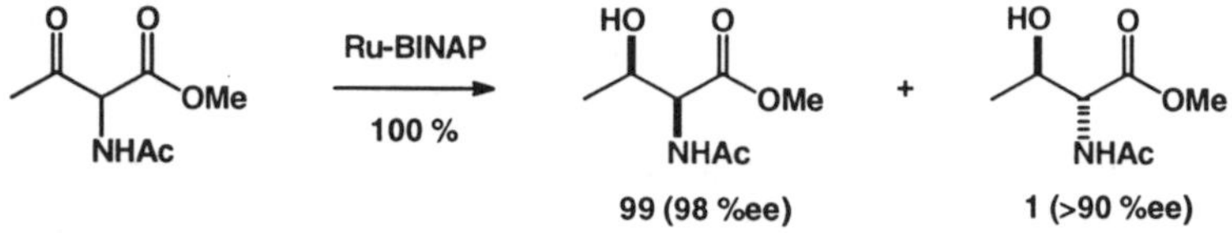

Scheme 5.23

Corey has developed a general catalytic method for the highly enantioselective synthesis of amino acids utilizing the CBS reduction. Trichloromethyl ketones are reduced with high enantioselectivity with catecholborane and the (S)-oxazaborolidine catalyst to the R secondary alcohols.[36] Treatment of these with a basic solution of NaN$_3$ gives the α-azido carboxylic acids with clean inversion of configuration at the α-centre. The azido group is finally converted to an amino group in a standard manner (Scheme 5.24). The method has been applied to the synthesis of widely varying R groups (including *tert*-butyl), and the observed enantioselectivities are high (>92 % ee). The overall conversion from the carbonyl is typically 70–80 %.[37]

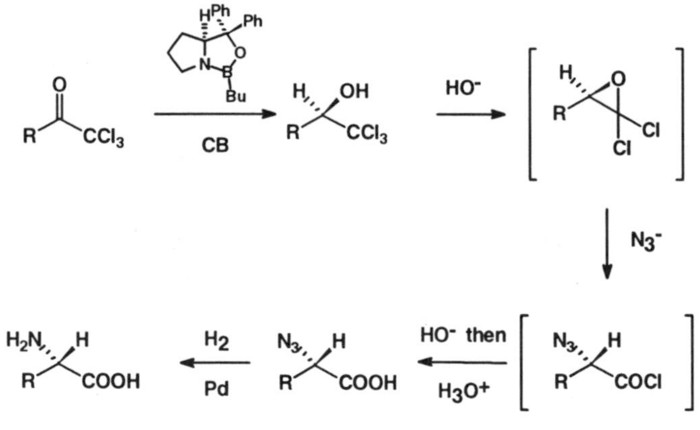

Scheme 5.24

Serine derived cyclic sulfamidates and sulfamidites function as alanyl β-cation equivalents.[38] The sulfamidates are synthesized from serine in five steps in about 50 % yield, and undergo highly regioselective nucleophilic ring opening at the β-carbon with a variety of soft nucleophiles (Scheme 5.25). It is interesting to note the similarity of this strategy to the one Nature adopts in the biosynthesis of tryptophan: in both cases the β-carbon of serine is activated towards reaction with a soft nucleophile.

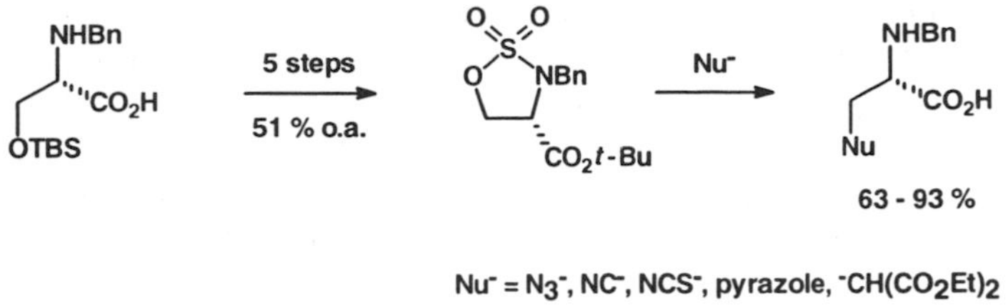

Scheme 5.25

β-Amino acids can be synthesized efficiently from achiral imines utilizing the chiral boron ester enolate developed by Corey.[39] S-*tert*-Butyl thiopropionate (Scheme 5.26), on reaction with the chiral diazaborolidine, gives rise to the $E(O)$ enolate which undergoes rapid addition to imines to give the *anti* products. The diazaborolidine catalyst secures high levels of enantioselectivity (>90 % ee).

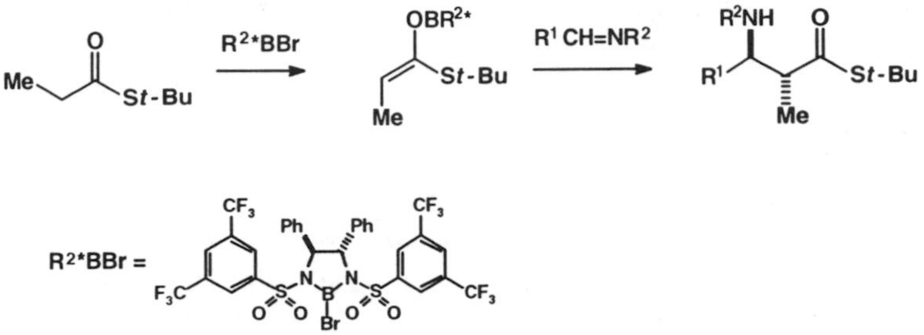

Scheme 5.26

The enantioselectivity is explained as arising from three factors: i) for steric reasons, the thermodynamically less favourable Z-aldimine complexes preferentially with the boron enolate; ii) the addition proceeds via a Zimmermann–Traxler type, six membered transition state; and iii) the transition state structure depicted in Figure 5.17 is favoured for steric reasons.

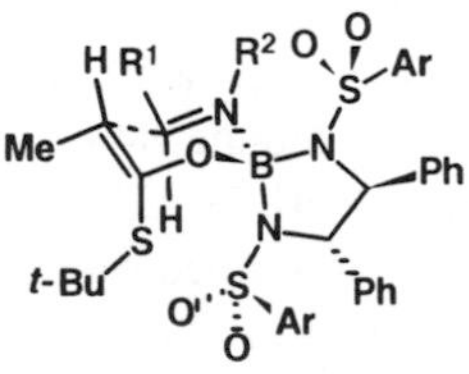

Figure 5.17

With the discovery of the widely applicable acid protease inhibitor pepstatin, a naturally occurring hexapeptide which incorporates a γ-amino-β-hydroxy acid moiety, much interest has been generated towards the synthesis of compounds bearing this structural feature. The most generally accepted synthetic strategy relies on the reaction of an ester enolate with a chiral α-amino aldehyde. The chemistry of α-amino aldehydes has been reviewed recently,[40] and we shall only take a few representative examples of the syntheses of γ-amino-β-hydroxy acids.

The diastereoselectivities of additions onto amino aldehydes are generally very low. The Felkin–Anh transition state model (Scheme 5.27) predicts the formation of the *syn* product, whereas the chelation controlled transition state model favours the formation of the *anti* product.

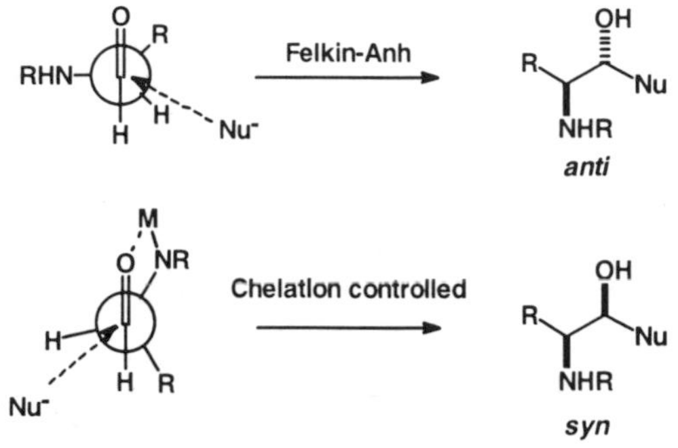

Scheme 5.27

The open transition state usually correlates with the results obtained, but by employing methods favouring the chelation controlled addition the *syn* products (Scheme 5.28) can usually be formed with acceptable diastereoselectivity (*ca* 4:1).[41]

Scheme 5.28

The addition of O-methyl O-trimethylsilyl ketene acetal to carbamate protected amino aldehydes under Lewis acidic conditions (Scheme 5.29) leads to the formation of the *syn* products in diastereomer ratios of around 94:6.[42] The high diastereoselectivity is achieved by using titanium tetrachloride as the Lewis acid. Other chelating Lewis acids also give the *syn* product, albeit in a diminished diastereoselectivity.

In the case of amino acids not capable of chelation, the open Felkin–Anh model can predict the outcome of the reaction quite favourably. This was beneficially utilized in the synthesis of dolaproine, a rare amino acid forming a part of the structure of the reportedly antineoplastic pentapeptide dolastatin 10.[43] Reaction of N-BOC-(S)-prolinal with the Z(O) boron enolate of S-phenyl thiopropionate gave the *syn.anti* product in 64 % yield (Scheme 5.30). The *syn.syn* and *anti.syn* isomers were formed in 10 % and 1 % yields, respectively. The aldol selectivity is thus extremely high (74:1), and the Felkin–Anh model is followed with 64:11 selectivity.

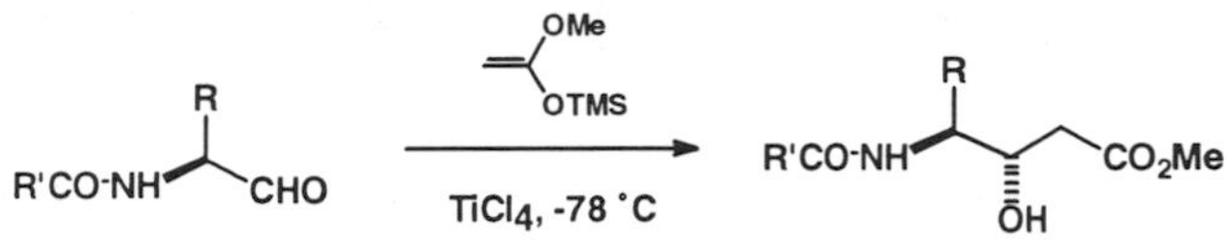

Scheme 5.29

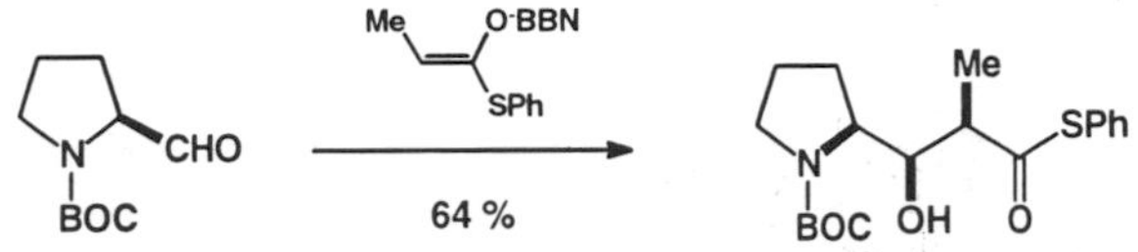

Scheme 5.30

The titanium–carbohydrate complex mediated addition of a glycine enolate to aldehydes gives efficient and economical access to *threo-3*-hydroxy-α-amino acids (Scheme 5.31).[44] The high *syn* selectivity and highly preferred addition of the enolate from the *Re* face of the aldehyde (typical diastereoselectivities >96 % de, enantioselectivities >96 % ee), coupled with the availability of the chiral ligand and the recoverability of the reagents, makes this method very attractive in many applications.

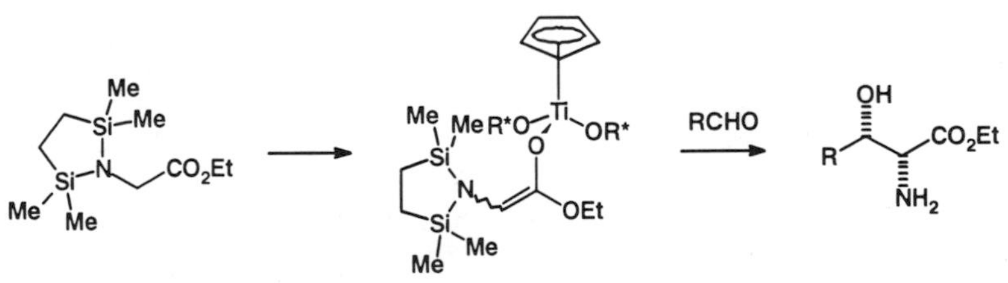

Scheme 5.31

The gold(I) catalysed, aldol type reaction of aldehydes and α-isocyanoacetate esters in the presence of chiral ferrocenylamine ligands possessing both central and planar chirality gives rise to optically active oxazolines in high enantiomeric purity (Scheme 5.32).[45]

The mechanism of the reaction has been studied and a transition state model (Figure 5.18) to account for the observed stereoselectivity has been advanced.[46]

A chiral version of the Cope ene–iminium cyclization has been utilized in the synthesis of substituted pipecolic acid derivatives.[47] The homoallylamine derived from phenylglycinol reacts with glyoxal in aqueous solution to furnish an intermediate iminium species which is directly converted through an intramolecular cyclization to the bicyclic hemiacetal (Scheme 5.33). This can be oxidized, the morpholinone ring cleaved and the benzylamine hydrogenolysed to give the pipecolic acid in enantiopure form.

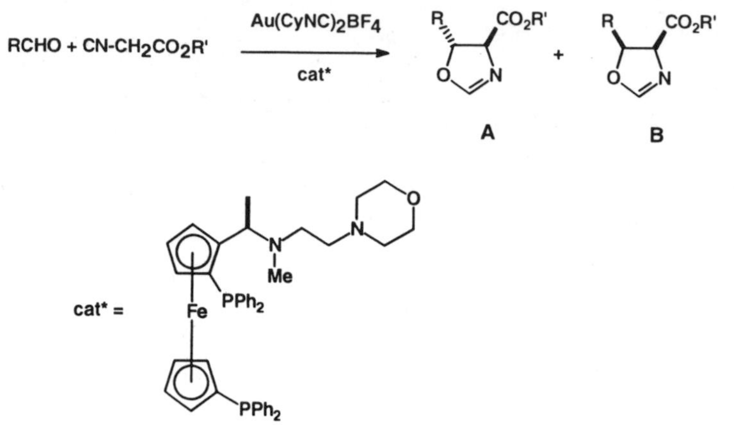

Scheme 5.32

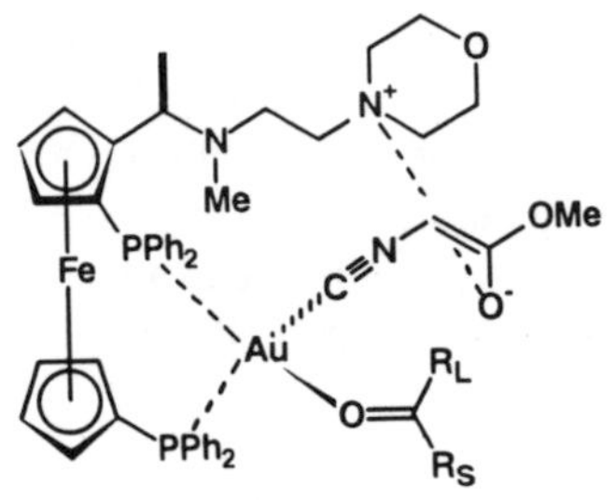

Figure 5.18

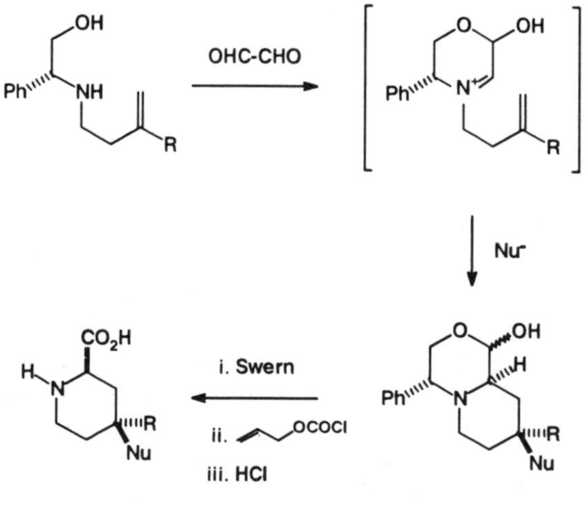

Scheme 5.33

The stereoselectivity in the cyclization is explained by stereoelectronic effects. The ene double bond undergoes stereospecific *anti* addition of the external nucleophile and of the electrophilic iminium double bond. Of the four alternative modes for the addition to the electrophilic iminium double bond (see Scheme 5.34), approaches A and D lead to a boat-like, six membered transition state. On the other hand, axial attack of the ene double bond on the iminium double bond generates the chair form of the incipient morpholine ring. Of the two alternative axial attack pathways, the one arising from the same side as the existing phenyl (path B) is disfavoured because of steric repulsion and path C is the favoured one, as observed.

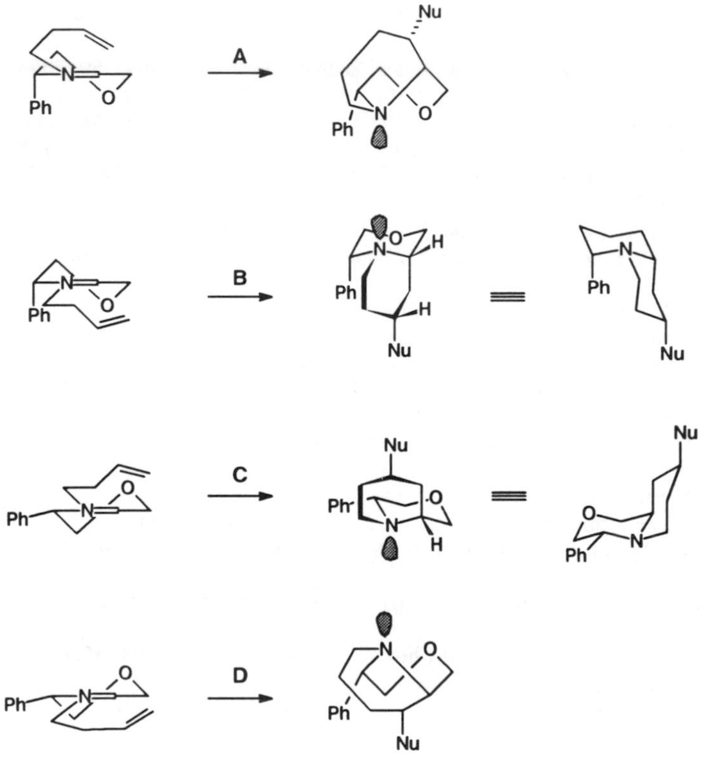

Scheme 5.34

References

1. Wagner, I. and Musso, H. *Angew. Chem., Int. Ed. Engl.* **22**, 816–828 (1983).
2. Bodanszky, M. *Peptide Chemistry, A Practical Textbook* Springer-Verlag: Berlin, 1988.
3. Pauling, L., Corey, R.B., and Branson, H.R. *Proc. Natl. Acad. Sci. USA* **37**, 205–211 (1951).
4. Lesczynski, J.F. and Rose, G.D. *Science* **234**, 849–855 (1986).
5. Milner-White, E.J. *Biochim. Biophys. Acta* **911**, 261–265 (1987).
6. Ollis, D.L. and White, S.W. *Chem. Rev.* **87**, 981–996 (1987).
7. Venkatachalam, C.M. *Biopolymers* **6**, 1425–1436 (1968).

8. Santaniello, E., Ferraboschi, P., Grisenti, P., and Manzocchi, A. *Chem. Rev.* **92**, 1071–1140 (1992).
9. Pelech, S.L. and Sanghera, J.S. *Science* **257**, 1355–1356 (1992).
10. Schöllkopf, U., Hartwig, W., and Groth, U. *Angew. Chem.* **91**, 863–864 (1979).
11. Groth, U. and Schöllkopf, U. *Synthesis* 37–38 (1983).
12. Schöllkopf, U. *Pure Appl. Chem.* **55**, 1799–1806 (1983).
13. Pettig, D., Schöllkopf, U. *Synthesis* 173–175 (1988).
14. Seebach, D., Boes, M., Naef, R., and Schweizer, W.B. *J. Am. Chem. Soc.* **105**, 5390–5398 (1983).
15. Hinds, M.G., Welsh, J.H., Brennand, D.M., Fisher, J., Blennie, M.J., Richards, N.G.J., Turner, D.J., and Robinson, J.A. *J. Med. Chem.* **34**, 1777–1789 (1991).
16. Williams, R.M., Glinka, T., Kwast, E., Coffman, H., and Stille, J.K. *J. Am. Chem. Soc.* **112**, 808–821 (1990).
17. Frater, G., Muller, U., and Gunther, W. *Tetrahedron Lett.* **22**, 4221–4224 (1981).
18. Seebach, D., Juaristi, E., Miller, D.D., Schickli, C., and Weber, T. *Helv. Chim. Acta* **70**, 237–261 (1987).
19. Juaristi, E., Quintana, D., Lamatsch, B., and Seebach, D. *J. Org. Chem.* **56**, 2553–2557 (1991).
20. Seebach, D., Lamatsch, B., Amstutz, R., Beck, A.K., Dobler, M., Egli, M., Fitzi, R., Gautschi, M., Herradon, B., Hidber, P.C., Irwin, J.J., Locher, R., Maestro, M., Maetzke, T., Mourino, A., Pfammatter, E., Plattner, D.A., Schickli, C., Schweizer, W.B., Seiler, P., Stucky, G., Petter, W., Escalante, J., Juaristi, E., Quintana, D., Miravitlles, C., and Molins, E. *Helv. Chim. Acta* **75**, 913–934 (1992).
21. Ojima, I., Qiu, X. *J. Am. Chem. Soc.* **109**, 6537–6538 (1987).
22. Williams, R.M., Sinclair, P.J., Zhai, D., and Chen, D. *J. Am. Chem. Soc.* **110**, 1547–1557 (1988).
23. Williams, R.M. and Im, M.-N. *J. Am. Chem. Soc.* **113**, 9276–9286 (1991).
24. Evans, D.A. and Weber, A.E. *J. Am. Chem. Soc.* **108**, 6757–6761 (1986).
25. Blaser, D., Ko, S.Y., and Seebach, D. *J. Org. Chem.* **56**, 6230–6233 (1991).
26. Evans, D.A., Evrard, D.A., Rychnovsky, S.D., Fruh, T., Whittingham, W.G., and DeVries, K.M. *Tetrahedron Lett.* **33**, 1189–1192 (1992).
27. Evans, D.A., Britton, T.C., Ellman, J.A., and Dorow, R.L. *J. Am. Chem. Soc.* **112**, 4011–4033 (1990).
28. Evans, D.A. and Lundy, K.M. *J. Am. Chem. Soc.* **114**, 1495–1496 (1992).
29. Oppolzer, W. and Tamura, O. *Tetrahedron Lett.* **31**, 991–994 (1990).
30. Oppolzer, W., Tamura, O., Sundarababu, G., and Signer, M. *J. Am. Chem. Soc.* **114**, 5900–5902 (1992).
31. Heathcock, C.H. In *Asymmetric Synthesis* (Morrison, J.D., Ed.) Academic Press: New York, 1984, Vol. 3, 111–212.
32. Masamune, S., Ellinghoe, J.W., and Choy, W. *J. Am. Chem. Soc.* **104**, 5526–5528 (1982).
33. Dang, T.P. and Kagan, H.B *J. Chem. Soc., Chem. Commun.* 481 (1971).
34. Vineyard, B.D., Knowles, W.S., Sabacky, M.J., Bachman, G.L., and Weinkauf, D.J. *J. Am. Chem. Soc.* **99**, 5946–5952 (1977).
35. Noyori, R. *Science* **248**, 1194–1199 (1990).
36. Corey, E.J. and Bakshi, R.K. *Tetrahedron Lett.* **31**, 611–614 (1990).
37. Corey, E.J. and Link, J.O. *J. Am. Chem. Soc.* **112**, 1906–1908 (1992).
38. Baldwin, J.E., Spivey, A.C., and Schofield, C.J. *Tetrahedron: Asymm* **1**, 881–884 (1990).
39. Corey, E.J., Decicco, C.P., and Newbold, R.C. *Tetrahedron Lett.* **32**, 5287–5290 (1991).
40. Jurczak, J. and Golebiowksi, A. *Chem. Rev.* **89**, 149–164 (1989).
41. Holladay, M.W. and Rich, D.H. *Tetrahedron Lett.* **24**, 4401–4404 (1983).
42. Takemoto, Y., Matsumoto, T., Ito, Y., and Terashima, S. *Tetrahedron Lett.* **31**, 217–218 (1990).
43. Tomioka, K., Kanai, M., and Koga, K. *Tetrahedron Lett.* **32**, 2395–2398 (1991).
44. Bold, G., Duthaler, R.O., and Riediger, M. *Angew. Chem., Int. Ed. Engl.* **28**, 497–498 (1989).
45. Ito, Y., Sawamura, M., and Hayashi, T. *J. Am. Chem. Soc.* **108**, 6405–6406 (1986).
46. Togni, A. and Pastor, S.D. *J. Org. Chem.* **55**, 1649–1664 (1990).
47. Agami, C., Couty, F., Poursoulis, M., and Vaissermann, J. *Tetrahedron* **48**, 431–442 (1992).

6 Nucleosides, Nucleotides and Nucleic Acids

Nucleic acids play an important role in biochemistry. They are the constituent parts of deoxyribonucleic acid (DNA) and ribonucleic acid (RNA). The former is the carrier of the cellular and organism's hereditary information, and the latter functions in signal transduction molecules for the synthesis of proteins; the information embedded in DNA is read and transcribed into RNA within the nucleus of the cell, and this information is then passed on to other parts of the cell where the sequence information in the messenger RNA is translated into a protein sequence.

In this chapter we shall not study the necessary functions of DNA or RNA, fascinating as they are in their chemistry, but these are left to the realms of biochemistry and molecular biology. Instead, we shall inspect molecules that are structurally related to the nucleic acid constituents that have medicinal importance.

The nucleic acids are formed from three structural building blocks: the base, the sugar and phosphoric acid. The base is a heterocyclic compound which belongs either to the purine or the pyrimidine bases. The purine bases adenine and guanine are common to both DNA and RNA, but there is variation in the pyrimidine bases: cytosine and thymine are found in DNA, and in RNA thymine is replaced with its demethylated analogue, uracil (Figure 6.1).

The sugar units in DNA and RNA form the foundation of the structural and, to some extent, functional differences between the two types of nucleic acid heteropolymers. In DNA, the carbohydrate is 2-deoxyribose, whereas in RNA it is ribose (Figure 6.2).

When a nucleic acid base (one of the five bases) is connected to either one of the sugars, a nucleoside is formed in the DNA series and a deoxynucleoside in the RNA series. Esterification with phosphoric acid gives the five nucleoside phosphates, or nucleotides, which are the basic building blocks of DNA and RNA. The nucleic acids differ both in their structure and function. The three-dimensional structures of DNA and RNA are different, and this most plausibly is also the structural reason for their

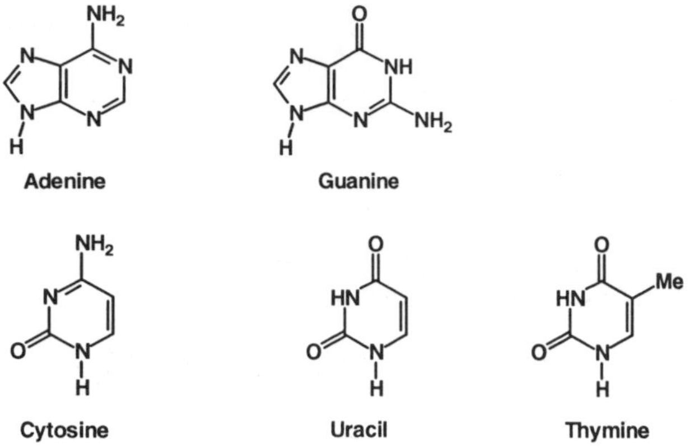

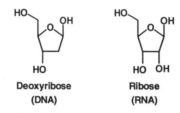

Adenine   **Guanine**

Cytosine   **Uracil**   **Thymine**

Figure 6.1

Deoxyribose   **Ribose**
(DNA)   **(RNA)**

Figure 6.2

different functions in cells. DNA is, basically, the information storage molecule in cells; the genetic information retained from one generation of cells to another is stored in the architecture of DNA. When the synthesis of a particular protein is needed, the information content of the DNA segment that codes the protein is transcribed into the RNA language—a messenger RNA (mRNA) is synthesized. The mRNA molecule is equipped with a number of signalling devices which relay the requisite information to the target site of the subsequent cellular events (transport from the nucleus to cytosol, transport to the different organelles in the cytosol, information regarding whether the peptide or protein that is synthesized is to be retained within the cell or excreted, etc.).

For the storage of genetic information, it is obviously important to have a relatively stable structure which is resistant to deleterious and random chemical transformations. At the same time, however, the structure must be accessible at the time when this information is processed, such as transcription to the RNA language. Since soluble RNA (such as mRNA) is processed relatively rapidly, and accumulation of different mRNA molecules is actually an unwanted process, this stability problem does not concern RNA as much as it concerns DNA. DNA must survive much longer periods of time under sometimes rather heavy bombardment from its environment. Through evolution, Nature has developed a finely tuned system capable of conserving the information (and even repairing some of the damages caused by external agents), at the same time retaining an efficient machinery for unravelling the information rapidly and accurately as it is needed. We shall not discuss the enzymatic basis for this machinery, nor shall we discuss the various ways of forming the physical forms or the ways of packing and unpacking the genetic information into the complex structures that form the chromosomes and other protein–nucleic acid complexes. We shall, however, take a brief look at the structural basis for the interactions and basic recognition events that form the foundations of molecular biology, i.e. the structure of DNA.

A closer look at the nucleic acid bases shows that all the bases contain functional groups capable of hydrogen bonding, either participating as hydrogen bond donors or acceptors. Such considerations led James Watson and Francis Crick to propose that DNA can form structures where one strand of the polynucleotide is matched with a complementary strand to form the maximum number of such interstrand hydrogen bonds which will impart stability to the complex of two strands relative to the isolated strands. This proposal was put forth some four decades ago, and it quickly gained experimental evidence to support it. The Watson–Crick base pairing rules are now widely accepted, and accordingly the base pairing is highly regular: adenosine (A) only forms stable pairs with thymine (T), and similarly guanine (G) pairs with cytidine (C). Watson and Crick also suggested how these base pairs are formed, and these are shown in Figure 6.3.

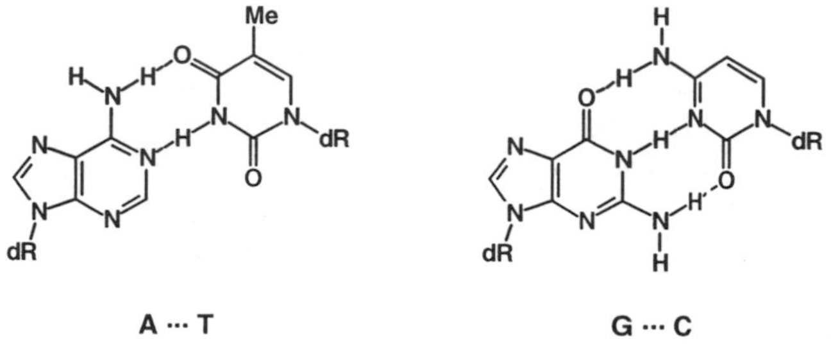

A ··· T   G ··· C

Figure 6.3

An alternative hydrogen bonding pattern, the Hoogsteen base pairing, is also possible. This is illustrated in Figure 6.4. The formation of a triple helix occurs by binding either a thymidine to an AT pair or a cytidine to a GC pair.

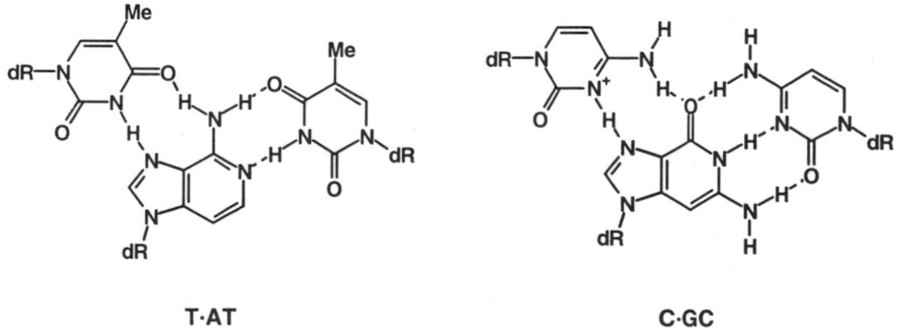

T·AT   C·GC

Figure 6.4

Another structural feature must be also taken into account, and that is the backbone of the sugar chain, i.e. the deoxyribose units connected through phosphodiester linkages. It has been suggested, and indeed verified by both X-ray crystallographic determinations and molecular mechanics and dynamics calculations, that the most stable conformation of the sugar phosphodiester backbone consists of two salient structural features. Firstly, the phosphodiester units adopt the energetically favourable gauche–gauche conformation, and secondly the O—C—C—O torsion angle in the deoxyribose unit is approximately 80°. Together these two structural features impart helicity to the DNA strand which is further strengthened by the interstrand hydrogen bonds. With relatively small changes in these three structural parameters, one can arrive at the three basic structural motifs of double-stranded DNA, which are known as the A-, B- and Z-form double helices.

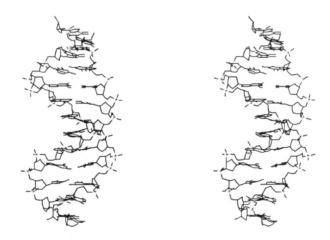

Figure 6.5

The structure of the B-form DNA, shown in Figure 6.5, exhibits two further structural features which are of importance in recognition events. The clefts between the two carbohydrate backbone strands, the major and minor grooves, provide different environments for the interacting species. Larger proteins, and occasionally a third DNA strand, can only bind to the more hydrophilic major groove, but the more hydrophobic minor groove provides a very snug fit for 'flat' hydrophobic molecules such as the DNA binding antibiotic distamycin A (Figure 6.6). The minor groove also allows close contact between the nucleic acid bases and the binding molecule, which is manifested in the higher sequence selectivity of recognition. Mitomycins are another class of DNA binding agents that bind through the minor groove and alkylate the DNA strands.

The minor groove is also the entry site for smaller, typically aromatic molecules which insert between two consecutive 'layers' of the bases. This *intercalation* causes a structural change in the double helix leading to lower stability and higher susceptibility to strand cleavage. The ene–diyne antibiotics (e.g. calicheamicin γ_1^I, see Chapter 1) and many antitumour compounds such as the anthracyclinones (e.g. daunomycin and adriamycin, Section 4.5) exert their action in this way.

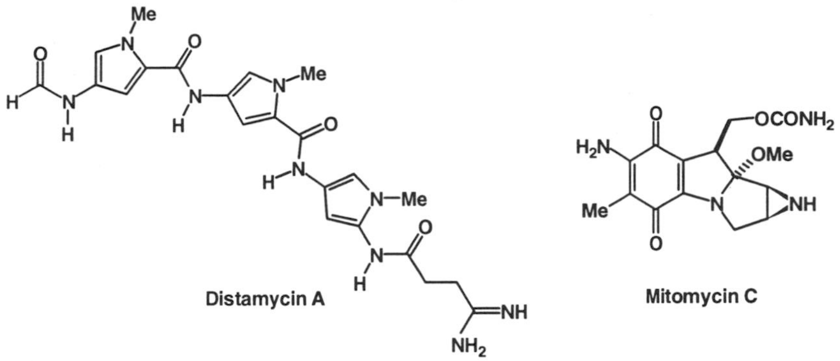

Figure 6.6

The intracellular energy storage system is provided by adenosine triphosphate, ATP. In this compound adenosine is esterified with a high energy triphosphate unit whose hydrolysis liberates a considerable amount of energy (ca 31 kJmol^{-1}). This energy can be used in, for example, protein synthesis, where the formation of each peptide bond consumes ca 2 kJmol^{-1}.

In biological systems, one often finds formal hydride reductions which resemble the usual sodium borohydride or lithium aluminium hydride reductions. One biological reductant capable of hydride delivery is nicotinamide adenine dinucleotide hydride, NADH (Figure 6.7), which functions as a coenzyme in such reactions.

In chemical signal transduction, many transmitters function by coupling their action with the hydrolysis of cyclic AMP (cAMP; Figure 6.7). This compound is a typical secondary messenger.

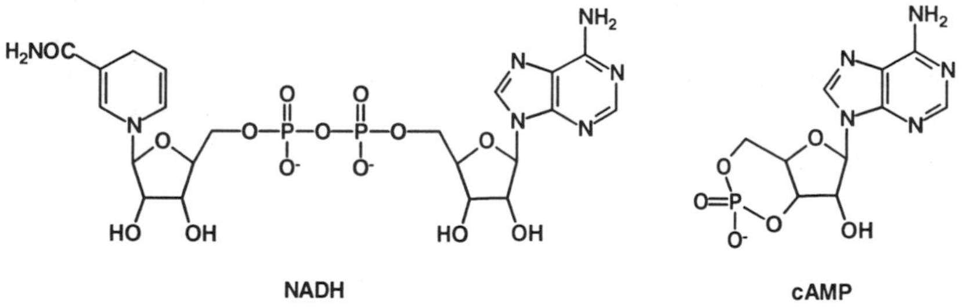

Figure 6.7

8-Azaadenosine and 1-β-D-arabinofuranosyladenine (araA) are typical examples of nucleoside antibiotics (Figure 6.8). Many nucleoside antibiotics have recently gained interest in the treatment of forms of cancer. The uridine derivative doxyfluridine (Figure 6.8) has recently been brought to market, and it shows excellent efficiency against forms of breast, stomach and intestinal cancer. Also the first AIDS drug, 3'-azidothymidine (AZT), is a nucleoside antibiotic.

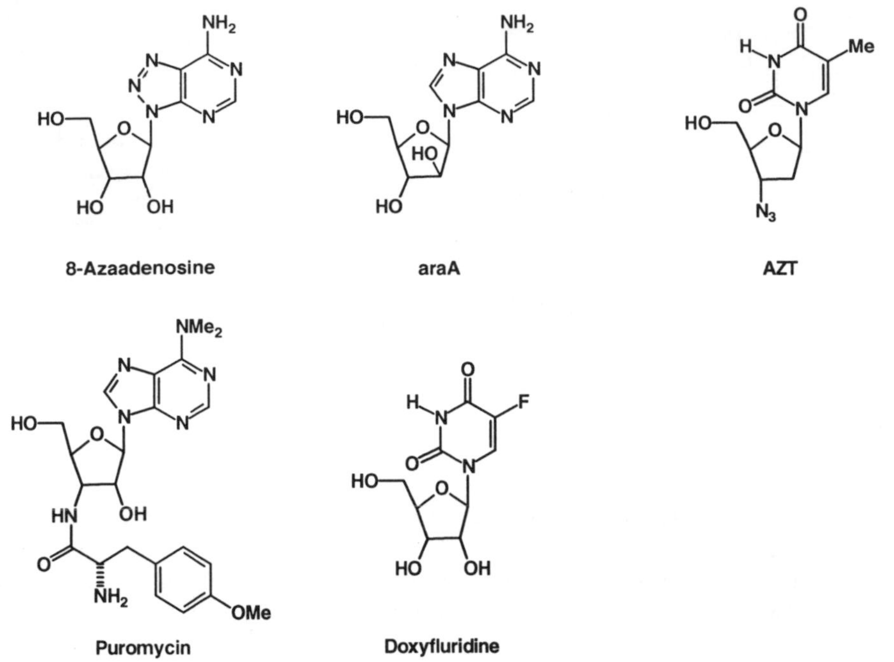

8-Azaadenosine   araA   AZT

Puromycin   Doxyfluridine

Figure 6.8

Puromycin is an adenosine derivative which inhibits the translation of RNA in protein synthesis. The compound is an effective antibiotic against trypanosomiasis and amaebosis, diseases typical of tropical regions (trypanosomiasis accounts for roughly 25 % of deaths in these regions). The compound, however, is very toxic, and therefore cannot be widely used in preventive medicine.

Cytokines are adenine derivatives which function as growth hormones in plants. Cytokines combine with messenger RNA (mRNA) and thereby regulate its function. A few dozen cytokines are known, and zeatine (Figure 6.9) is a typical example. Besides the growth promoting effect on plants, cytokines have been shown to have similar effects on the cells of animals, bacteria and fungi.

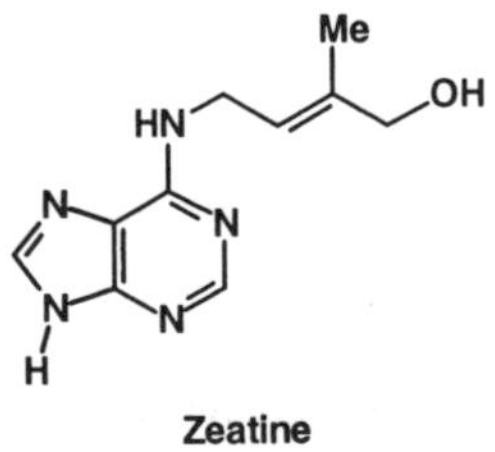

Zeatine

Figure 6.9

7 Polyketides

The lengthening of carbon chains of biological molecules occurs by adding a two-carbon unit into the molecule. Those natural products whose biosynthesis involves few other transformations, and which are distinguishably derived from the two-carbon fragments by straightforward chain extensions and reductions/oxidations, are typically grouped into polyketides. As the name implies, acetate units can be replaced with longer carbon chains. The substituted two-carbon units form the main chain of these natural products, as shown for the structure of erythronolide B (Figure 7.1), a biogenetic precursor of all the erythromycins. The bold lines indicate the propionate units which come intact from propionyl-CoA.

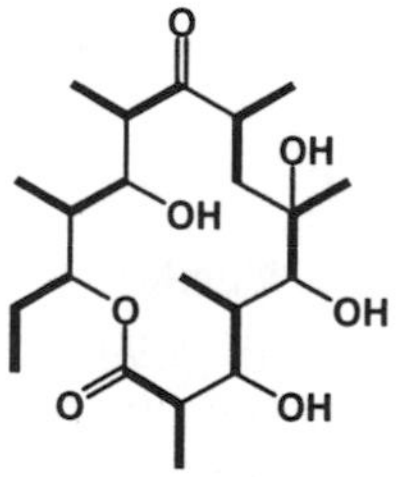

Erythronolide B

Figure 7.1

Polyketides are formally synthesized by aldol reactions, and therefore it is no surprise that over the last couple of decades this class of natural products has provided most of the driving force to the development of new synthetic methods. The need for stereospecific generation of 1,2- and 1,3-dioxygenated systems anticipates thoroughly developed 1,2-hydroxylation methods (epoxidation or dihydroxylation); 1,3-dioxygenated systems similarly require the methodology for the aldol and related reactions. These reactions provided the cornerstones in the era of racemic synthesis, and they also seem to play a similarly important role in the era of asymmetric synthesis.

7.1 Biosynthesis

As we have already seen, the elongation of carbon chains usually occurs through the addition of two carbon atoms at a time. In Chapter 1, we saw the biosynthesis of acetic acid from carbon dioxide and water. The activation of acetic acid towards both electrophiles and nucleophiles is achieved by the reaction of the acetate with coenzyme A (CoA; Figure 7.2) to form a thioester. The thioester is much more acidic at the α-position. It is also relatively stable but still easily cleaved under physiological conditions. Substituents are acceptable, and this facilitates the biosynthesis of, for example, polypropionates. These enzymatic reactions require the participation of a derivatizing agent, another molecule in addition to the substrate. Coenzyme A exerts its function by carrying a molecule or a part of it into the reaction. The coenzyme functions to activate the acyl group towards nucleophilic attack, just like in syntheses we try to achieve this by means of a variety of activating groups. Since acetate (or thioacetate) α-protons still have relatively low acidities, further activation is often needed. This can be brought about by further carboxylation to give a malonate unit. Malonyl-CoA is a much more powerful nucleophile than the corresponding acetyl-CoA.

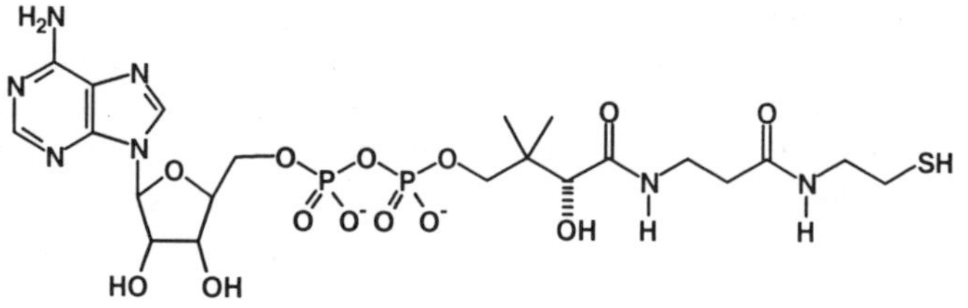

Coenzyme A

Figure 7.2

Because of the nature of the chain lengthening process, natural carbon chains most often contain an even number of carbon atoms; both of the acetate carbon atoms are joined into the growing carbon chain. This process is shown in Scheme 7.1.

Instead of an acetyl group, CoA and the corresponding acyl carrier protein (ACP) can also carry an acetoacetyl or a propionyl group. Scheme 7.2 outlines the participation of an acetoacetyl-CoA in the biosynthesis of mevalonic acid, the ubiquitous intermediate in the biosynthesis of the isoprene unit needed for the terpenoid synthesis.

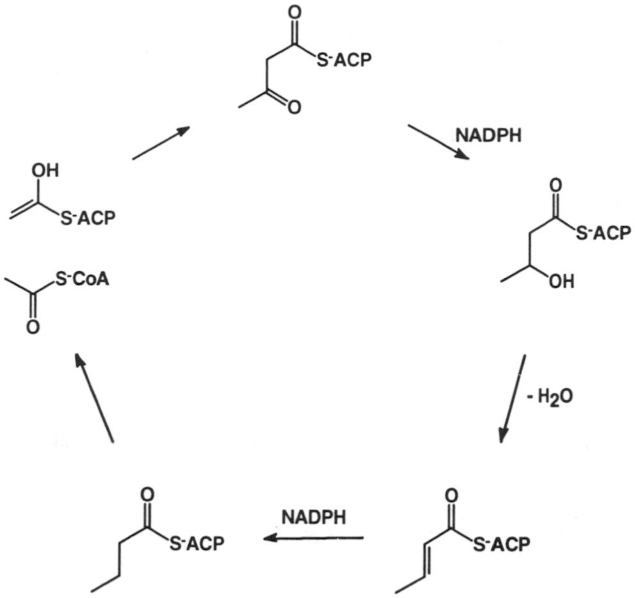

Scheme 7.1

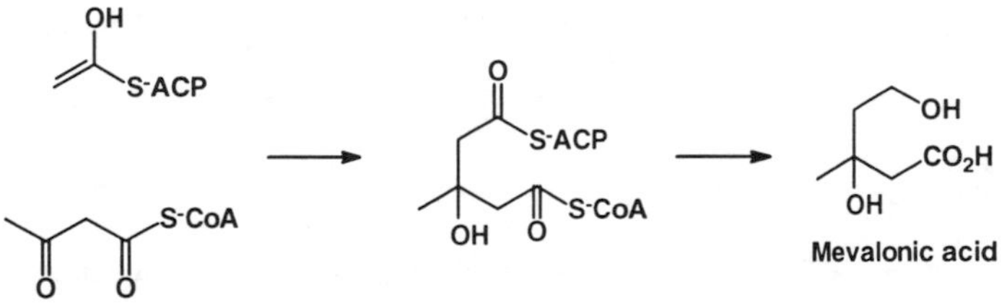

Scheme 7.2

The natural C-acylation, a relative of the Claisen condensation, leads to the incorporation of two carbon atoms at a time. This is clearly reflected in the composition of fatty acids (lipids); by far the most common carbon chains contain an even number of atoms.

Polyketides are commonly grouped according to the number of acetate units they contain: four acetates form a tetraketide, five a pentaketide, etc. The initial product in the condensation is a β-keto ester. This is usually then reduced to the corresponding alcohol or deoxygenated to the fully saturated compound. The alcohol intermediate can further eliminate a molecule of water to give rise to the unsaturated polyketides.

7.2 Fatty Acids

Fatty acids, fatty alcohols and phospholipids belong to the polyketides. Typical fatty acids contain a long alkyl chain with an even number of carbon atoms in the main chain.

The chain is seldom branched, but unsaturation is often encountered. For instance, the 18-carbon straight chain acids palmitic, oleic, linoleic and linolenic acids (Figure 7.3) are all known and found, for instance, in soy bean lecithin. The polyunsaturated C_{20} acids form the basis of the arachidonic acid cascade, which will be discussed in connection with prostaglandins, thromboxanes and leukotrienes. The unsaturation is typically near the centre of the chain (Δ^9 is the common site, regardless of the chain length), and the double bond geometry is usually *cis*. The polyunsaturated fatty acids typically contain skipped polyene units, where the two olefinic bonds are separated by a methylene unit.

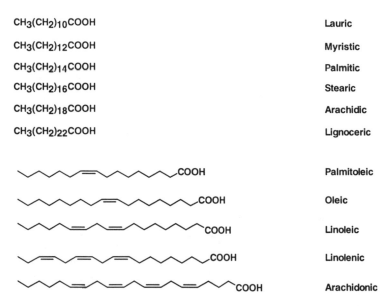

$CH_3(CH_2)_{10}COOH$	**Lauric**
$CH_3(CH_2)_{12}COOH$	**Myristic**
$CH_3(CH_2)_{14}COOH$	**Palmitic**
$CH_3(CH_2)_{16}COOH$	**Stearic**
$CH_3(CH_2)_{18}COOH$	**Arachidic**
$CH_3(CH_2)_{22}COOH$	**Lignoceric**
	Palmitoleic
	Oleic
	Linoleic
	Linolenic
	Arachidonic

Figure 7.3

Fatty acids usually occur in cell membranes, bound to glycerol phosphate through an ester bond. These phospholipids function as the storage sites for fatty acids, which are released through secondary enzymatic processes as needed (e.g. during nerve excitation).

Terpenoids will be discussed separately in the following chapter, but it is wise to remind ourselves here of the fact that, for example, cholesterol binds to the phospholipid bilayer membrane, and thus rigidifies its structure. Since double bonds are also more rigid that simple single bonds, unsaturation has a similar rigidifying effect on the lipid membrane.

The fatty acids in phospholipids can be exchanged rather easily through a transesterification process. This facilitates the intermediary storage function of the membranes. A polyunsaturated fatty acid, arachidonic acid, is particularly important in this respect. After its release from the phospholipid, arachidonic acid participates in controlling several biological phenomena through a series of chemical reactions.

7.2.1 Prostaglandins, Thromboxanes and Leukotrienes

In 1935 von Euler isolated a fraction from sperm which had both blood pressure lowering and smooth muscle contracting properties. Two decades later the Swedish group led by Bergström and Samuelsson identified several compounds from this prostaglandin fraction, and they showed that the compounds were derivatives of arachidonic acid. The total syntheses of these compounds were achieved at a rapid pace, mainly through the efforts of Corey at Harvard, and the biosynthetic connections between arachidonic acid, prostaglandins, leukotrienes (slow reacting substances) and thromboxanes have been elucidated.

Upon excitation of the respective cellular receptors, arachidonic acid is released from its phospholipid conjugate through the action of phospholipase A, which leads to the initiation of the arachidonic acid cascade (Scheme 7.3). Depending on the stimulus, cell type and desired function, arachidonic acid is ultimately converted into either prostaglandins, leukotrienes or thromboxanes. 5-Lipoxygenase oxidizes arachidonic acid to 5-hydroperoxyeicosatetraenoic acid (5-HPETE) which is dehydrated to leukotriene A_4 (LTA$_4$). Coupling with glutathione-S-transferase gives the peptidoleukotriene LTC$_4$. An alternative oxidation of arachidonic acid by cyclo-oxygenase gives an endoperoxide which is the precursor for both prostaglandins and thromboxanes.

Arachidonic acid can be oxidized through the action of two different enzymes, thus triggering the biosynthesis of either prostaglandins and thromboxanes (prostaglandin

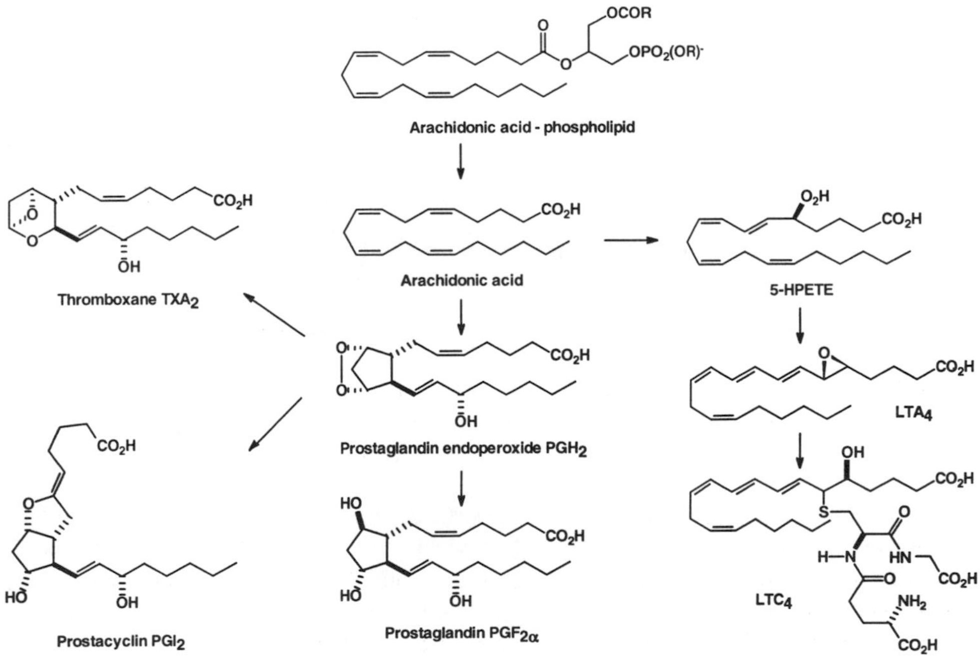

Scheme 7.3

endoperoxidase or cyclo-oxygenase) or leukotrienes (lipoxygenase). Prostaglandins have several physiological functions, including blood pressure lowering effects and contraction of smooth muscles. Some of these effects are used to induce labour.

Thromboxanes accelerate the coagulation of blood. Leukotrienes play a significant role in immediate hypersensitivity reactions and they also possess pronounced pro-inflammatory effects. Many of these compounds have very specific tissue-dependent effects. Of the anti-inflammatory agents the cortisone derivatives, for example, inhibit the release of arachidonic acid from phospholipids, and many other non-steroidal, anti-inflammatory agents (NSAIDs) inhibit cyclo-oxygenase and thus also the synthesis of prostaglandins.

7.2.2 Sphingolipids

Sphingosine (Figure 7.4) provides another extremely abundant structural class of polyketides which are prevalent in cell membranes. It has been estimated that up to 70 % of the dry weight of the myelin sheath (the protective layer surrounding the nerve cell) is sphingosine. The glycosphingolipids contain sphingosine as their lipid portion, whereas the cerebrosides have a different amino alcohol unit, as well as differently constructed N-acyl and glycoside fragments.

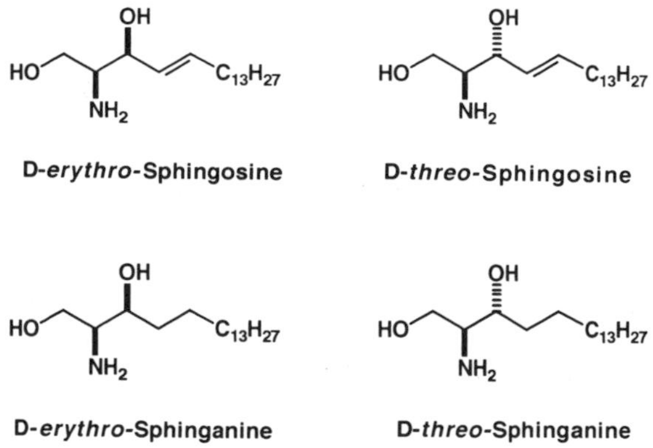

D-*erythro*-Sphingosine   D-*threo*-Sphingosine

D-*erythro*-Sphinganine   D-*threo*-Sphinganine

Figure 7.4

In mammals, the sphingolipids are constructed from sphingosine or its saturated analogue sphinganine, whereas in higher plants and yeasts the major base is 4-hydroxysphinganine. In marine vertebrates it is not uncommon to find more highly functionalized sphingosine derivatives, as exemplified in Figure 7.5.

Symbioramide (Figure 7.6), an antileukaemic ceramide isolated from the cultured dinoflagellate *Symbiodinium* species,[1] has been synthesized and its absolute and relative stereochemistry have been confirmed.[2]

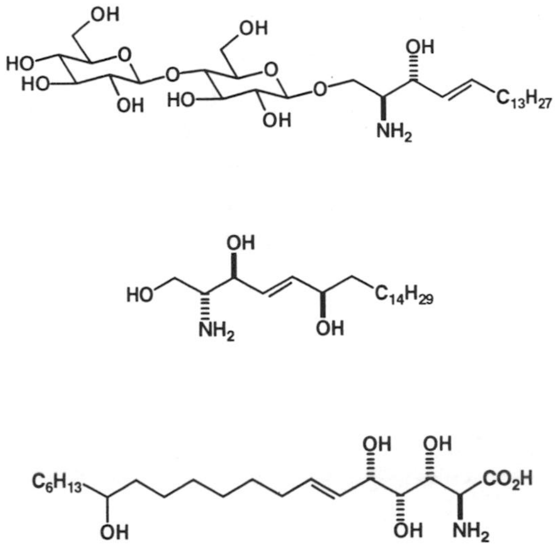

Figure 7.5

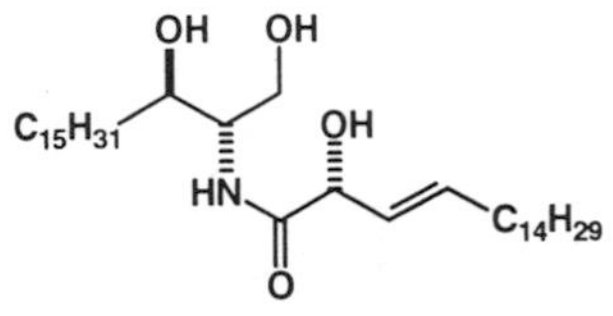

Symbioramide

Figure 7.6

Marine organisms are very rich in natural products, providing a variety of exciting and novel structures often with challenging substitution patterns and atom contents. It is not uncommon to find polyhalogenated compounds among these natural products. The seastars of the *Astropecten* and *Acanthaster* species produce structurally novel glycosphingolipids, asterocerebrosides and acanthagangliosides (Figure 7.7), respectively, which have truncated sphinganine units as well as novel types of carbohydrate residues.

7.3 Polypropionates: Polyether Antibiotics

Polyether antibiotics,[3] isolated from the *Streptomyces* organisms, have prompted much interest in the development of novel synthetic avenues for the construction of polyhydroxylated carbon arrays. These compounds usually contain a mixture of 1,2- and 1,3-diols, the alcohol groups of which are often etherified, esterified or bound into spiroketal arrays. Typical examples of the structures of polyether antibiotics include Ionomycin C, calcimycin, ionomycin and lasalocid A (Figure 7.8), all of which also contain a carboxylic acid moiety. Many of these compounds chelate cations in the living

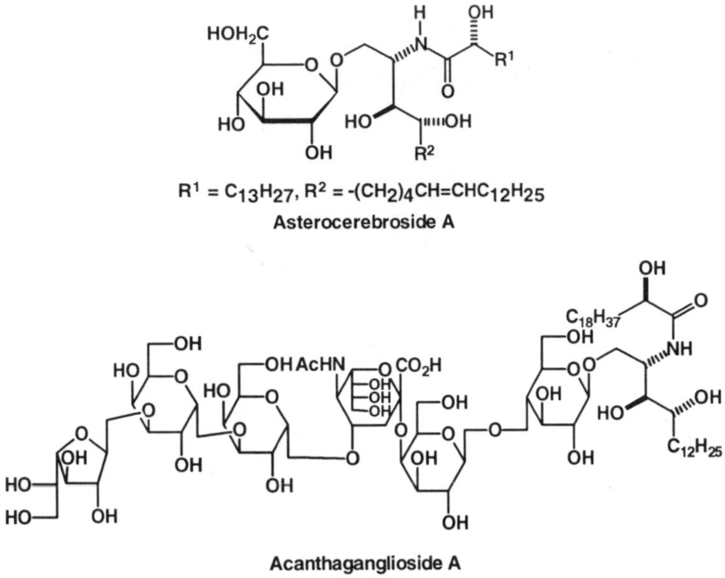

R[1] = C₁₃H₂₇, R[2] = -(CH₂)₄CH=CHC₁₂H₂₅

$R^1 = C_{13}H_{27}, R^2 = -(CH_2)_4CH=CHC_{12}H_{25}$

Asterocerebroside A

Acanthaganglioside A

Figure 7.7

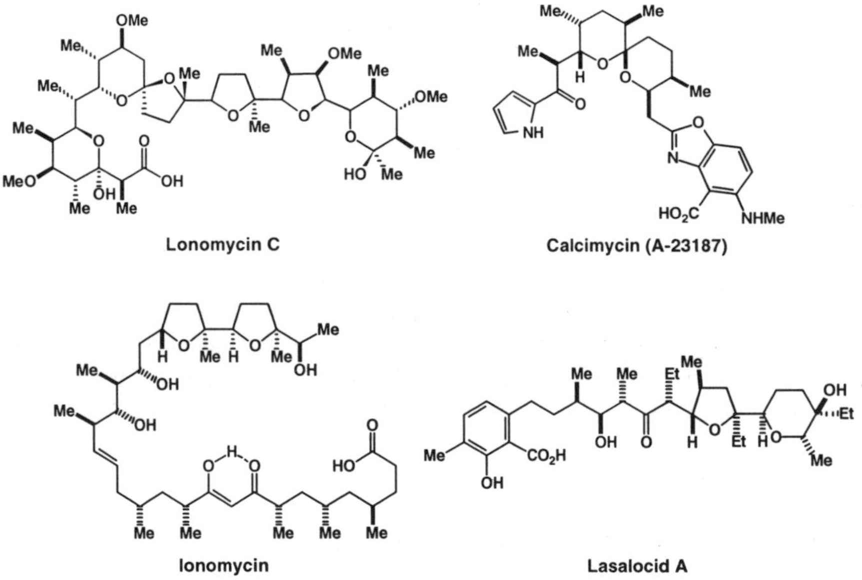

Lonomycin C

Calcimycin (A-23187)

Ionomycin

Lasalocid A

Figure 7.8

cell and can thereby transport cations into and out of living cells, thus functioning as ionophores. By binding to the ionophore the metal is masked in a hydrophobic shell, thus facilitating the passage of cations through the lipid bilayer. The acidic ionophores capable of forming a neutral complex with monovalent cations have proven particularly useful in veterinary medicine.

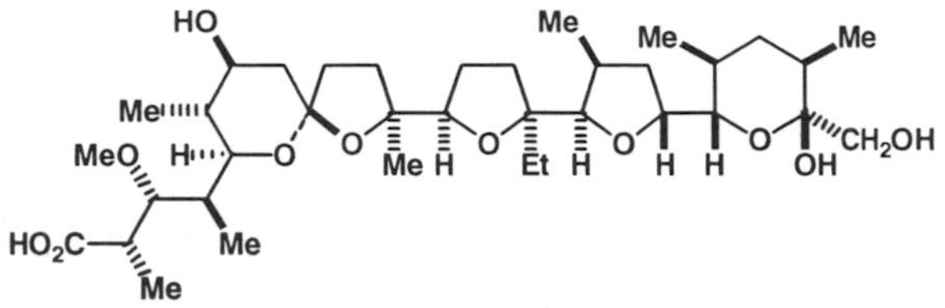

Monensin

Figure 7.9

Monensin (Figure 7.9) is widely used, especially in the US, as a coccidiostat in poultry. After administration of monensin, the transport of Na^+ outside the organism increases the intracellular osmotic pressure, causing the coccidia to explode.

7.3.1 Macrolides

Macrolide antibiotics are a relatively large group of polyketides, including well over 100 metabolites, all characterized by the macrocyclic lactone moiety incorporated in their structures.[4] Within the macrolides, one can distinguish four main types of structures: the polyoxo, polyene, ionophore and ansamycin macrolides. Nearly all macrolides exhibit antibacterial activity, and research into their structures and chemistry has provided many compounds which are in medicinal use.

The polyoxo macrolides typically contain either a 12, 14 or 16 membered lactone ring which is usually also oxygenated at several sites. Unsaturation is not uncommon, and the main carbon chain is derived typically of a mixture of acetate and propionate units. In the case of the 16-membered macrolides, one butyrate unit is also incorporated. The macrolide ring is also always connected to one or more carbohydrate units, which are often of the amino sugar type. The compound with the sugar residue detached is called the corresponding aglycone (e.g. tylonolide is the aglycone of tylosin).

The first macrolide to be discovered was pikromycin (Figure 7.10).[5] By the end of the 1950s, the structures of methymycin, erythromycins A and B and carbomycin A (magnamycin) were elucidated through classical chemical degradation reactions. The structural and conformational variations in these natural products have provided several examples where X-ray crystallography, NMR and mass spectrometry have been used ingeniously in the structure elucidation.

Characteristic to polyene macrolides are the low degree of alkylation on the lactone ring and the presence of a conjugated polyene moiety. The ring size is correspondingly larger than in the polyoxo macrolides. Most of the polyene macrolides exhibit antifungal activity with no antibacterial activity. Nystatin A_1 and amphotericin B (Figure 7.11) are but two examples of clinically widely used congeners. Amphotericin B, isolated from *Streptomyces nodosus*, is used against systemic fungal infections (especially histoplasmosis and blastmycosis), and its mode of action involves binding with the cell membrane sterols, especially ergosterol.

Ionophore macrolides are composed of two or more ω-hydroxycarboxylic acids through the formation of an oligolactone. Thus, nonactin (Figure 7.12) is formed through

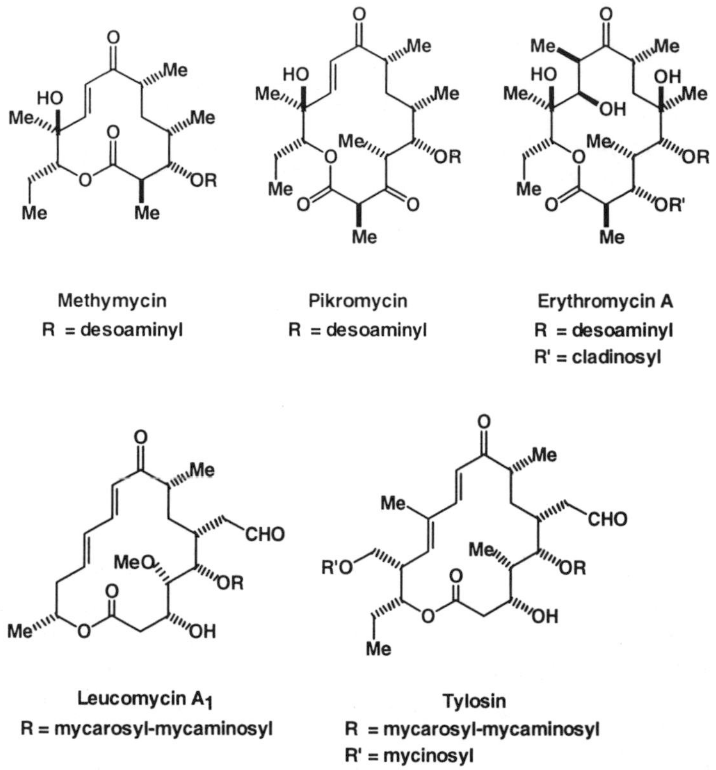

Methymycin
R = desoaminyl

Pikromycin
R = desoaminyl

Erythromycin A
R = desoaminyl
R' = cladinosyl

Leucomycin A₁
R = mycarosyl-mycaminosyl

Tylosin
R = mycarosyl-mycaminosyl
R' = mycinosyl

Figure 7.10

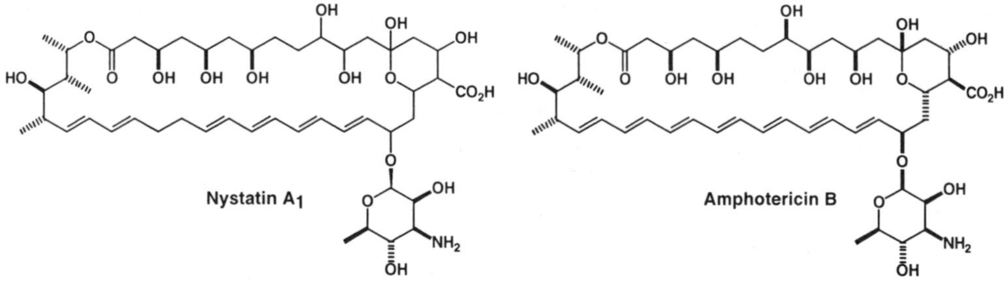

Nystatin A₁

Amphotericin B

Figure 7.11

cyclotetramerization of nonactic acid. As the name implies, the ionophore antibiotics can transport alkali metal cations in biological systems quite efficiently.

Ansamycins are not typical macrolides but contain a large-ring lactam moiety. Also typical to them is an aromatic nucleus bridged through non-adjacent carbon atoms by the ansa ring. Most of the ansamycins exhibit broad antibacterial and powerful antitumour activity. Rifamycin S and maytansine (Figure 7.13) are used clinically.

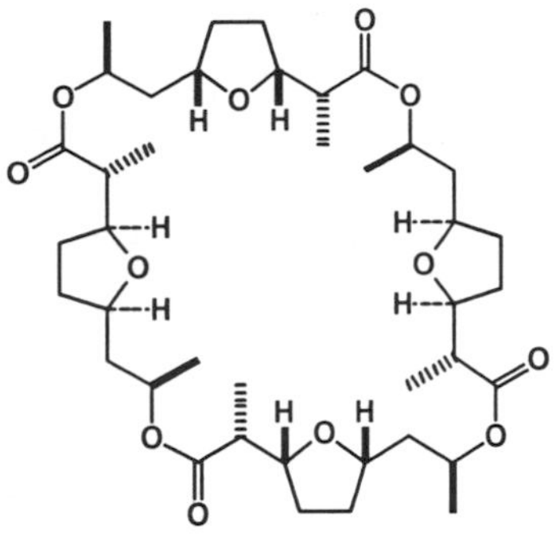

Nonactin

Figure 7.12

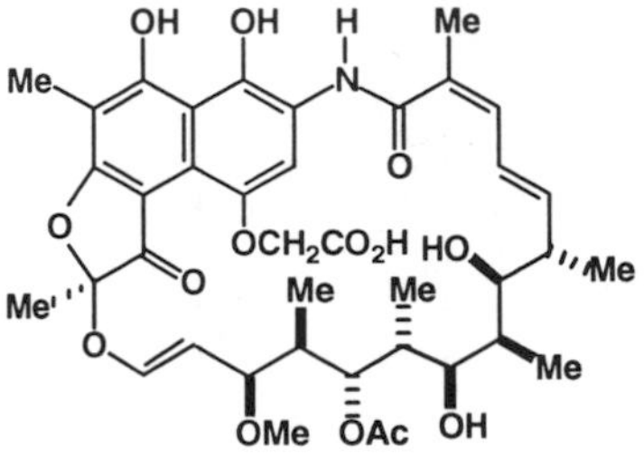

Rifamycin S

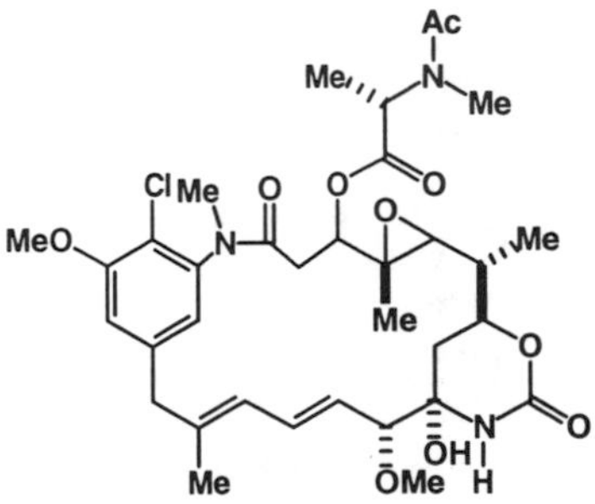

Maytansine

Figure 7.13

Other compounds, such as zearalenone, pyrenophorin and brefeldin A (Figure 7.14), also contain a macrolide structure. Zearalenone has growth inducing properties and is widely used in animal breeding. Brefeldin specifically blocks the transport of proteins from the endoplasmic reticulum to the Golgi apparatus, which accounts for its antiviral activity.

The stereochemistry and conformation of the polyoxo macrolides are surprisingly free of variations. Although different ring sizes can be found, and the oxidation patterns vary, the relative stereochemistry of each chiral centre can often be ascertained from the Celmer model (Figure 7.15).[6] Drawn as the Fischer projection, the chiral centres of methymycin (12 membered ring), erythromycin A (14) and tylosin (16), as well as the ansa chain of rifamycin, match with the Celmer model. This invariability has been used in some cases to correct the originally proposed (wrong) stereostructure of an isolated macrolide.

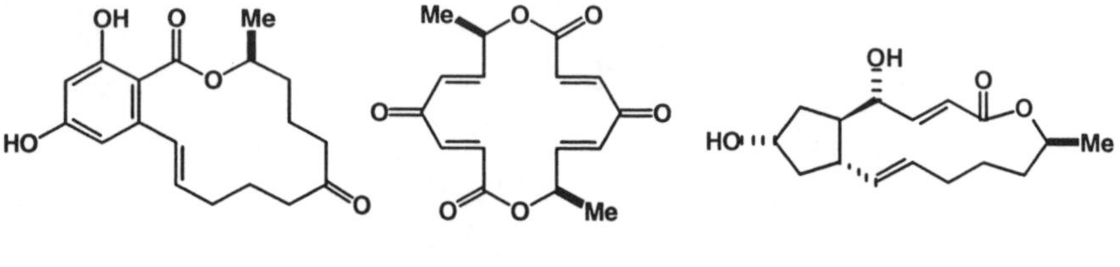

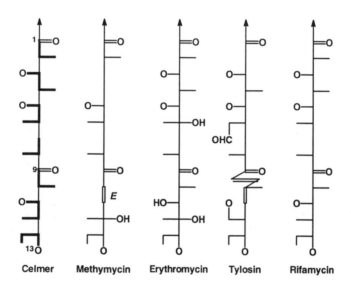

Zearalenone   Pyrenophorin   Brefeldin A

Figure 7.14

Celmer   Methymycin   Erythromycin   Tylosin   Rifamycin

Figure 7.15

7.3.2 Spiroketals

The spiroketal unit enjoys a widespread occurrence in polyketide structures from fungi, insects, microbes, plants and marine organisms. These natural products are also of wide pharmacological importance.[7] The spiroketal unit was originally identified in steroidal structures such as the saponins (glycosides), in which the aglycone (sapogenin) contains a steroid nucleus fused through its D-ring to a spiroketal moiety. Glycosylation usually occurred in the A-ring unit of the molecule. Tomatidine and hecogenin are typical steroidal spiroketals (Figure 7.16). Both of these can be used as starting materials in steroid synthesis.

Okadaic acid (Figure 7.17) is produced by the marine sponge *Halichondria okadaii*, the organism that has been implicated in ciguaterig fish poisoning. Okadaic acid is one of those compounds which exhibit strong protein phosphatase inhibitory activity, especially towards the acid phosphatases A1 and A2. Calyculins (e.g. Figure 7.18) are polyketides isolated from another marine sponge, *Discodermia calyx*, which also possess potent protein phosphatase inhibitory activity as well as antitumour activity.

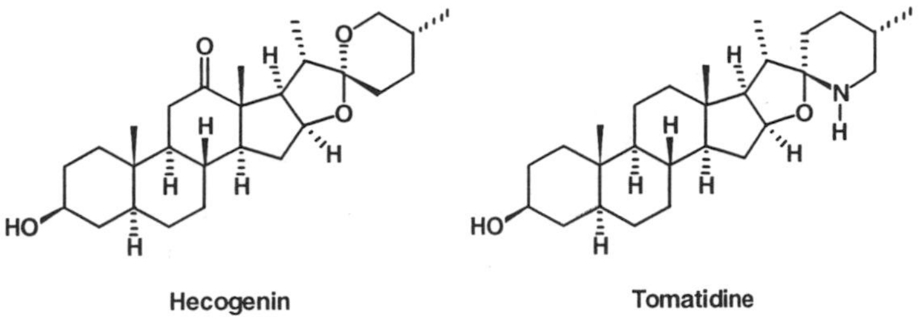

Hecogenin

Tomatidine

Figure 7.16

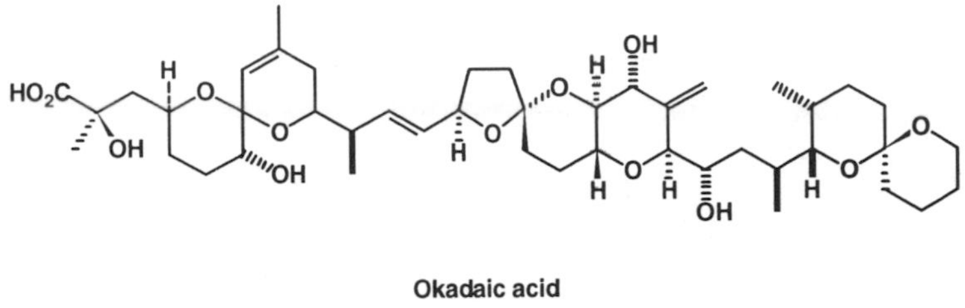

Okadaic acid

Figure 7.17

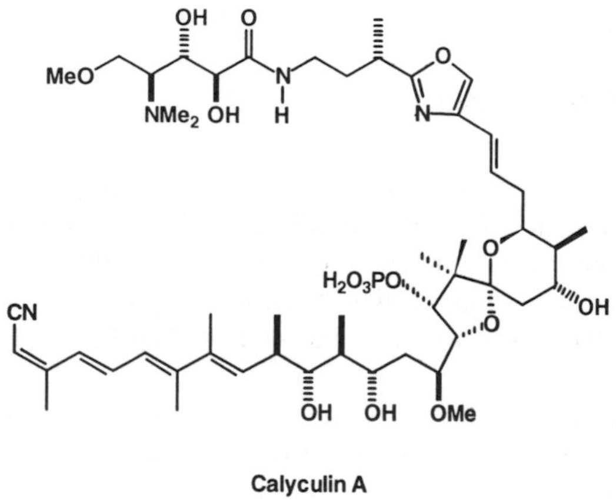

Calyculin A

Figure 7.18

Milbemycins and avermectins (e.g. Figure 7.19) are 16-membered macrolide antibiotics which also contain a spiroketal moiety. These compounds possess significant insecticidal and acaricidal activity. Coupled with their low mammalian toxicity, these compounds hold promising potential for the treatment of parasitic infections. In particular, 22,23-dihydroavermectin B_1, or ivermectin, is effective against the transmission of *Onchocerca volvulus* microfilariae by the blackfly *Simulium yahense*. Females of this species spread onchocerciasis, a parasitic disease ultimately leading to blindness which affects some 20–40 million people worldwide.

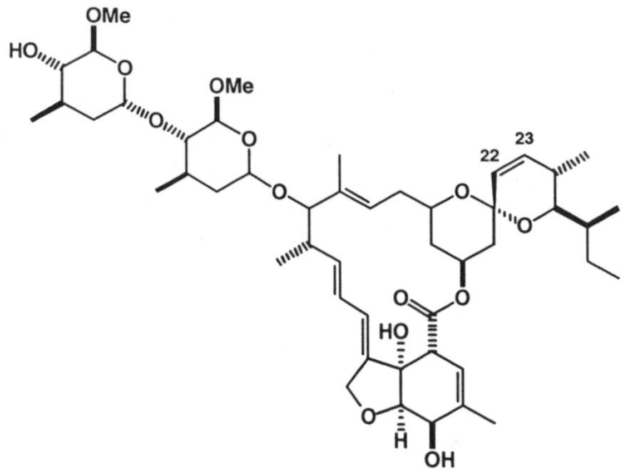

Avermectin B_{1a}

Figure 7.19

The formation of spiroketals from the corresponding open chain dihydroxy ketones is a thermodynamically favoured process, and it is often nearly impossible to prevent the keto diol from cyclizing, especially under acidic conditions.[7]

7.4 Aromatic Polyketides

The bright colours of several flowers and fruits are given by aromatic compounds, which are derived from polyketides. Flavones usually give rise to a yellow or orange colour, and anthocyanins produce a red or blue tinge. Aromatic polyketides (e.g. Figure 7.20) arise by simple cyclocondensation reactions from acyclic precursors. After the cyclization many other chemical transformations may ensue, including halogenation, alkylation and reduction.

Many very potent antibiotics belong to the class of aromatic polyketides. Some of these are in medicinal use in the chemotherapy of cancer. These compounds inhibit nucleic acid synthesis by intercalating between the DNA bases, thereby blocking the transcription of the DNA. Daunomycin is in use only for the treatment of acute leukaemia, and its close congener adriamycin and other derivatives are being tested for the treatment of solid tumours (for structures, see Figure 4.18).

Aflatoxins are mycotoxins produced by certain strains of *Aspergillus* which can contaminate many commodities, including peanuts, Brazil nuts, pistachio nuts and corn and grain sorghum during growth, harvesting, processing, storage and shipment. Aflatoxins are potent carcinogens, mutagens and teratogens, thereby potentially causing major health and economic problems. Aflatoxin B₁ (Figure 7.21) is the most potent of these compounds, all of which contain an angularly fused bisdihydrofuran system. Sterigmatocystin and versicolorin A (Figure 7.21) also contain a similar structural unit, and these compounds are also active in mutagenicity tests.

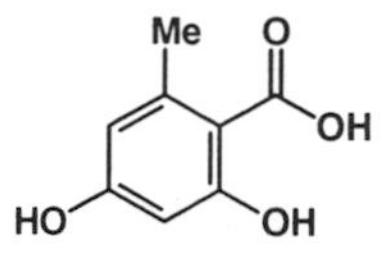

Orsellinic acid 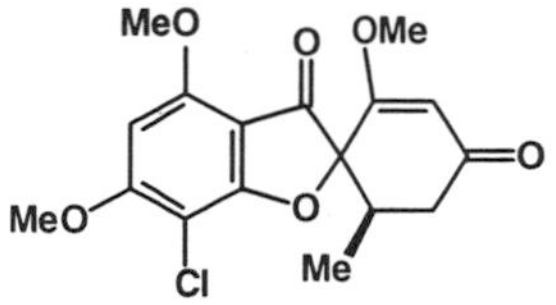   Griseofulvin

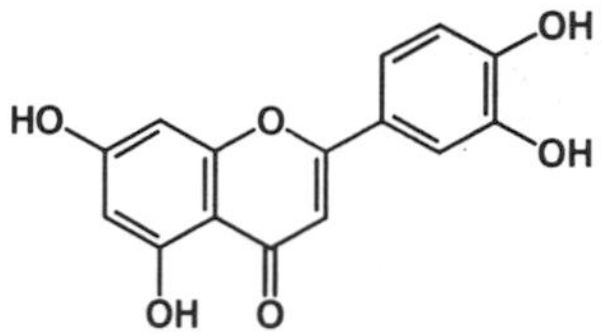

Luteoline

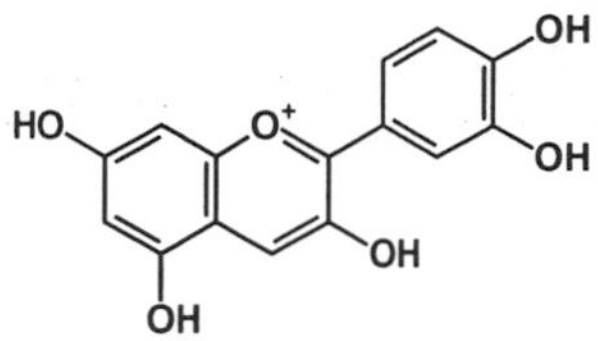

Cyanidine

Figure 7.20

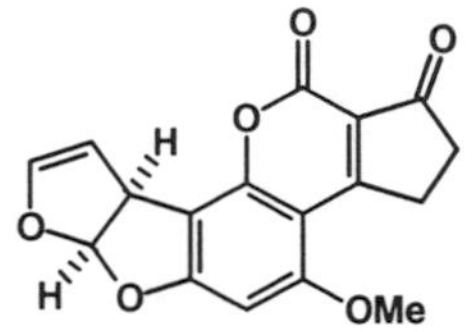

Aflatoxin B$_1$ 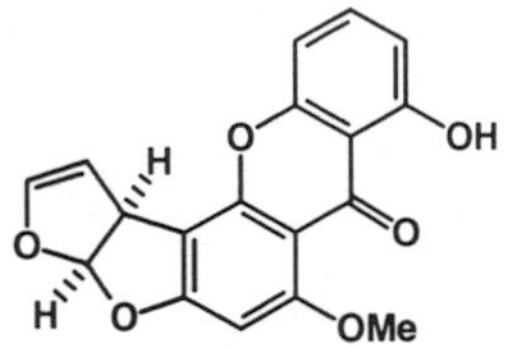   Sterigmatocystin 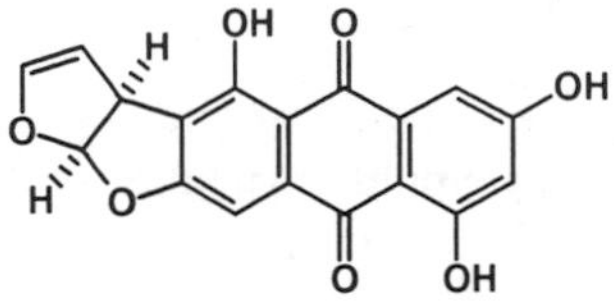   Versicolorin A

Figure 7.21

References

1. Kobayashi, J., Ishibashi, M., Nakamura, H., Hirata, Y., Yamasu, T., Sasaki, T., and Ohizumi, Y. *Experientia*, **44**, 800–802 (1988).
2. Nakagawa, M., Yoshida, J., and Hino, T. *Chem. Lett.* 1407–1410 (1990).
3. Westley, J.W. (Ed.) *Polyether Antibiotics* Marcel Dekker: New York, 1982, vols 1–2.
4. Masamune, S., Bates, G.S., and Corcoran, J.W. *Angew. Chem.* **89**, 602–624 (1977).
5. Brockmann, H. and Henkel, W. *Naturwissenschaften* **37**, 138–139 (1950); *Chem. Ber.* **84**, 284–288 (1951).
6. Celmer, W.D. *Pure Appl. Chem.* **28**, 413–453 (1971).
7. Perron, F. and Albizati, K. *Chem. Rev.* **89**, 1617–1661 (1989).

8 Isoprenoids

In the early history of natural product chemistry, many strongly odorous plant compounds were observed to be formed from C_5 units called isopentenyl or isoprene units. These compounds were termed terpenes, the term derived from the terebinth tree, *Pistacia terebinthus*. Formally, terpenes are derived from isoprene by joining two or more units from either end, the head or the tail. Thus, for instance, limonene (Scheme 8.1) can be synthesized by a formal Diels–Alder reaction by joining the head of one isoprene unit with the tail of another.

Terpenoids (e.g. Figure 8.1) are classified according to the number of these units present in the molecule: monoterpenes, C_{10}; sesquiterpenes, C_{15}; diterpenes, C_{20}; sesterterpenes,

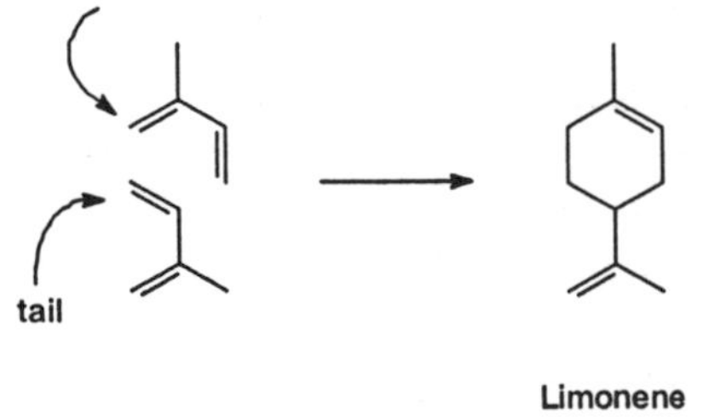

head

tail

Limonene

Scheme 8.1

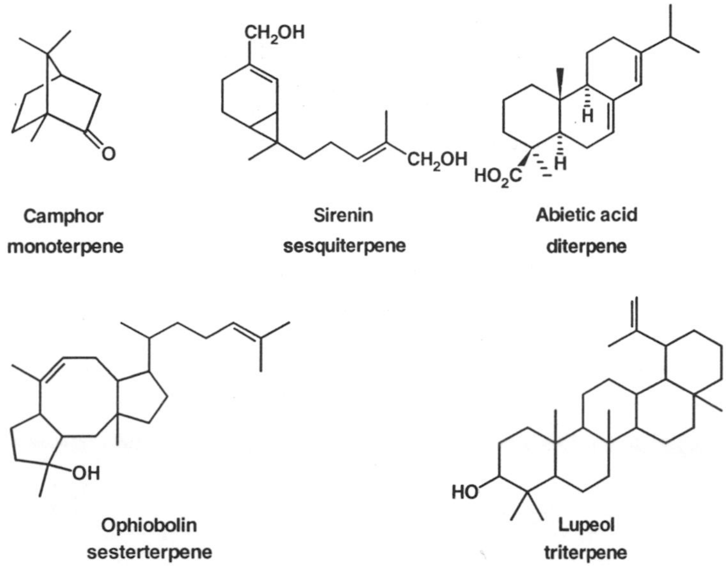

Figure 8.1

C_{25}; triterpenes, C_{30}, etc. Often one or more carbon atoms are excised from the molecule, and these terpenoids are indicated by the prefix nor (e.g. norditerpene contains 19 carbon atoms in its skeleton).

8.1 Terpenoids

Isoprene itself does not function as the reactive biogenetic species. Isopentenyl and dimethylallyl pyrophosphates are the reactive species involved in the formation of terpenes. These are formed from mevalonic acid (Scheme 8.2) by phosphorylation followed by ATP-assisted loss of water and carbon dioxide to give isopentenyl pyrophosphate (IPP). Isomerization of the double bond gives dimethylallyl pyrophosphate (DMAP).

Terpenes occupied a central role in the chemical research at the turn of the century. The first compounds to be studied were the smaller, volatile ones, and in the 1920s Ruzicka and his co-workers at the University of Zurich developed the methods needed for the study of larger terpenes. Ruzicka also systematized the then known concepts in the biosynthesis of terpenes in a theory which is known as the isoprene rule.[1] It states that terpenes are formed from isoprene (C_5) units linked together from head to tail. Although this generalization fails to be true in all cases, it has proven to be very useful in the majority of cases. The isolation of mevalonic acid in 1956 helped in the understanding of the various rearrangement, methylation and dealkylation reactions involved in the biogenesis of the more complex terpenes.

Terpenoids are highly prevalent in the plant kingdom. Many familiar fragrances are terpenes with relatively small size and high volatility. The odour typical to lemons is mainly owing to limonene, and the main component in dill oil is carvone.

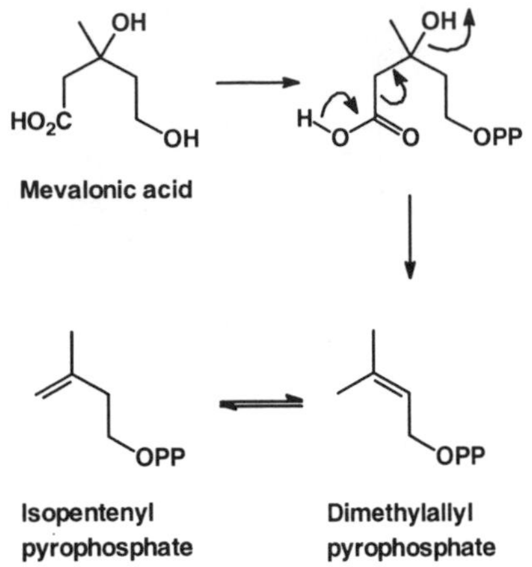

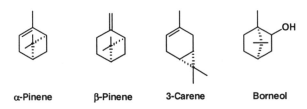

Scheme 8.2

α-Pinene   β-Pinene   3-Carene   Borneol

Figure 8.2

Furthermore, a large number of compounds derived from terpenes are classified as alkaloids, because their physical properties more closely resemble those of the alkaloids (see Chapter 10). Turpentine (distilled tall oil) has been used for centuries as a solvent. The main component of turpentine oil is α-pinene (up to 65 %, the rest mainly being β-pinene). α-Pinene (Figure 8.2) from European turpentine oils is laevorotatory, whereas in North American oils the dextrorotatory form is found. 3-Carene is also found in the *Pinus sylvestris* turpentine (up to 42 %). Lavoisier studied the behaviour and chemistry of turpentine, and the first studies of optical activity by Biot in the early 19th century were conducted on turpentine.

Camphor has been used for medicinal purposes for a long time, and it is still used in the plastics and photographic industries. Although the structure of camphor is relatively simple by modern standards, altogether nearly 30 structural proposals were presented before Bredt showed the correct structure for camphor in 1893. The Finnish chemist, Gustav Komppa, first prepared fully synthetic camphor in 1903, and this is regarded as the first total synthesis of a complex natural product. The Briton William Perkin presented the synthesis of terpineol in the following year, 1904.

Pine resin contains large quantities of abietic acid (Figure 8.3), and is thus one of the least expensive sources of optically active compounds. Abietic acid is a resin acid which is widely used in the production of plasticizers for the polymer industry.

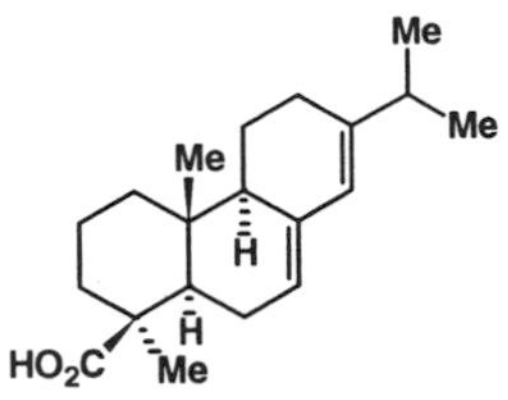

Abietic acid

Figure 8.3

8.1.1 Monoterpenes

The biosynthesis of monoterpenes involves the dimerization of two isoprene units, in a head-to-tail fashion, to form geranyl pyrophophate. Isomerization to the *cis* olefin, neryl pyrophosphate, sets the stage for cationic cyclization to give the menthane skeleton. Loss of a proton and hydration of the endocyclic olefin gives menthol (Scheme 8.3).

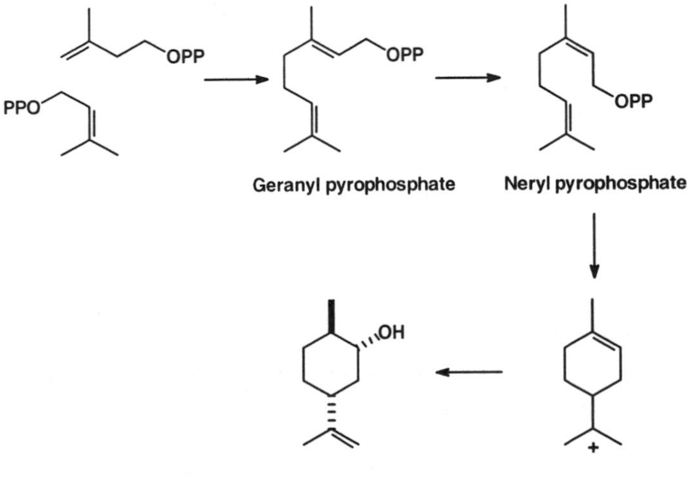

Geranyl pyrophosphate   Neryl pyrophosphate

Menthol

Scheme 8.3

Further cationic cyclizations of the menthane cation lead to the pinane, bornane and carane skeleta. A Wagner–Meerwein 1,2-shift of the menthane cation gives a rearranged cation which can be trapped by hydroxide to give terpinen-4-ol (Scheme 8.4). An alternative cationic cyclization with the participation of the ring double bond leads to the thujane skeleton.

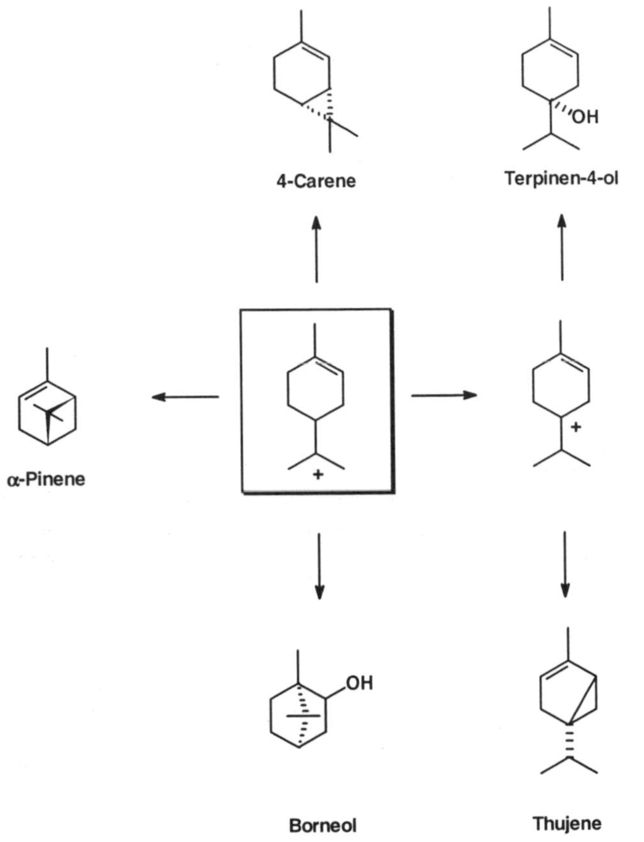

Scheme 8.4

8.1.2 Sesquiterpenes

Sesquiterpenes are formed from three isoprene units and thus contain 15 carbon atoms. We have already seen the structure of sirenin, a sperm attractant in the marine mould *Allomyces*. The sesquiterpenes are formed from *cis–trans*-farnesyl pyrophosphate (Scheme 8.5) through cationic cyclization similar to the formation of the menthane cation. The bisabolyl cation can undergo similar cyclization reactions as the menthane cation to give pinane, carane, bornane and menthane ring systems. Bergamotene is a constituent of bergamot oil, which is widely used in the perfume industry.

The additional C-10 double bond can also participate in the cyclization reactions, leading to new ring systems. Cationic rearrangements are common in terpene chemistry, and the biosynthesis of widdrol (Figure 8.4) exhibits a particularly intriguing example. A quick look at the structure of this sesquiterpene would suggest the incorporation of the isoprene units in accordance with the isoprene rule, i.e. head-to-tail, as shown in arrangement A. However, labelling studies have shown that this is not the case. Arrangement B represents the incorporation of the three isoprene units. Thus, the central unit must undergo a rearrangement at some stage of the biosynthesis.

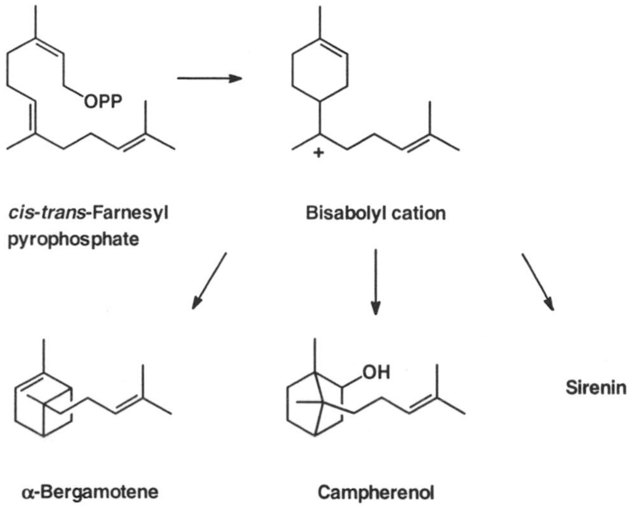

cis-trans-Farnesyl
pyrophosphate

Bisabolyl cation

α-Bergamotene

Campherenol

Sirenin

Scheme 8.5

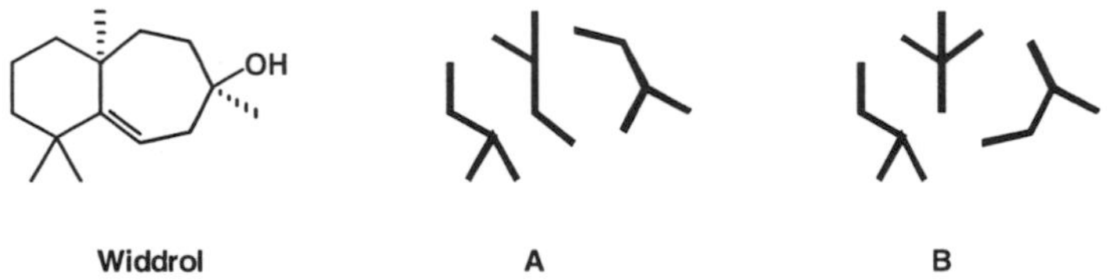

Widdrol

A

B

Figure 8.4

Scheme 8.6 shows the postulated biosynthesis of widdrol. Cyclization of the bisabolyl cation gives a new cation which can form the conjugated diene through loss of a proton (cuparenene). 1,2-Shift of the *cis* alkyl group leads to the spirocyclic chamigrane skeleton, which undergoes another alkyl shift to give a 6,7-fused ring system. The isomerization of the double bond finally occurs through a ring contraction–ring enlargement pathway involving two cyclopropanoid intermediates.

Aristolactone (Figure 8.5) is a typical germacrane, a sesquiterpene which contains a 10-membered carbocyclic ring. Further cyclization gives the bicyclic eudesmane skeleton (e.g. eudesmol).

The guaiane skelton contains a 5,7-fused ring system, which often contains a third ring, an unsaturated α-methylene-γ-lactone, fused to the seven membered ring. Several of the guaianes possess interesting pharmacological activities, including promising antineoplastic effects, and they occupied a pronounced position in the development of synthetic methods during the 1970s and early 1980s. Helenalin (Figure 8.6) is a bitter tasting compound which occurs in several *Helenium* species. It irritates the mucous membrane, causes indigestion and paralyses the heart muscle. It has also aroused interest because of its insecticidal and vermifugal activity.

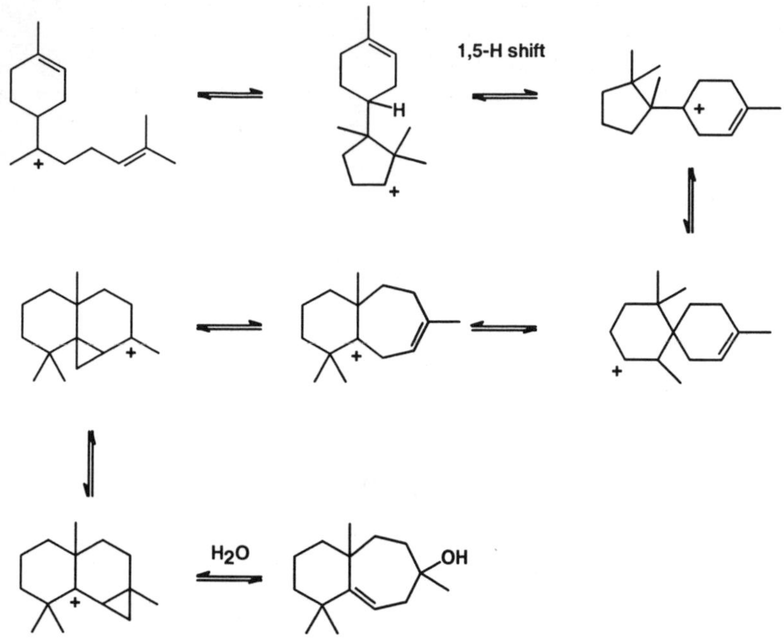

Scheme 8.6

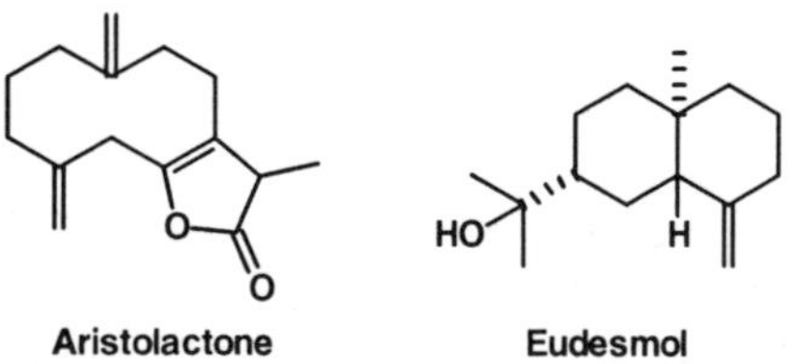

Aristolactone   Eudesmol

Figure 8.5

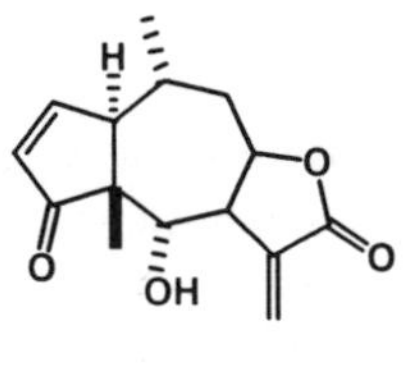

Helenalin

Figure 8.6

Polyquinane terpenes form a structurally intriguing class of natural products.[2] The linear and angular fusions of cyclopentane rings lead to a wide variation in their structures. Pentalenene (Figure 8.7), isolated from *Streptomyces griseochromogenes*, is the parent hydrocarbon of the pentalenolactone family of antibiotic fungal metabolites. The antibiotic and antitumoural compound coriolin (Figure 8.7) is a member of the hirsutane class of sesquiterpenes.

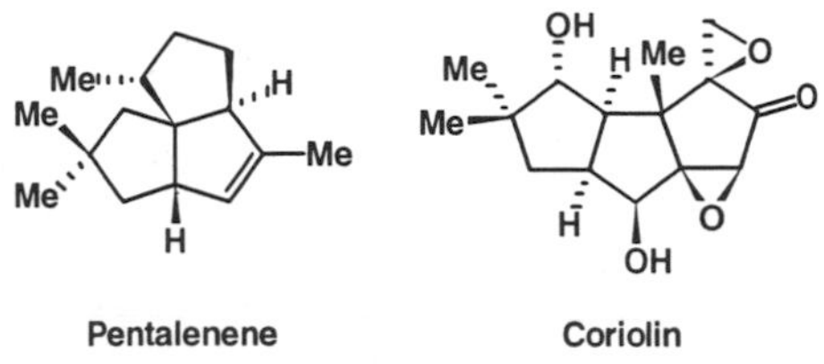

Pentalenene   Coriolin

Figure 8.7

8.1.3 Diterpenes

Diterpenes contain a C_{20} skeleton formed from four isoprene units. The common precursor is the linear *trans–trans*-geranylgeranyl pyrophosphate (Scheme 8.7), whose cyclization can be effected in many ways. The cyclization normally proceeds directly to a bicyclic *trans*-decalin system which then undergoes a variety of different transformations. Straightforward loss of a proton gives labdadienyl pyrophosphate, which functions as the intermediate to several structural types of diterpenes.

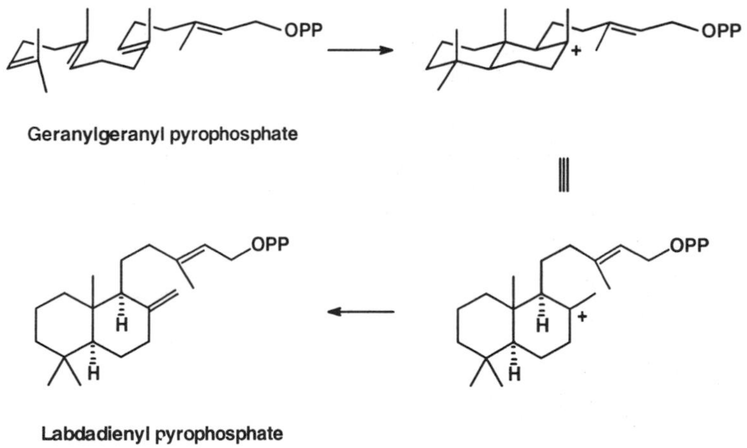

Geranylgeranyl pyrophosphate

Labdadienyl pyrophosphate

Scheme 8.7

Gibberellins are important plant growth hormones which control cell elongation. These compounds were first isolated from the fungus *Gibberella fujikuroi*, which is a parasite of rice that causes the straw cells to grow too long and thin, thereby making the straw less stiff. The infection is a very serious threat to the rice crops in the rice producing countries. Gibberellins have subsequently been found in several plants in small quantities, acting as plant growth hormones.[3]

The biosynthesis of gibberellins involves the cyclization of labdadienyl pyrophosphate to give the *ent*-8-pimarenyl cation through loss of pyrophosphate (Scheme 8.8). Loss of the proton from the angular methyl group is accompanied by contraction of the C-ring through an alkyl shift, which terminates in the closure of the bridge of *ent*-kaurene. Oxidation gives the hydroxycarboxylic acid which upon further oxidation at the 6-position undergoes a ring contractive rearrangement of the B-ring to furnish the gibberellin skeleton (gibberellin A_{12} aldehyde). Final oxidative transformations then lead to the various gibberellins.

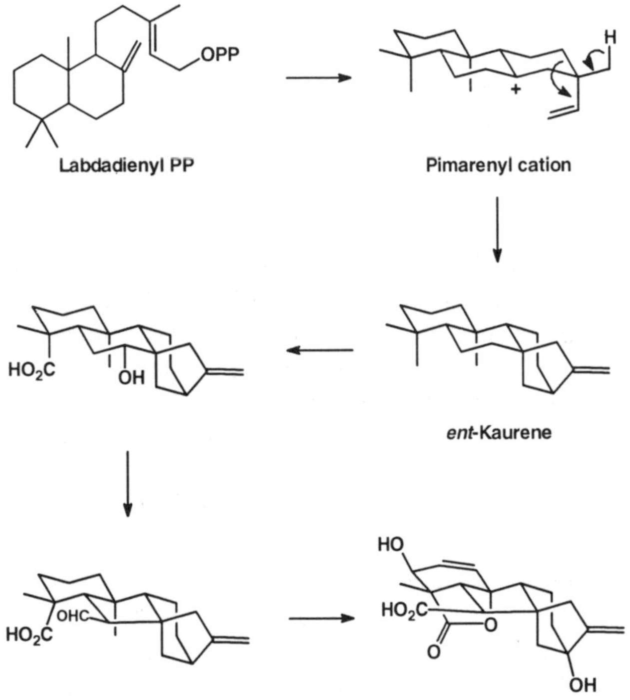

Labdadienyl PP

Pimarenyl cation

ent-Kaurene

Scheme 8.8

8.1.4 Higher Terpenes

Before turning our attention to two individual classes of terpenes, the carotenoids and steroids, let us examine the construction of a tail-to-tail joined C_{30} hydrocarbon, squalene (Figure 8.8). This functions as the intermediate to the steroids, and the formation of the tail-to-tail linkage is instructive of the mechanism for the formation of phytoene, the biogenetic precursor of the carotenoids.

Squalene itself was originally isolated from shark liver (*Squalus* species), but was later found to be widely distributed. Squalene consists of two farnesyl groups joined tail-to-tail. The mechanism of this transformation remained a challenging problem until the isolation of a cyclopropane containing intermediate, presqualene pyrophosphate.[4,5] The two farnesyl units are joined together, and the carbocation undergoes cyclization to

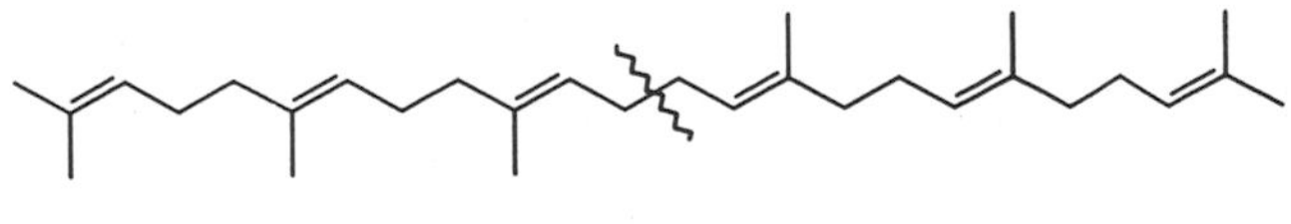

Squalene

Figure 8.8

the cyclopropane (presqualene pyrophosphate) through a stereospecific loss of the pro-*S* proton, as indicated in Scheme 8.9. The pyrophosphate in the cyclopropylmethanol functions as a powerful leaving group, giving rise to a rearranged carbocation with a cyclobutane skeleton. Being still a high-energy species, this undergoes rapid ring opening to the much more stable allylic cation which is finally trapped by NADPH, the biological hydride reductant, to give squalene.

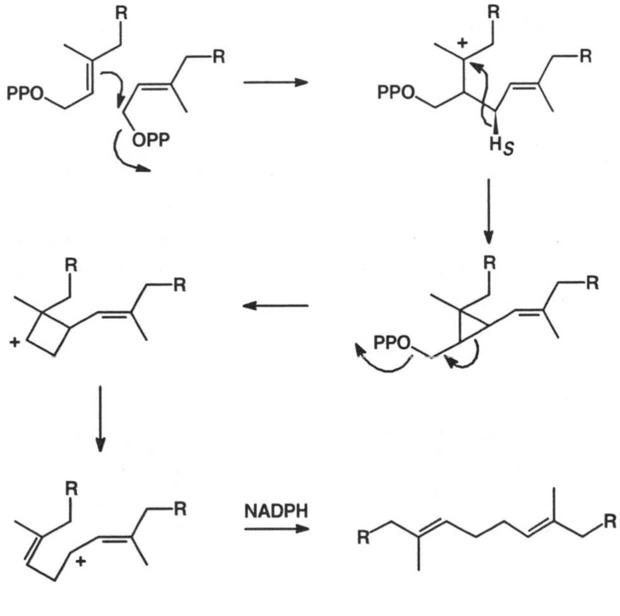

Scheme 8.9

The flexible squalene molecule can be folded in a number of ways. The enzymes that catalyse the cyclization of squalene exhibit remarkable specificity in folding the chain into the proper orientation before cyclization, and Scheme 8.10 shows the form incorporating mainly chair conformations This is the usual pathway observed in photosynthetic plants, which produce a number of triterpenes after a wealth of rearrangement steps. For an essentially cationic polycyclization, squalene needs activation, and this is achieved by epoxidation of the terminal double bond. Acid catalysed opening of the epoxide gives the protosterol carbocation, which is converted into a number of plant sterols through a cascade of cationic rearrangements.

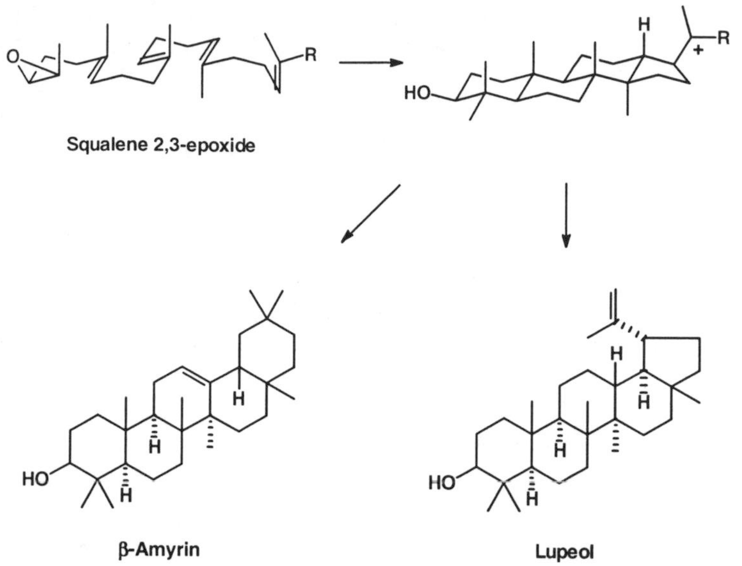

Scheme 8.10

Fernene (Figure 8.9), a fern triterpene lacking the C-3 hydroxyl group, and the protozoan metabolite tetrahymanol (which has a hydroxyl group at C-21) are produced from squalene by direct protonation rather than through squalene epoxide. Squalene adopts an all-chair conformation, which after protonation undergoes a number of Wagner–Meerwein 1,2-shifts to give the products.

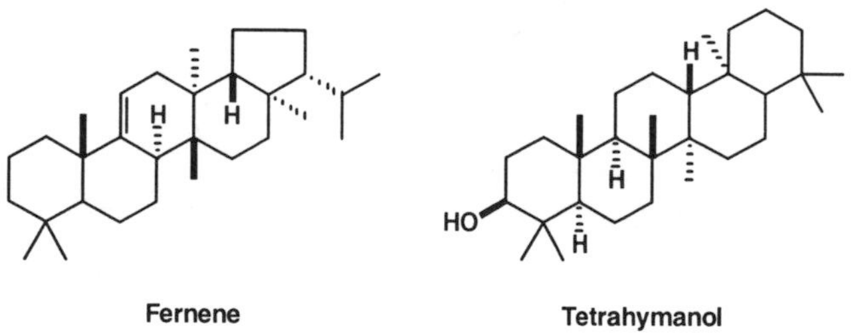

Figure 8.9

In animals, squalene adopts a chair–boat–chair arrangement for the forming ABC rings, and an oxidative cyclization via the epoxide was proposed for the mechanism of transformation of squalene to lanosterol (Scheme 8.11) in the 1950s by Eschenmoser (ETH, Switzerland) and Corey (Harvard, USA).[6,7] Since then, the epoxidase has been isolated, and is known to utilize molecular oxygen and NADPH.[8,9] The direct enzymatic formation of protosterol with the side-chain at C-17 β-oriented has recently been proved.[10]

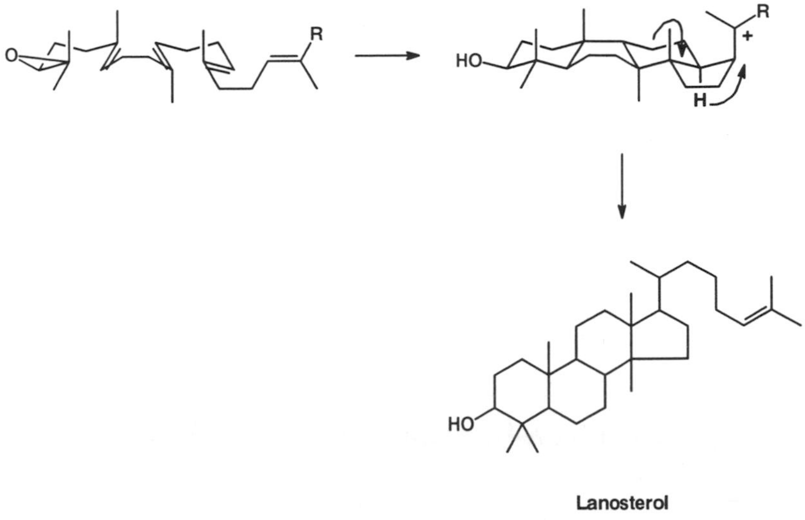

Lanosterol

Scheme 8.11

The triterpene nucleus can undergo several modifications after its synthesis. Oxidation of the existing hydroxyl groups, oxygenations to introduce further oxygen functionalities, cationic rearrangements involving Wagner–Meerwein shifts and S-adenosyl methionine mediated alkylations provide the possible routes to the large number of structurally varied triterpenes. Quassinoids (e.g. Figure 8.10), the bitter principles of *Simaroubaceae*, require extensive degradation and recyclization reactions for their formation, as do the limonoids, the bitter principles of citrus species and the bitter principles of the Mediterranean *Cneoria* species. Azadirachtin (Figure 8.10) is a highly oxygenated tetranortriterpene isolated from the neem tree *Azadirachta indica*. It is extremely potent as an insect antifeedant and growth regulator and has prompted extensive synthetic efforts towards simpler analogues that retain these desired properties.[11,12]

8.2 Carotenoids

The strong colours of carrots, egg yolk, tomatoes, yellow autumn leaves and algae are owing to carotenoids, tetraterpenes with long, conjugated systems. The carbon chain usually contains a long stretch of conjugated double bonds, giving rise to the coloured properties of these compounds. In photosynthetic organisms, the carotenoids function as supplemental, light absorbing molecules that pass the excitation energy to chlorophyll.

The most important tetraterpene is retinol (vitamin A_1), which is necessary for sight. Retinol is formed in the liver from β-carotene through oxidative degradation. Because the human organism cannot produce β-carotene itself, it must be ingested with food. The best sources are carrots, spinach and salad. The lycopene of tomatoes is an intermediate in the biosynthesis of β-carotene.

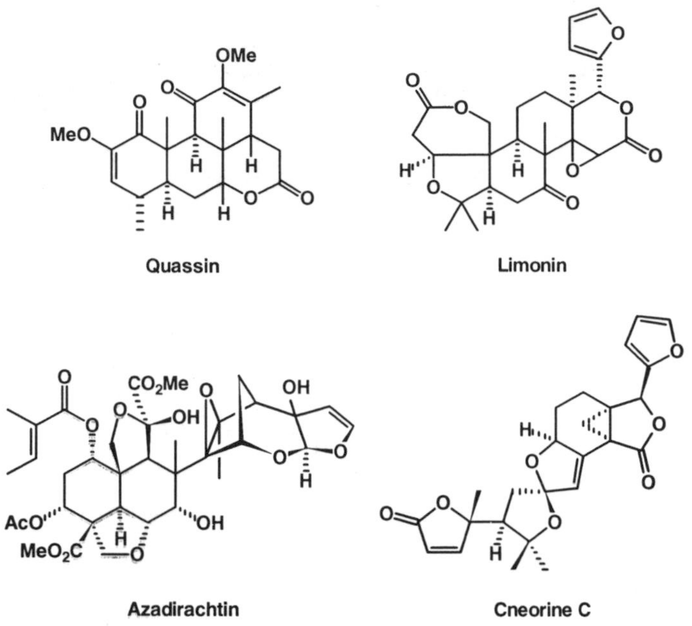

Quassin

Limonin

Azadirachtin

Cneorine C

Figure 8.10

In the biosynthesis of β-carotene, two geranylgeranyl pyrophosphates are joined tail-to-tail to give phytoene in a fashion not unlike the one we saw for the synthesis of squalene (i.e. involving a cyclopropane intermediate). Oxidation to the highly conjugated (11 conjugated double bonds) lycopene is followed by cationic cyclization at each end of the chain to form the carotene molecule (Scheme 8.12).

Dark vision is based on a photochemical reaction in which vitamin A plays a crucial role. As mentioned previously, β-carotene is transformed in the liver to retinol (vitamin A_1; Figure 8.11). This is further oxidized to the corresponding aldehyde (retinal), which reacts with the opsin protein present in the rod cells of the retina to form a covalent bond. This new protein complex, rhodopsin, contains the *cis* double bond form of retinal, which absorbs light at 500 nm. Light absorption causes isomerization of the double bond to the *trans* configuration, and *trans*-retinal is released from the opsin complex. Simultaneously, a nerve signal is transmitted to the centre of vision in the brain. *trans*-Retinal is enzymatically isomerized back to the *cis* form, and the cycle can begin again. Rhodopsin absorbs light with a very high sensitivity, the absorption of a few light quanta leading to the isomerization. The sensitivity of the rod cells towards blue–green light is also explained by the aforementioned events: light at 500 nm is blue–green.

Some bacteria can produce even longer carotenoids and terpenes, and continuation of the prenylation process in plants produces a polymeric latex. Rubber is obtained from the rubber tree, *Hevea brasiliensis*, and is a high molecular weight polymer of isoprene (*ca* 2000 units). Nearly all the double bonds in rubber are *cis*, whereas the harder gutta-percha (from *Palaquium* species) contains *trans* double bonds.

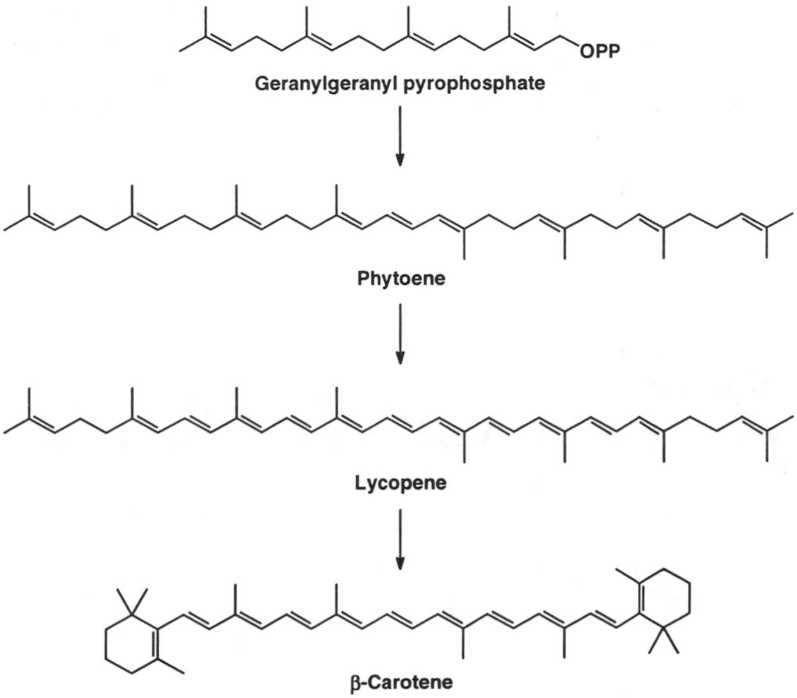

Geranylgeranyl pyrophosphate

Phytoene

Lycopene

β-Carotene

Scheme 8.12

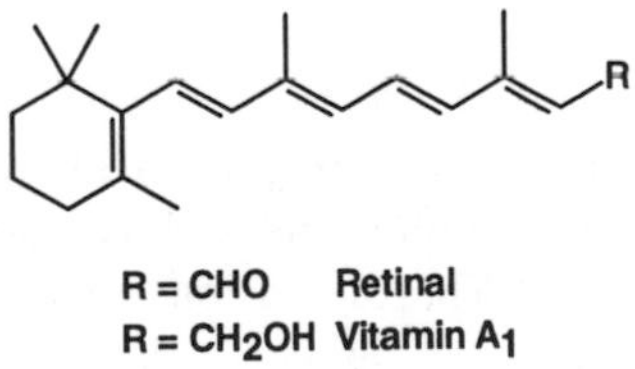

R = CHO   Retinal
R = CH$_2$OH   Vitamin A$_1$

Figure 8.11

8.3 Steroids

When plant or animal cells are extracted with ether, chloroform or benzene, one can isolate a mixture of several compounds containing lipids. This extract can be divided into two fractions based on the behaviour of the compounds towards basic hydrolysis. The hydrolysable lipids produce water soluble compounds, and the remaining fraction contains, among others, the steroids.

Steroids occur widely as building blocks for all cell membranes. For instance, the dry material of the human brain is 17 % free cholesterol, and gallstones are mainly composed of cholesterol. The structure of cholesterol is very rigid, and incorporation of it into the cell membrane has a rigidifying action. Cholesterol has molecular dimensions of *ca* 4 × 7 × 20 Å, which makes a nearly perfect match for the dimensions of the phospholipids, the cell membrane constituents forming the lipid bilayer (Figure 8.12). Cholesterol also displays a hydrophilic end (the hydroxyl group at C-3) and a hydrophobic end (the terminal alkyl side-chain at C-17).

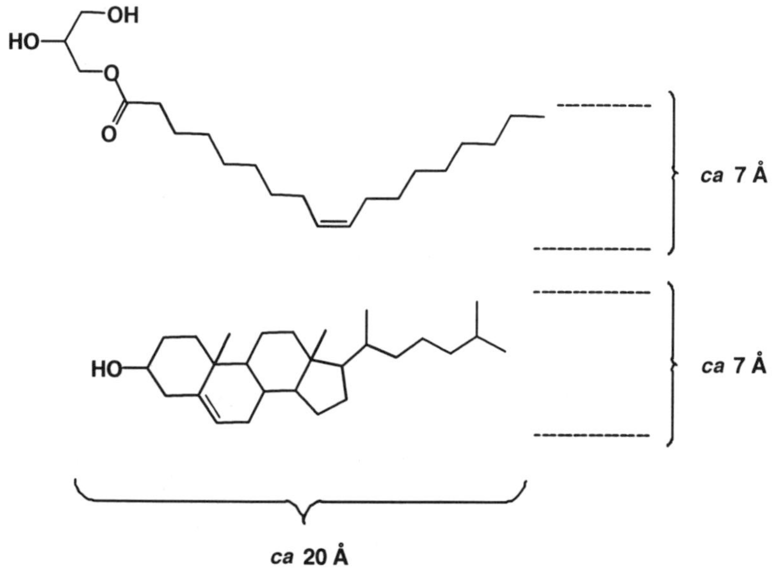

Figure 8.12

Whereas cholesterol is the rigidifying component in eukaryotic cell membranes, in prokaryotes a higher terpene, bacteriohopanetetrol (Figure 8.13), with a C_{35} carbon skeleton, occupies a similar role.[13] The dimensions of bacteriohopanetetrol (7.7 Å by 18.45 Å to the tetrol portion) are very similar to those of cholesterol, and it can thus easily be accommodated in a similar position in the lipid bilayer. Prokaryotes also contain a high proportion of longer (C_{40}) terpenes which cross-link the two lipophilic surfaces of the bilayer.

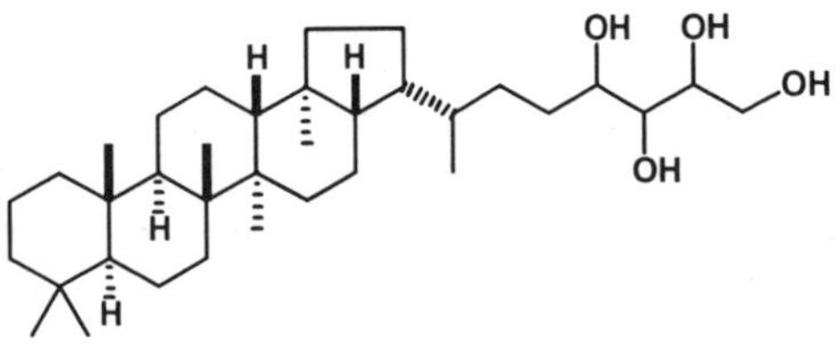

Bacteriohopanetetrol

Figure 8.13

The elucidation of the structures of the steroids took several decades, and in 1928 Wieland and Windaus obtained the Nobel prize in chemistry for their studies on the structures of these important compounds. However, the structure they proposed for cholesterol (Figure 8.14) was based on oxidative degradation studies and the application of Balc's rule, which states that pyrolysis of substituted glutaric acids will yield an anhydride, whereas adipic acids will be transformed to a cyclopentanone derivative. This rule, and thus also the structure of cholesterol, turned out to be wrong, and the structure was corrected with the aid of the first application of X-ray diffraction studies to natural product structure elucidation a few years later by Bernal.

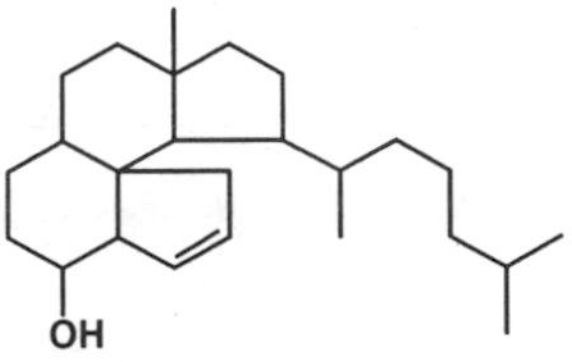

**The Wieland-Windaus structure
for cholesterol in 1928**

Figure 8.14

The isolation of and structural studies on the first steroid hormones give a good indication of the difficulties encountered by natural product chemists in the early part of this century. The first male sex hormone to be isolated and characterized was the excreted hormone androsterone, which is less active than the primary hormone testosterone (Figure 8.15). For the structure elucidation of the first, Butenandt extracted 15 000 L of urine to obtain only 15 mg of pure androsterone.

Steroids (e.g. Figure 8.15) still play an important role in the drug industry. The hormone preparations make up the widest group of these drugs, which are used mainly as contraceptives. The steroid hormones include the female sex hormones such as the oestrogens and the gestagens and the male sex hormones such as the androgens and the adrenocorticoids secreted by the cortex of the adrenal gland. The male hormones are further divided into the adrenocorticoids and the mineralocorticoids.

The cardiac glycosides are also steroid derivatives. These include the highly toxic strophantidine (from lily of the valley; Figure 8.16), the bile acids (e.g. cholic acid) and the sapogenin diosgenin.

A common structural feature of the steroids is the cyclopentanophenanthrene (androstane) in a skeleton (Figure 8.17) where the rings are joined together in a regular fashion. Only the fusion between the rings A and B varies, and even here regularities can be observed. In the bile acids and cardenolides the fusion is *cis*, whereas the standard steroids have either a *trans* fusion or ring A is aromatic. A third structural variation with *cis* A/B, *trans* B/C and *cis* C/D ring junctions occurs in the cardiac glycosides.

The methyl groups at C-10 and C-13 in natural steroids are always β (above the plane of the paper), except of course in those cases where the A-ring is aromatic and the angular methyl group is thus missing. Natural steroids also often have a side chain at C-17 which is β-oriented, as shown for cholesterol in Figure 8.18. The numbering scheme for the atoms in steroids is also shown.

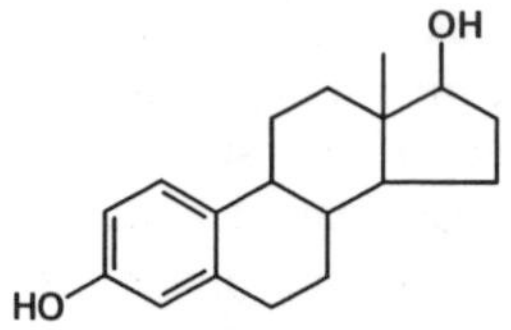

Oestradiol
(an estrogene)

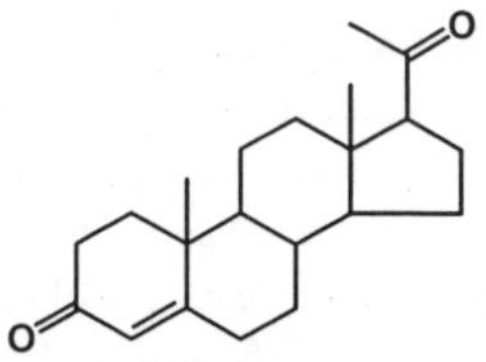

Progesterone
(a progestine)

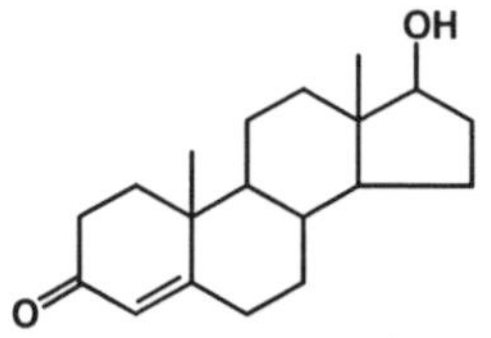

Testosterone
(an androgen)

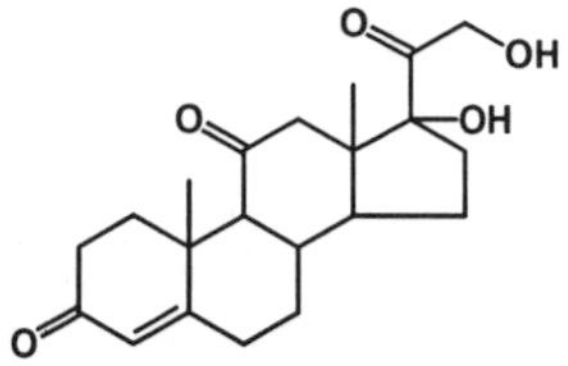

Cortisone
(a corticoid)

Figure 8.15

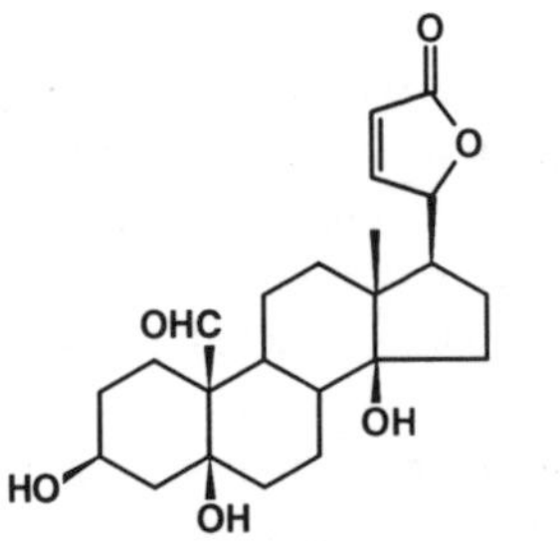

Strophantidine
(a cardenolide)

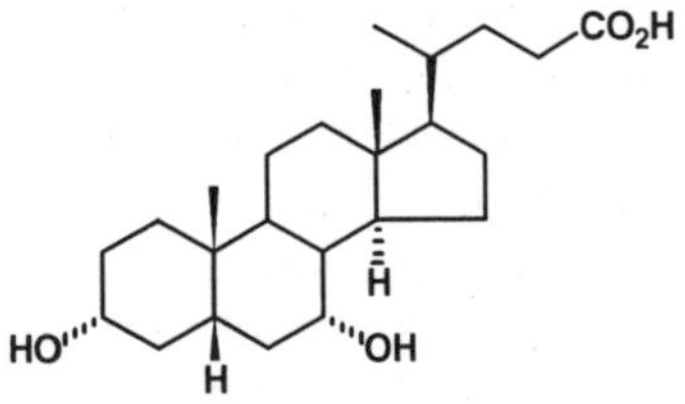

Cholic acid
(a bile acid)

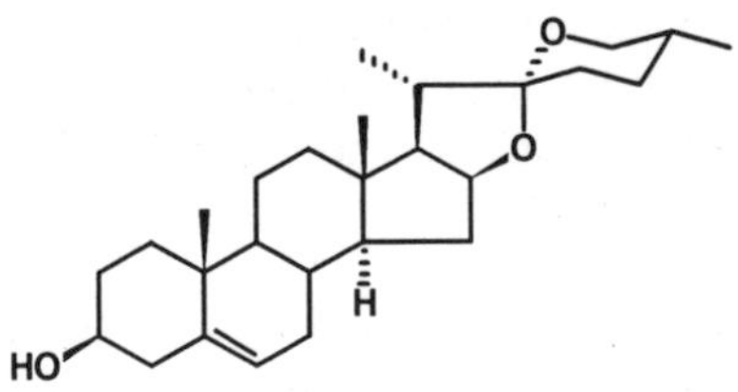

Digitoxigenin
(a sapogenin)

Figure 8.16

The steroid hormones are biosynthesized from cholesterol. After liberation from its storage site in the cell walls, cholesterol is transformed into pregnenolone, which functions as the starting material for the remaining steroid hormones. The ovary gland produces oestrogens which regulate the release of the egg (ovulation) and the development of the external sexual properties of females. Through the response to adenohypophysis,

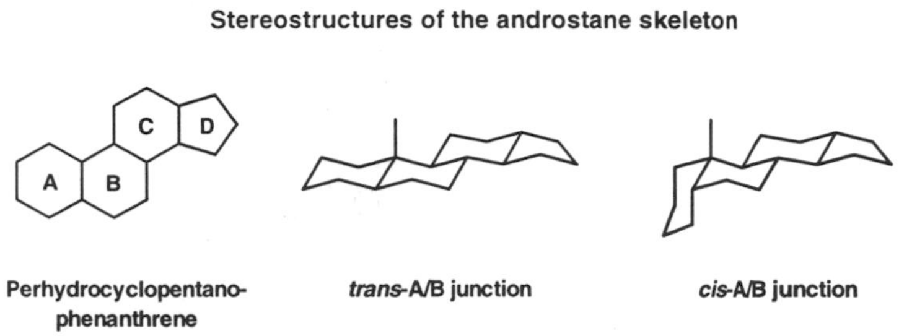

Stereostructures of the androstane skeleton

Perhydrocyclopentano-
phenanthrene

trans-A/B junction

cis-A/B junction

Figure 8.17

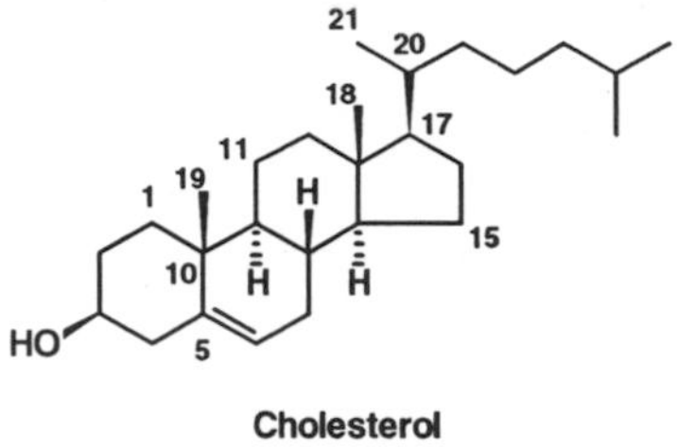

Cholesterol

Figure 8.18

oestrogen inhibits the secretion of follicle stimulating hormone (FSH) which results in the lack of ovulation when the oestrogen effect is strong. Oestrone (Figure 8.19) is the oldest known female sex hormone, and oestradiol has the strongest effects. Several molecular modifications of these compounds are in medicinal use, such as ethinyloestradiol (contraceptive) and various stilbene derivatives which also bind to the oestrogen receptor. Administration of diethylstilboestrol to pregnant women has been found to induce tumour development, and its use has been discontinued.

When the release of oestrogens has lowered the levels of FSH, the ovary gland (follicle) is turned into an endocrinous gland (*corpus luteum*), which during the first third of pregnancy secretes progesterone. After this period the placenta takes care of the excretion of both progesterone and oestrogen. In later phases of pregnancy, progesterone prohibits ovulation, and this action has been used in the development of oral contraceptives since the early 1950s. Progesterone has an inhibitory effect on the secretion of luteinizing hormone, and it also prepares the membranes of the womb to support fertilized pregnancy. Progesterone itself is rapidly metabolized in the liver, and it has little practical medicinal value. Gestagens are hormones which function as progesterone does, but they are slowly metabolized in the liver and can be used in oral formulations. Examples of progesterones and gestagens are given in Figure 8.20.

The male sex hormones, androgens, have effects on the development of the external sexual properties of males, the production of sperm (spermatogenesis) and the growth of muscle tissues (anabolic effect). Androgens are used in the treatment of male sterility, impotency and female breast and genital cancers.

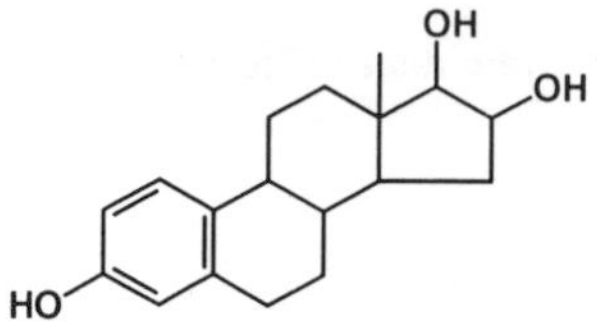

Oestriol

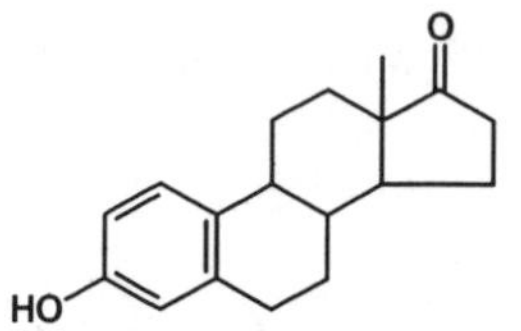

Oestrone

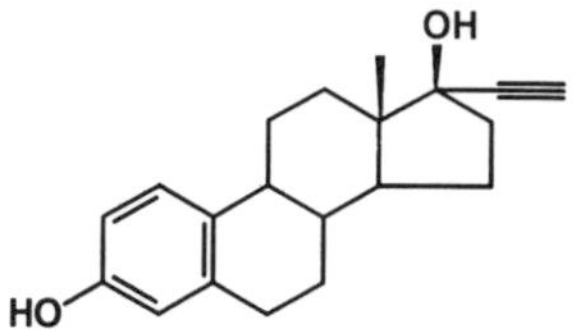

Ethinyloestradiol

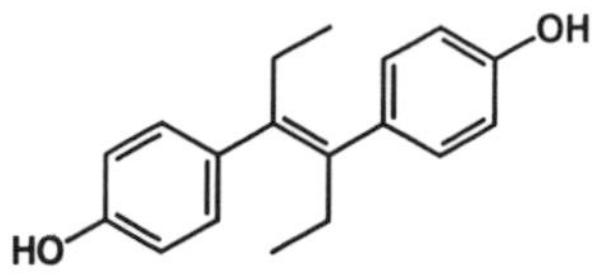

Diethylstilboestrol

Figure 8.19

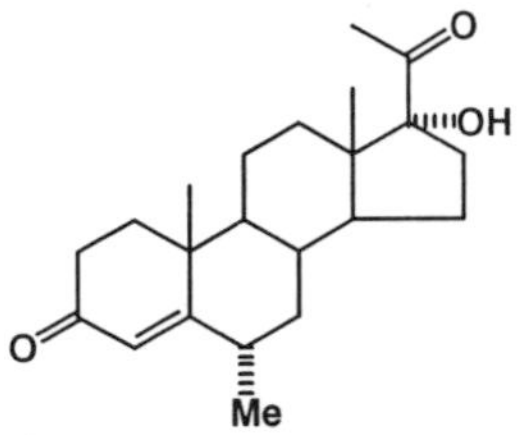

Medroxyprogesterone

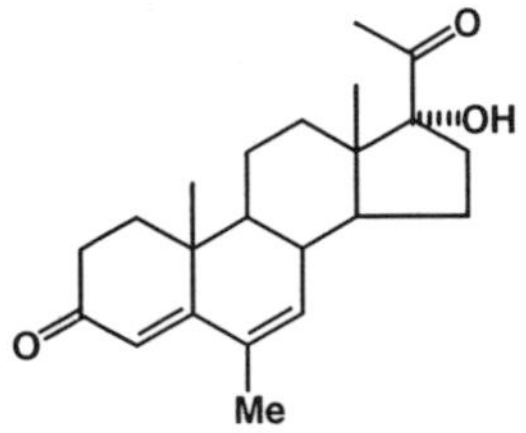

Megestrol

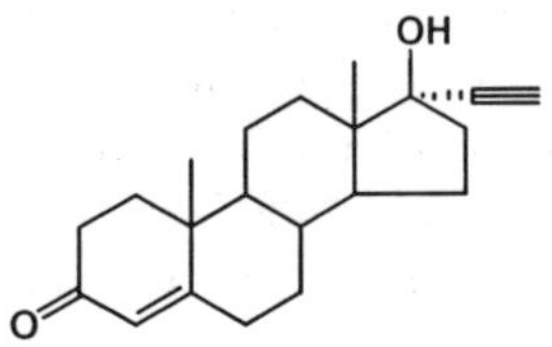

Norethisterone

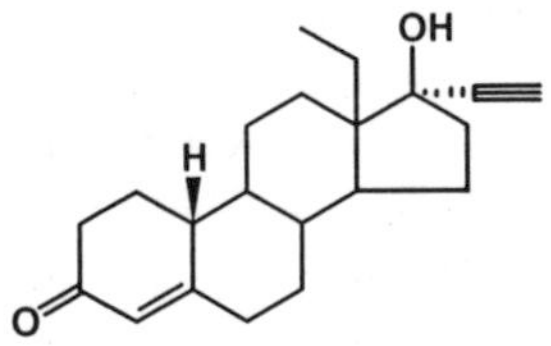

Norgestrel

Figure 8.20

In the anabolic effect, the excretion of nitrogenous compounds is decreased and the biosynthesis of structural proteins is increased. This is externally manifested as the growth of muscles. With molecular modifications the androgenic component can be eliminated, and the anabolic steroids are used to speed up tissue growth after operations, to treat growth abnormalities in muscles and bones and also in the notorious doping cases in sports. Androstanone (Figure 8.21) is an example of an androgen, and metelonone is an anabolic steroid.

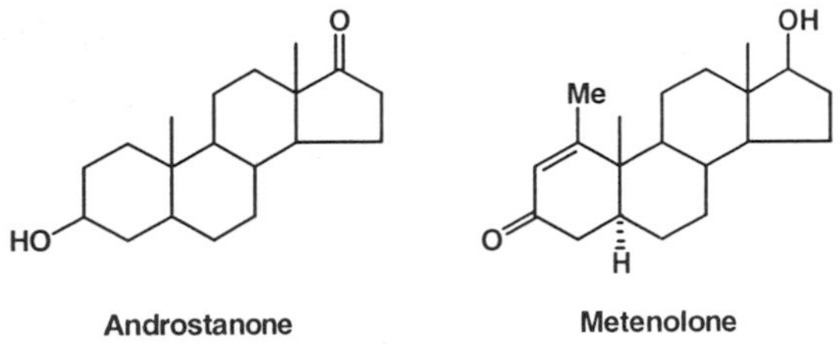

Androstanone   Metenolone

Figure 8.21

The external regions of the kidneys secrete corticosteroids which are of two main classes. The glucocorticoids act on the carbohydrate, lipid and protein metabolism, and the mineralocorticoids regulate the secretion of Na^+ ions in the kidneys. The main medicinal effect of the glucocorticoids is on the irritation and prevention of rheumatic effects. The use of the steroidal anti-inflammatory drugs is declining, because new, more effective drugs with less severe side effects are being developed. Of the natural hormones, only cortisone and hydrocortisone (Figure 8.22) are medicinally used. Synthetic molecular modifications include prednisone (five times as effective as hydrocortisone) and dexamethasone (30 times). The mineralocorticoids regulate the Na^+/K^+ equilibrium. The only hormone with this ability is aldosterone (Figure 8.22), but it is not used as such in medical practice. Desoxycortone and fludrocortisone are synthetic mineralocorticoids used in the treatment of Addison's disease.

8.3.1 Biosynthesis of Steroids

As we have already seen, the steroids are biosynthesized from squalene. The intermediate for the steroid biosynthesis is lanosterol, which still has to undergo a number of degradative steps to reach cholesterol, the common precursor to the rest of the steroid hormones. The side-chain double bond needs to be reduced, the endocyclic double bond has to be isomerized and altogether three methyl groups must be removed. The hydrogenation of the side-chain double bond is the most straightforward: an NADPH mediated *cis* hydrogenation gives the fully saturated precursor.

The next event is the oxidative removal of the angular C-14 methyl group. This occurs through sequential oxidation of the methyl group. The methyl carbon is lost as formic acid (Scheme 8.13). It is also known that the 15α proton is selectively lost and replaced during this oxidation. The emerging double bond is then hydrogenated (NADPH) to give the demethylated compound.

The demethylation of the geminal dimethyl grouping at C-4 is one of the last events to occur in the biosynthesis of cholesterol. This occurs via oxidation of the methyl on the lower face of the ring system (α) to a carboxylic acid and oxidation of the C-3 hydroxyl to a ketone (Scheme 8.14). The β-keto ester undergoes decarboxylation and equilibration of the remaining methyl group to the α-face. Repetition of the oxidation–decarboxylation cycle is followed by NADPH mediated reduction to give the demethylated C-3 alcohol.

The mechanism (and indeed the timing) of the isomerization of the C-8 double bond is not as clear. Whether it occurs relatively early (even before the oxidative removal of the C-14 methyl group) or late in the biosynthesis is still under debate. A plausible

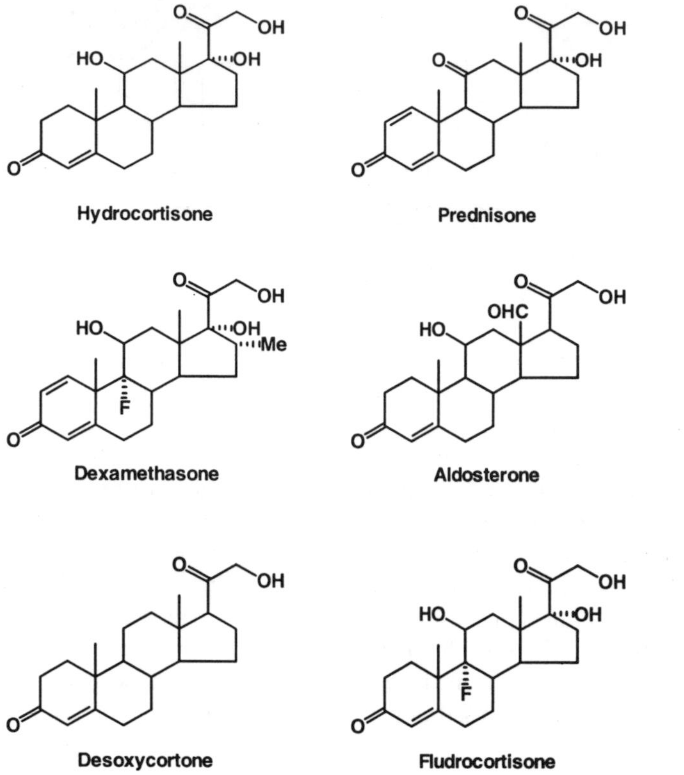

Figure 8.22

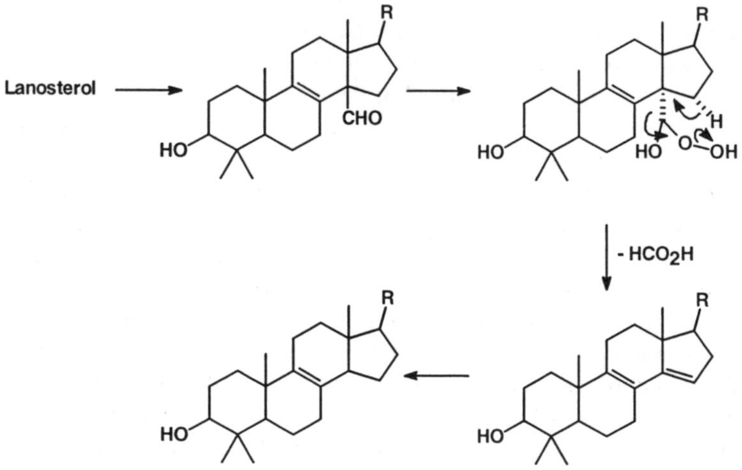

Scheme 8.13

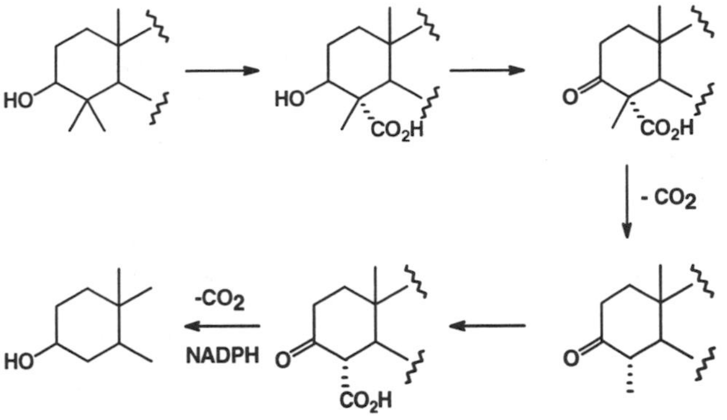

Scheme 8.14

suggestion has been advanced,[14] based on the occurrence of certain natural products, that before elimination of the C-14 methyl group the original double bond is isomerized to give the C-7 olefin. A 5,7-diene (7-dehydrocholesterol) is the next proposed intermediate, in line with the existence of ergosterol, the precursor of vitamin D_2.

Ergosterol was first isolated from ergot, but is also readily available in yeast. Ergosterol is converted to precalciferol by a light-induced retrocyclization of the B-ring, and further thermal isomerization leads to calciferol (vitamin D_2; Scheme 8.15). Deficiency of vitamin D causes rickets, a weakening of the bone structure, which can be prevented by supplementing the diet with vitamin D rich nutrients. Fish liver oil is a traditional source of vitamin D.

8.3.2 Asymmetric Synthesis of Steroids

Because of their importance in medicinal use, steroids have enjoyed a tremendous amount of synthetic activity, and much of this is already incorporated into the basic organic chemistry texts. Some recent developments in the asymmetric synthesis methodology have facilitated the development of new strategies for the elaboration of the tetracyclic nucleus. The common routes depend on either the formation of a hydrindanone system (C and D rings) or the construction of the B and C rings through a stereoselective cyclization.

In the early 1970s, chemists at Roche and Schering nearly simultaneously developed a remarkably facile catalytic aldolization process to construct the hydrindanone in practically enantiopure form.[15,16] Starting from the C_2 symmetric cyclopentanedione, catalysis with natural L-proline gives the CD ring fragment with the natural stereochemistry (Scheme 8.16).

The intramolecular [4+2] cycloaddition reaction utilizing a benzocyclobutene precursor for a quinodimethane is a route used widely for the construction of the B-ring of steroids. Recently a modification of this strategy has facilitated the first total synthesis of (+)-cortisone.[17] The chiral benzocyclobutene was induced to undergo thermal ring opening, and the *o*-quinodimethane underwent a completely *exo* selective, intramolecular [4+2]-cycloaddition to give the BCD ring system (Scheme 8.17). Final elaboration of the A-ring with standard Robinson annulation type chemistry and implementation of the correct functionality gave cortisone in enantiopure form.

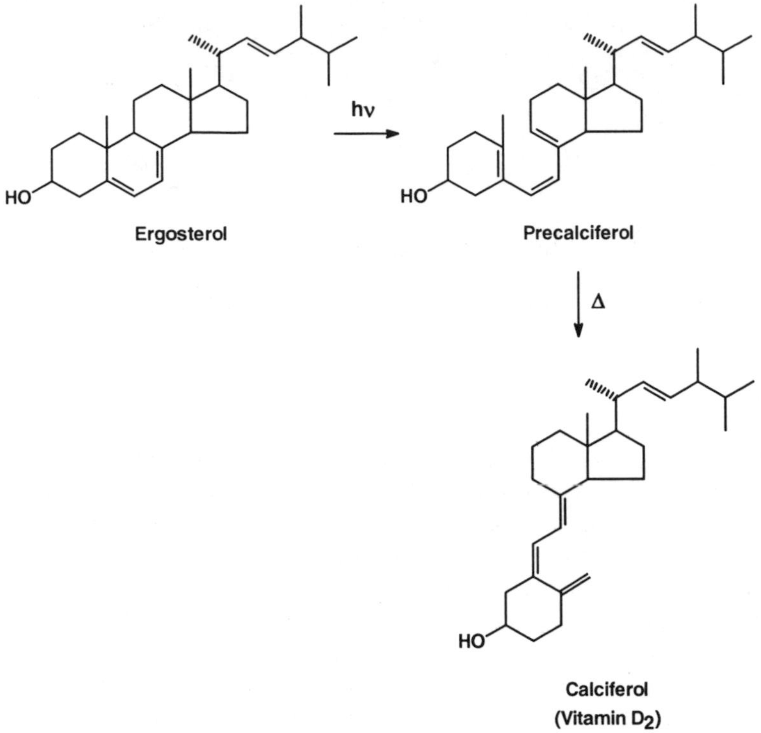

Scheme 8.15

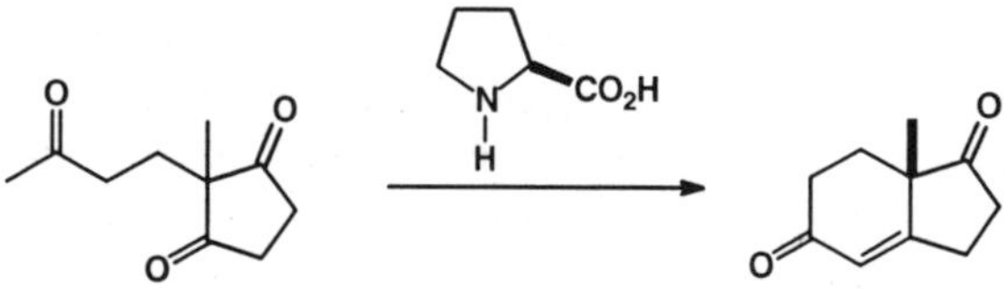

Scheme 8.16

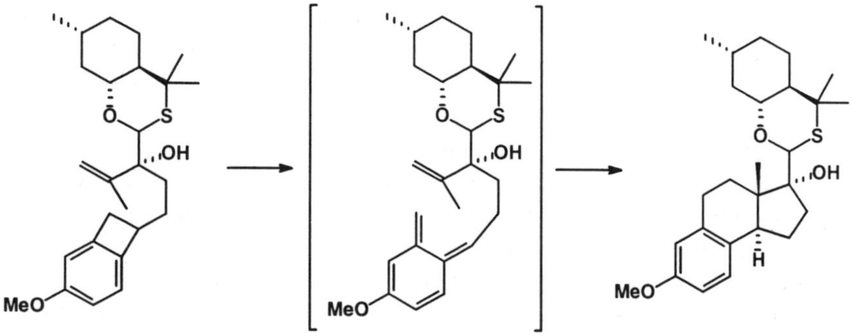

Scheme 8.17

References

1. Ruzicka, L. *Experientia*, **9**, 357–367 (1953).
2. Paquette, L.A. and Doherty, A.M. *Polyquinane Chemistry: Syntheses and Reactions* Springer Verlag: Berlin, 1987.
3. Mander, L.N. *Chem. Rev.* **92**, 573–612 (1992).
4. Epstein, W.W. and Rilling, H.C. *J. Biol. Chem.* **245**, 4597–4605 (1970).
5. Popjak, G., Edmond, J., and Wong, S.-M. *J. Am. Chem. Soc.* **95**, 2713–2714 (1973).
6. Corey, E.J., Russey, W.E., and de Montellano, P.P.O. *J. Am. Chem. Soc.* **88**, 4750–4751 (1966).
7. van Tamelen, E.E., Willett, J.D., Clayton, R.B., and Lord, K.E. *J. Am. Chem. Soc.* **88**, 4752–4754 (1966).
8. Yamamoto, S. and Bloch, K. *J. Biol. Chem.* **245**, 1668–1672 (1970).
9. Ebersole, R.C., Godtfredsen, W.O., Vangedal, S., and Caspi, E. *J. Am. Chem. Soc.* **95**, 8133–8140 (1973).
10. Corey, E.J. and Virgil, S.C. *J. Am. Chem. Soc.* **113**, 4025–4026 (1991).
11. Jones, P.S., Ley, S.V., Morgan, E.D., and Santafianos, D. *The Chemistry of the Neem Tree* in *1988 Focus on Phytochemical Pesticides: Vol. 1, The Neem Tree* (Jacobson, M., Ed.) CRC Press Inc.: Boca Raton, FL 19 (1989).
12. Kolb, H. and Ley, S.V. *Tetrahedron Lett.* **32**, 6187–6190 (1991).
13. Ourisson, G., Rohmer, M. and Poralla, K. *Annu. Rev. Microbiol.* **41**, 301–333 (1987).
14. Schroepfer, G.J., Jr., Lutsky, B.N., Martin, J.A., Huntoon, S., Fourcans, B., Lee, W.-H., and Vervilion, J. *Proc. R. Soc. London, Ser. B* **180**, 125–146 (1972).
15. Hajos, Z.G. and Parrish, D.R. *J. Org. Chem.* **39**, 1615–1621 (1974).
16. Eder, U., Sauer, G., and Wiechert, R. *Angew. Chem.* **83**, 492–493 (1971).
17. Nemoto, H., Matsuhashi, N., Imaizumi, M., Nagai, M., and Fukumoto, K. *J. Org. Chem.* **55**, 5625–5631 (1990).

9 Shikimic Acid Derivatives

A large number of aromatic natural products belong to a biogenetically homogeneous group, the so-called shikimates. These are formed from a common intermediate, shikimic acid, which was originally isolated from the Japanese plant *Illicium religiosum*, (in Japanese *shikimi-no-ki*). Shikimic acid (Figure 9.1) functions in many plants as the starting material for aromatic amino acids (phenylalanine, tyrosine and tryptophan), and is also the origin of a wide variety of other aromatic products. Shikimic acid is derived from pyruvate via reaction with D-erythrose-4-phosphate, as discussed in Chapter 5 in connection with the biosynthesis of the aromatic amino acids.

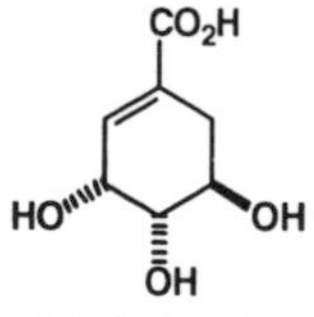

Shikimic acid

Figure 9.1

Salicylic and benzoic acids occur in many berries. These compounds are bacteriostatic (inhibit the growth of bacteria), and thus, for example, cloudberry can be preserved without added preservatives. These acids are formed from phenylalanine via elimination of ammonia to give cinnamic acid. Oxidative cleavage of the double bond then gives benzoic acid. Cinnamic acid can also be oxidized in the aromatic ring, first at the *para* position, then again at the *meta* position to give caffeic acid. Caffeic acid and ferulic acid (Figure 9.2) occur in the tannic acids of coffee as their quinic acid esters. 3-Caffeylquinic acid was isolated as early as 1846 by Payen from coffee. The composition and relative abundances of the glyconjugates vary depending on the different coffee bean types.

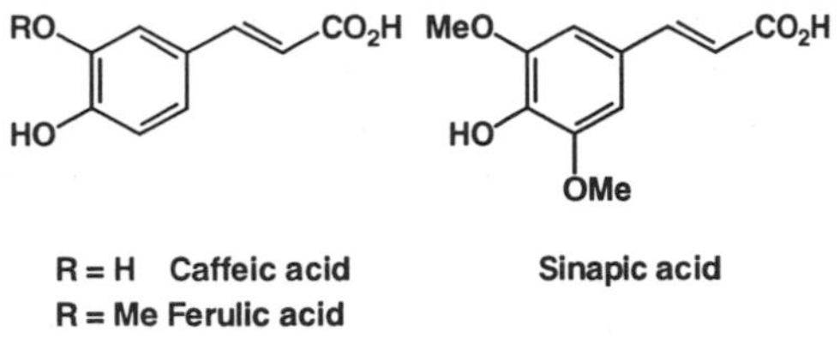

R = H   Caffeic acid
R = Me Ferulic acid

Sinapic acid

Figure 9.2

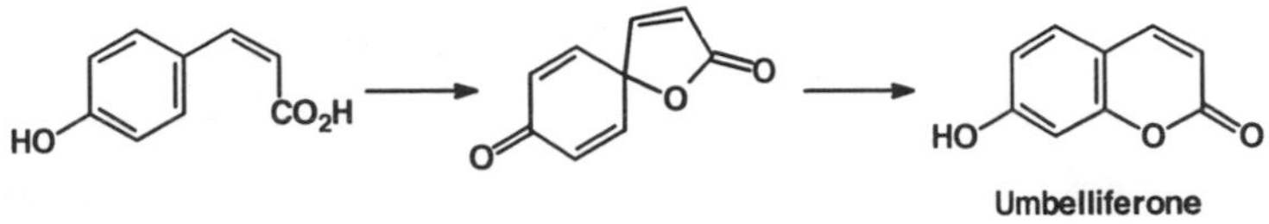

Umbelliferone

Scheme 9.1

Coumarins and quinones are also derived from shikimic acid through cinnamic acid. Most natural coumarins contain a hydroxyl group at C-7. The *o*-hydroxylation is formulated to occur by way of a spirolactone (Scheme 9.1) followed by a rearrangement step. This has been shown by tracer experiments utilizing an ^{18}O labelled *p*-hydroxycoumarin acid.

Psoralen is derived from umbelliferone through alkylation (prenylation) with dimethylallyl pyrophosphate followed by epoxidation, cyclization and cleavage of the side-chain (Scheme 9.2). It is used for the treatment of psoriasis, mainly in the form known as PUVA treatment (psoralen-UV). Psoralen is an example of a furanocoumarin, and its action is directed to DNA. The psoralen molecule intercalates with DNA and then binds with the two DNA strands upon UV irradiation, which triggers a [2+2] photochemical cycloaddition the results in the formation of covalent bonds with both of the DNA strands. Thus, replication of the DNA is prohibited.

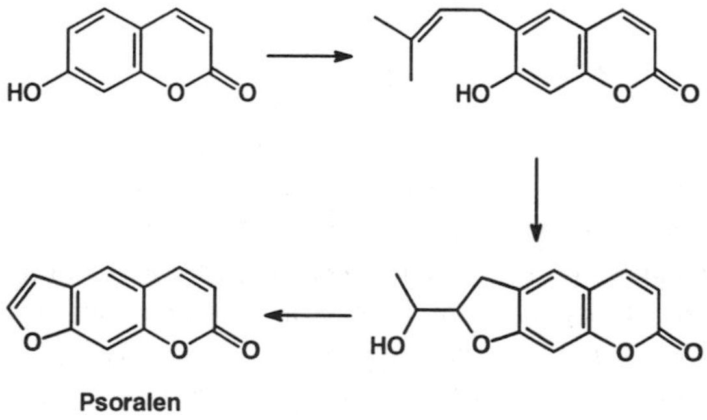

Psoralen

Scheme 9.2

Quinones are formed as oxidation products of quinic acid (hence their name), and the quinone moiety occurs in a number of natural products, including antibiotics (rifamycin, see macrolide antibiotics, Section 7.3.1) and vitamin K. Ubiquinones and plastoquinones (Figure 9.3) are coenzymes involved in the one-electron transport in living systems.

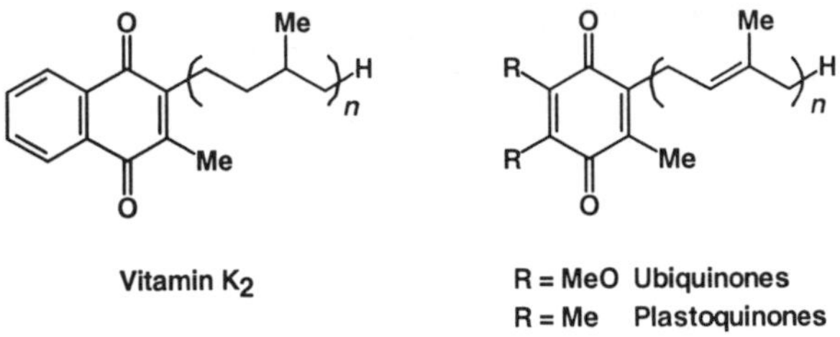

Vitamin K$_2$

R = MeO Ubiquinones
R = Me   Plastoquinones

Figure 9.3

The usually highly coloured flavones (from Greek *flavius* = yellow) form the largest group of oxygen heterocycles found in plants. The flavonoids show a mixed biogenesis from shikimates and polyketides, and their structures vary very widely. Their biological role is to interfere with insects pollinating or feeding on plants. Some flavonoids also have a characteristically bitter taste, which makes them repel caterpillars. Anthocyanins (e.g. Figure 9.4) are derived from flavonoids, and they usually have a strong red, violet or blue colour. The magnificent colours seen in autumn leaves are due to flavonoids, certain isoprenoids (yellow) and anthocyanins.

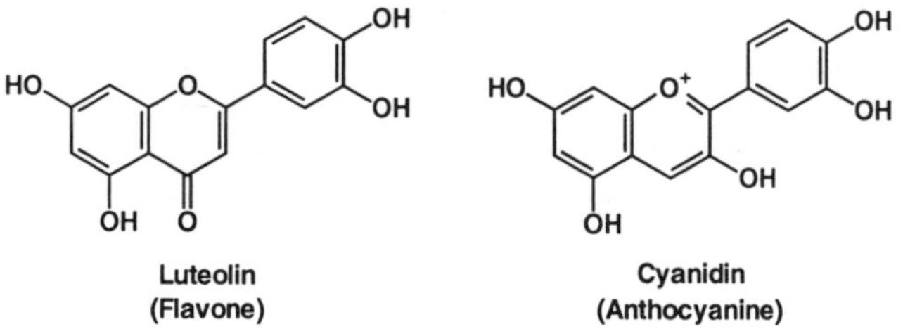

Luteolin
(Flavone)

Cyanidin
(Anthocyanine)

Figure 9.4

Compounds structurally related to the shikimates have an important role in the formation of lignin, the binding material of wood. Lignin is mainly formed by radical polymerization of phenylpropionyl units, especially those derived from coniferyl alcohol, the alcohol corresponding to ferulic acid. Because of the several possible precursor alcohols, and the possibility of various radical combinations, lignin has a highly heterogeneous structure. Aromatic compounds have a relatively high value, and therefore the chemical refining of lignin into smaller, industrially useful molecules is an important research topic. The structure in Figure 9.5 is a lignin model, showing several possible ways of combining the radical species.

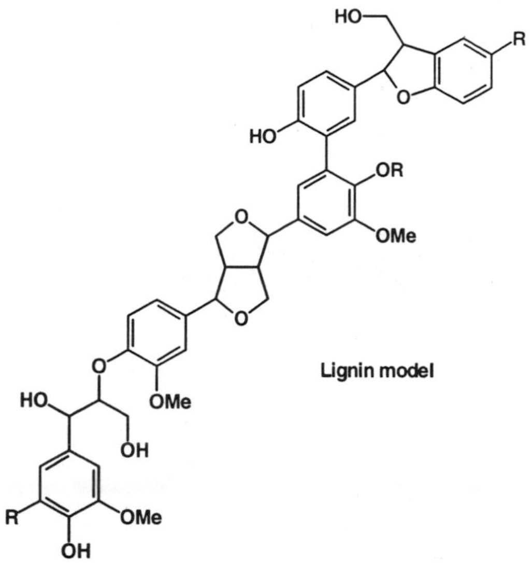

Lignin model

Figure 9.5

10 Alkaloids

The term alkaloid was originally coined by Meissner in 1818 to cover compounds similar in behaviour to alkalis, in other words, basic compounds. The word alkali derives from the Arabic *al qalay* = to roast. Many alkaloids are not, however, notably basic in character. It is often very difficult to distinguish naturally occurring compounds from non-natural ones. This definition of alkaloids has been doubted for a long time. Hesse has presented a new definition for alkaloids: nitrogen containing compounds derived from plants or animals.[1] Even this definition has its shortcomings, as for instance DNA, RNA and peptides are not considered to be alkaloids. On the other hand, we shall see that there exists a definite group of peptide alkaloids, a group of compounds containing the ergot alkaloids and cyclopeptide alkaloids, both of which are medicinally important.

At present, some 6000 natural alkaloids are known. These can be subdivided according to their chemical structures into the following groups: i) heterocyclic alkaloids; ii) alkaloids with an exocyclic nitrogen atom; iii) polyamines; iv) peptide alkaloids; and v) terpene alkaloids. Typical examples of each group will be discussed in the following in terms of their structures, biosynthesis and synthesis.

10.1 Heterocyclic Alkaloids

In common use, the term alkaloids refers to heterocyclic alkaloids. Many of these have had and still have an important role to play in the healing practices of most cultures.

Typical structural groups are indole alkaloids (e.g. reserpine, which possesses blood pressure reducing properties), pyrrolidine alkaloids (e.g. mesembrine), tropane alkaloids (cocaine and its relatives, such as atropine and scopolamine; these are anticholinergic agents), quinoline alkaloids (the malaria drug quinine and morphine for pain relief) and the izidine alkaloids (securinine is an indolizidine). Also pilocarpine, a histidine derivative which is used for the treatment of glaucoma, belongs to this broad class of natural products (Figure 10.1).

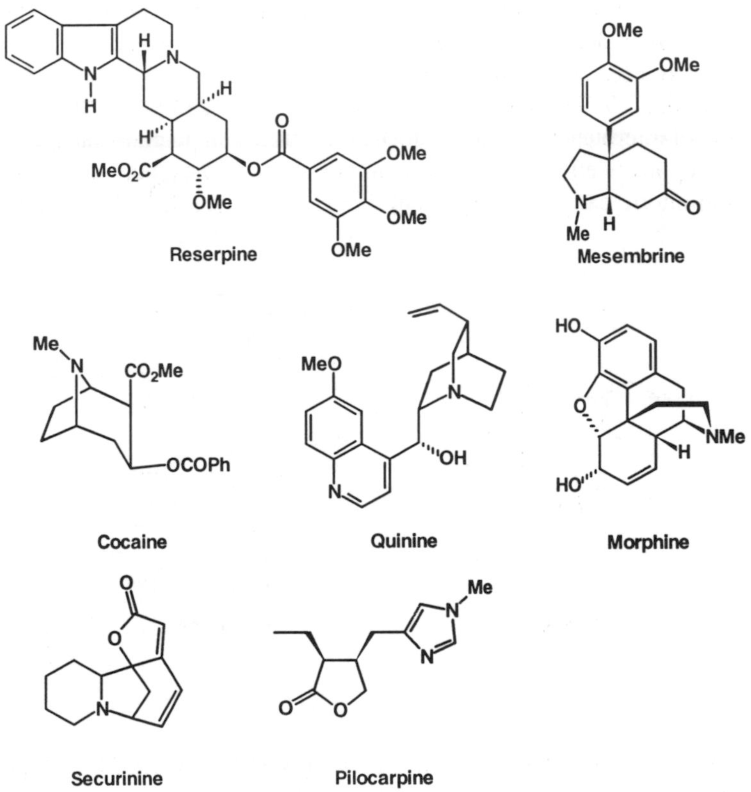

Figure 10.1

Medicinally, the alkaloids have played a key role for milennia, and even today some 25 % of commercial drugs are either alkaloids or their structural modifications or analogues. In drug design, the search for new chemical entities is still heavily dependent on natural compounds, and as long as new structures can be found from the plant and animal kingdoms this process can continue. A major threat is the destruction of the rain forests; it has been estimated that there are still tens, perhaps hundreds of thousands of unclassified and unexplored plant varieties which are under serious threat owing to the exploitation of these areas. We shall take some examples of the pharmacologically active alkaloids in the following.

10.1.1 Indole Alkaloids

Many pharmacologically important compounds belong to the indole alkaloids. Vincamine (Figure 10.2), a member of the *Aspidosperma–Hunteria* alkaloids, promotes blood circulation in the brain, and is being used for the treatment of stroke in many European countries. The commercial production of vincamine is based on both fermentation methods as well as total synthesis.

Another indole alkaloid with remarkable medicinal impact is vinblastine (Figure 10.2). This was the first compound observed to heal Hodgkin's disease, a form of lymphoid cancer. Vincristine, a close structural relative of vinblastine, is claimed to cure up to 70 % of acute lymphocytic leukaemia cases in children. Vinblastine is produced by the Madagascan periwinkle, *Catharanthus roseus*. The plant is related to the *Vinca* plants which are widely used as ornamental plants, and these alkaloids are often referred to as the *Vinca* alkaloids. Vinblastine and vincristine inhibit the cell propagation by inhibiting the formation of microtubules during mitosis. It is interesting to note that another alkaloid, taxol, has an opposite effect on the formation of microtubules, but the compound is still an active antimitotic cancer therapy agent. We shall return to the case of taxol shortly.

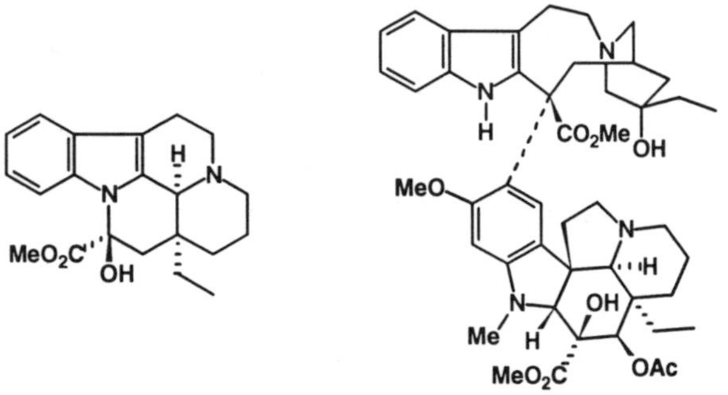

Vincamine   Vinblastine

Figure 10.2

The chemistry of the indole alkaloids is strongly influenced by the unique reactions occurring to the indole moiety. Two of these are related to the biosynthesis in terms of the formation of the polycyclic alkaloid skeleton, namely the condensation reactions of indole with aldehydes and amides. These two reactions form the basis of many biogenetic reactions, and they have also been very efficiently utilized in the so-called biomimetic syntheses of several indole alkaloids. We shall briefly take a look at these two important reactions.

The condensation of an indolylethylamine (tryptamine) with an aldehyde provides an example of the Pictet–Spengler cyclization (Scheme 10.1). The reaction is initiated by the formation of the Schiff base, which after protonation (or the action of a Lewis acid) gives an electrophilic iminium ion prone to cyclization with the indole unit. Much work was expended on finding out whether it was the 3- or 2-position of the indole that would perform as the nucleophile. The end result is 2-alkylation, although similar reactions with simple pyrroles tend to be directed to the 3-position. It was finally Jackson's labelling

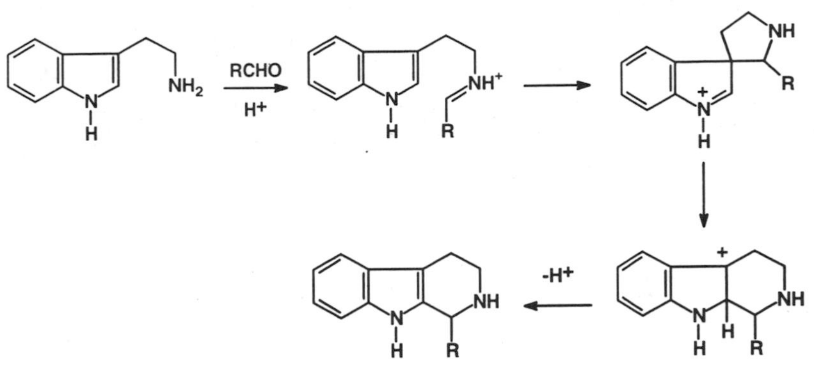

Scheme 10.1

studies which confirmed that the reaction occurs by initial alkylation at C-3, followed by a Wagner–Meerwein type 1,2-shift.[2]

If the newly forming C-ring contains a substituent at the carbon bearing the amino group (e.g. tryptophan), there exists the possibility for the formation of either the *cis* or *trans* substituted products. The stereochemical aspects of the cyclization have been studied.[3–5] Under kinetic conditions, the attack of the indole nucleophile to form the spiroindolenine intermediate would favour the formation of the *cis* substituted product (both substituents equatorial in the final six–membered ring). Substitution of the indole nitrogen or the N_α nitrogen (R^3 and R^2, respectively) will increase the allylic $A^{1,2}$ strain in the transition state leading to the *cis* product, and therefore the *trans* cyclization is observed (Scheme 10.2).

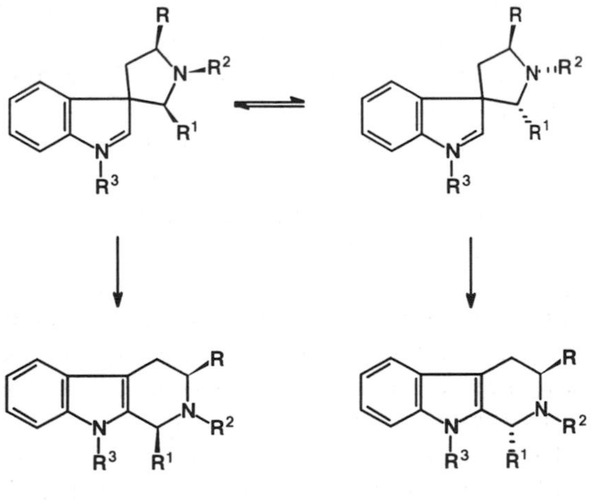

Scheme 10.2

The amide derived from tryptamine and a carboxylic acid can be induced to cyclize by using dehydrating conditions. This is known as the Bischler–Napieralski reaction. Usually one employs phosphorus oxychloride, tosyl chloride or similar (acidic) dehydrants, and the mechanism has been rationalized as involving a chloroiminium intermediate. The initial product is the corresponding unsaturated tricycle, which can be hydrogenated to the saturated compound (Scheme 10.3).

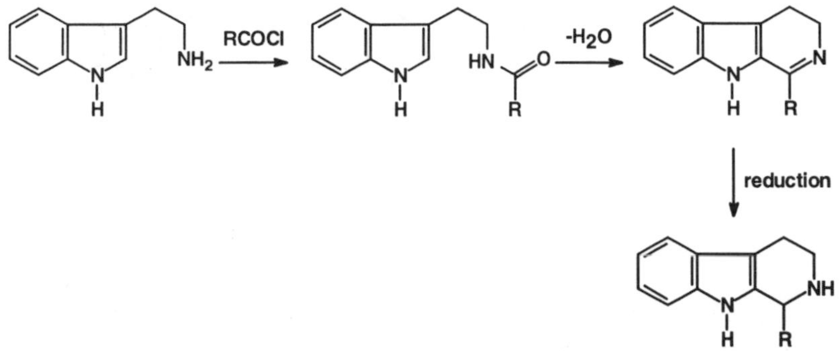

Scheme 10.3

10.1.1.1 Biosynthesis of indole alkaloids

Biogenetically, indole alkaloids are derived from tryptophan and secologanin. The earliest proposals for the origins of the indole alkaloids were put forth by Perkin and Robinson in 1919,[6] when they postulated the role of tryptophan as the source of the indole moiety in these alkaloids. The monoterpene unit secologanin was the subject of much debate until the late 1970s. The Barger–Hahn hypothesis suggested that the carbon skeleton of many of the indole alkaloids is derived from tryptophan, phenylalanine and formaldehyde.[7,8] Woodward refined this model and substituted 3,4-dihydroxyphenylalanine for the aromatic precursor.[9] According to this theory, the o-hydroquinone moiety would facilitate the cleavage and further processing of the carbon structure to produce the *Strychnos* and other indole alkaloids. The similarity of the non-tryptophan moiety of the indole alkaloids to several non-alkaloidal glucosides led to the suggestion that the C_{10} unit arises from two mevalonate units.[10,11] This Thomas–Wenkert hypothesis was much disputed when first presented, but it has been shown by radioactive labelling studies to be a correct representation of the events. In recent years, plant cell cultures and isolated enzyme preparations have enabled the identification of several new intermediates in the biogenetic pathways, and thus our understanding of the biosynthesis of alkaloids has increased considerably in detail.[12] According to their biogenetic origins, the indole alkaloids can be divided into the following classes (see Scheme 10.4):

   (i) secologanin unrearranged (e.g. ajmalicine);
  (ii) secologanin rearranged (route **a**, e.g. vincadifformine);
 (iii) secologanin rearranged (route **b**, e.g. catharanthine);
 (iv) indole alkaloids not derived from secologanin;
  (v) bisindole alkaloids.

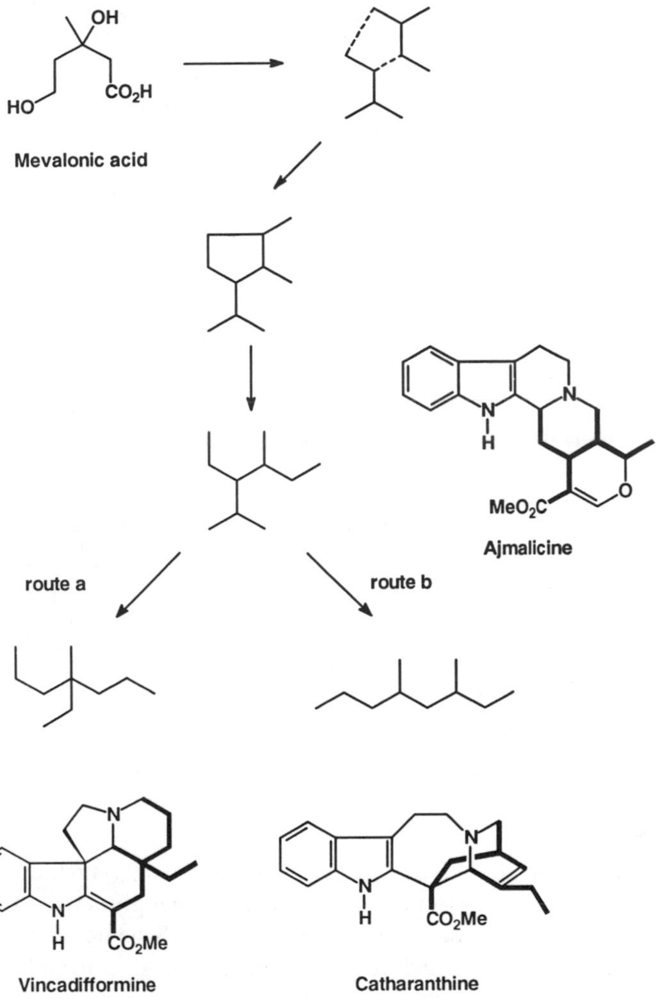
OH
HO
CO2H
Mevalonic acid
N
H
MeO2C
O
Ajmalicine
route a
route b
N
H
CO2Me
Vincadifformine
N
H
CO2Me
Catharanthine

Scheme 10.4

The indole alkaloids comprise a group of compounds which represents some 1500 natural products. Their value lies in their varied and efficiently utilized applicability as medicinal agents. Ajmaline (Figure 10.3), a member of the sarpagine–ajmaline alkaloids, is used for the treatment of cardiac arrhythmia. Vincamine, an *Eburna* alkaloid, finds its use in the treatment of hypertension. The dimeric *Catharanthus* alkaloids, exemplified by vinblastine and vincristine, enjoy their position as efficient medicaments for the treatment of various forms of cancer.

Structurally, the indole alkaloids are classified into three main groups: the *Corynanthe–Strychnos*, the *Aspidosperma–Hunteria*, and the *Iboga* alkaloids (e.g. Figure 10.4). The classification is based on the complexity of the biosynthetic pathway, the first class arising by relatively simple chemical and enzymatic transformations, and the latter two requiring more thorough rearrangement steps. We shall first consider the common pathway for all these alkaloids, and then take a brief look at the biogenesis of the individual classes.

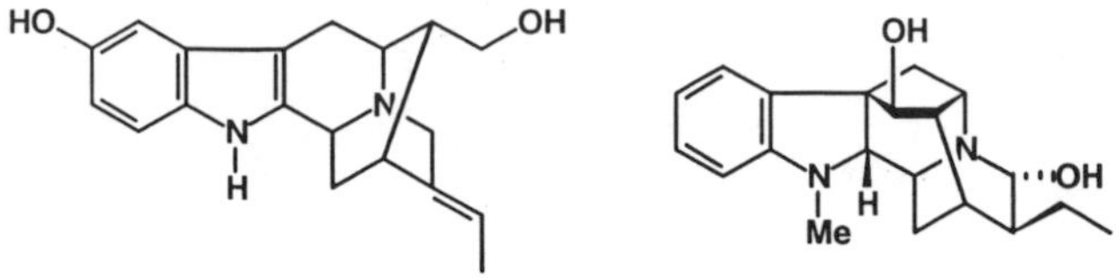

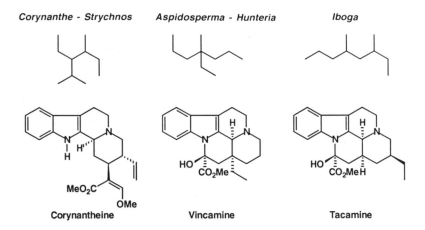

Sarpagine **Ajmaline**

Figure 10.3

Corynanthe - Strychnos *Aspidosperma - Hunteria* *Iboga*

Corynantheine **Vincamine** **Tacamine**

Figure 10.4

The indole alkaloids are derived from tryptophan and a C_{10} unit, a monoterpene. The monoterpene unit itself arises from mevalonic acid, which gives rise to the monoterpene geraniol. A number of oxidative transformations (see Scheme 10.5) are followed by glucosylation and cyclization and further oxidation to give loganin. This is further oxidatively cleaved to give secologanin, the ultimate terpenoid intermediate in the biosynthesis of the indole alkaloids.

Secologanin is then coupled with tryptophan (tryptamine) with the loss of the carboxyl function to give the common intermediate for all indole alkaloids, strictosidine (Scheme 10.6). Some two decades ago, the roles of strictosidine and its C-3 epimer, vincoside, were heavily disputed, and finally strictosidine was found to be the actual biogenetic intermediate.

Different modes of cyclization give rise to the various structural classes of the indole alkaloids. Vallesiachotamine and the heteroyohimbanes (e.g. tetrahydroalstonine, ajmalicine and cathenamine; Scheme 10.6) arise, and these function as further relay points for the more elaborate *Corynanthe* alkaloids.

The pentacyclic *Corynanthe–Strychnos* alkaloids can be grouped into structurally related classes. Yohimbanes contain a carbocyclic E-ring, whereas in the heteroyohimbanes the E-ring is heterocyclic. The E-ring cleaved alkaloids are secoyohimbanes (corynanes). The basic ring structures are shown in Figure 10.5, as well as the biogenetic ring numbering which is commonly used in alkaloid chemistry.[13]

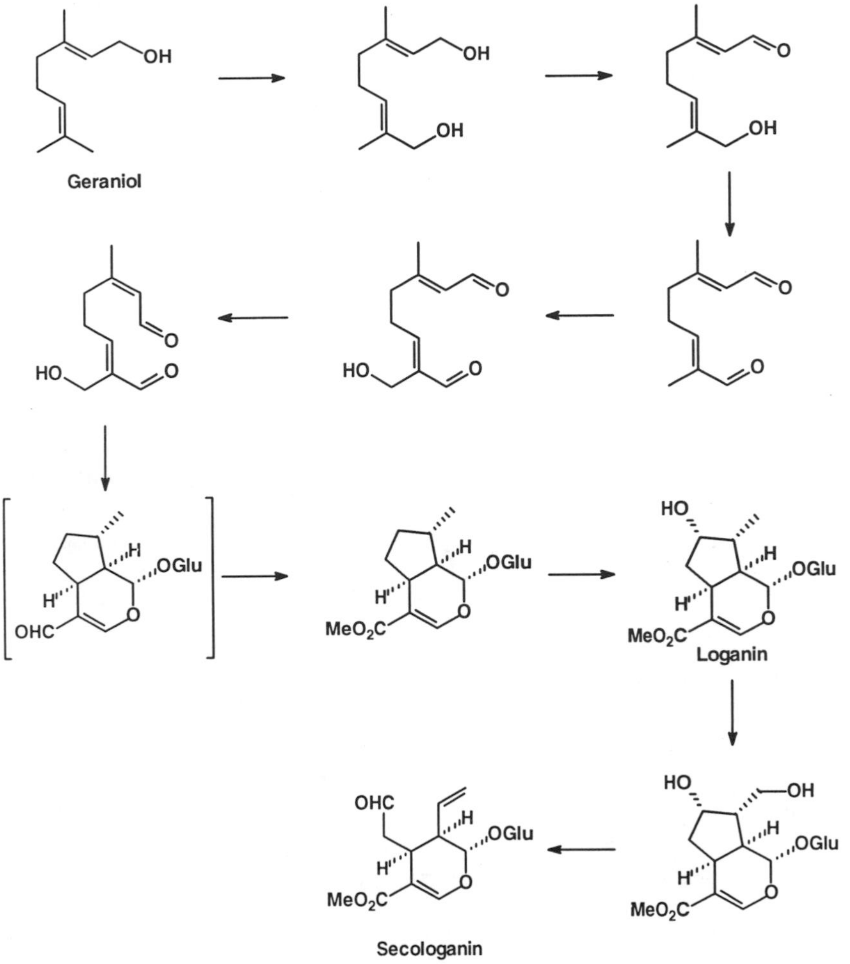

Scheme 10.5

In yohimbanes, there are three chiral centres in the ring carbon atoms, giving rise to the possibility of eight stereoisomers. The hydrogen atom at C-15 is always down (α), reducing the number of naturally occurring isomers to four. All four structures, yohimbane ($3\alpha,20\beta$), pseudoyohimbane ($3\beta,20\beta$), alloyohimbane ($3\alpha,20\alpha$) and epialloyohimbane ($3\beta,20\alpha$), have been found in naturally occurring alkaloids (Figure 10.6).

In the biosynthesis, strictosidine is converted by a glucosidase enzyme into the hemiacetal, which undergoes a sequence of steps to geissoschizine and ajmalicine (Scheme 10.7).

The biogenesis of both the *Aspidosperma–Hunteria* and the *Iboga* alkaloids requires more deep-seated rearrangements. There is overwhelming evidence that *Strychnos* alkaloids are formed from strictosidine through the *Corynanthe* alkaloids. These are the precursors for the *Aspidosperma* and *Iboga* alkaloids. The biogenesis proceeds via geissoschizine which, through oxidation of the indole ring, is cleaved and recyclized

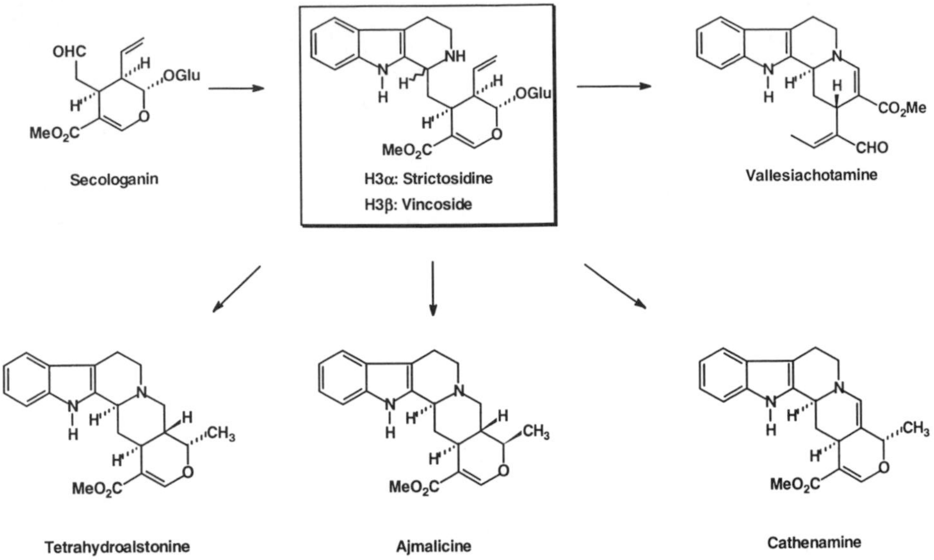

Scheme 10.6

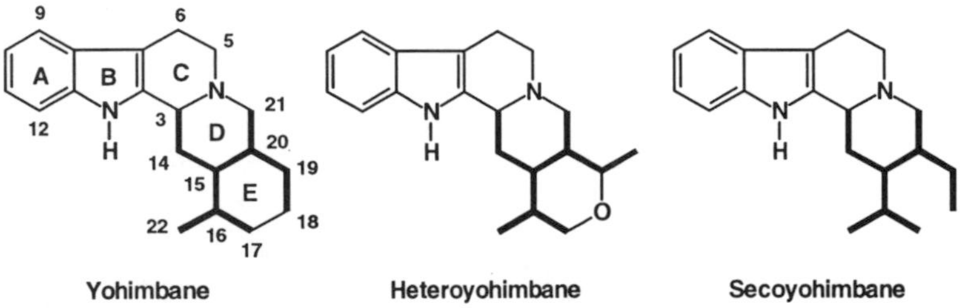

Figure 10.5

to preakuammicine. This in turn functions as the intermediate for both akuammicine and stemmadenine, the former leading to the *Strychnos* alkaloid skeleton and the latter leading to the *Aspidosperma–Hunteria* alkaloids (Scheme 10.8).

The formation of the *Aspidosperma–Hunteria* alkaloids from stemmadenine is depicted in Scheme 10.9. Isomerization of the ethylidene double bond is followed by fragmentation with participation of the nitrogen lone pair (1,6-Grob type fragmentation) to give secodine, which is the common intermediate for both types of alkaloids. A formal Diels–Alder reaction leads to the *Iboga* alkaloids. An alternative cyclization mode of the enamine portion onto the acrylate unit gives rise to an iminium ion, whose further cyclization gives the *Aspidosperma* skeleton.

10.1.1.2 Asymmetric synthesis of indole alkaloids

Since the Pictet–Spengler cyclization reaction is highly stereoselective, most of the generally used strategies for the asymmetric syntheses of indole alkaloids rely on this powerful process. The asymmetric information can be introduced either through the use of tryptophan or via an aldehyde containing the desired chirality. We shall look at one example of each of these strategies, as well as a strategy utilizing a chiral auxiliary.

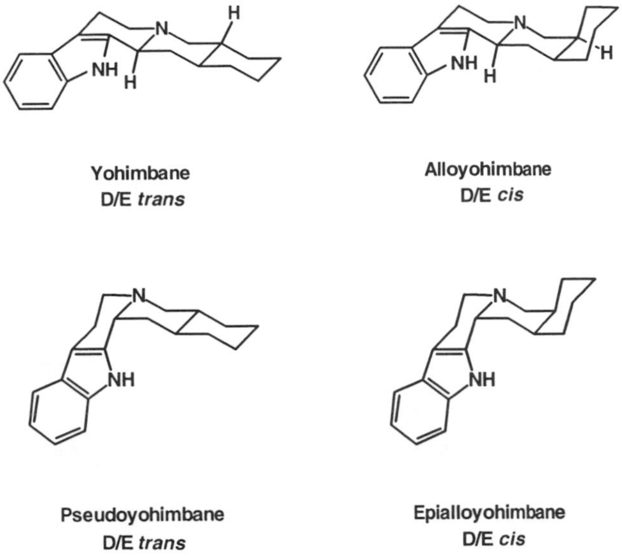

Yohimbane
D/E *trans*

Alloyohimbane
D/E *cis*

Pseudoyohimbane
D/E *trans*

Epialloyohimbane
D/E *cis*

Figure 10.6

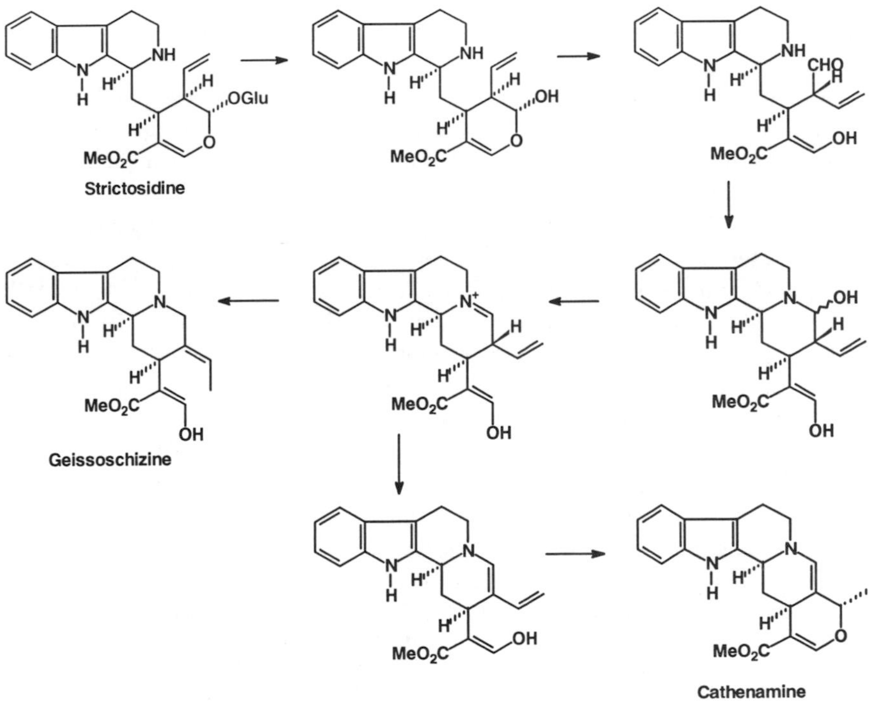

Strictosidine

Geissoschizine

Cathenamine

Scheme 10.7

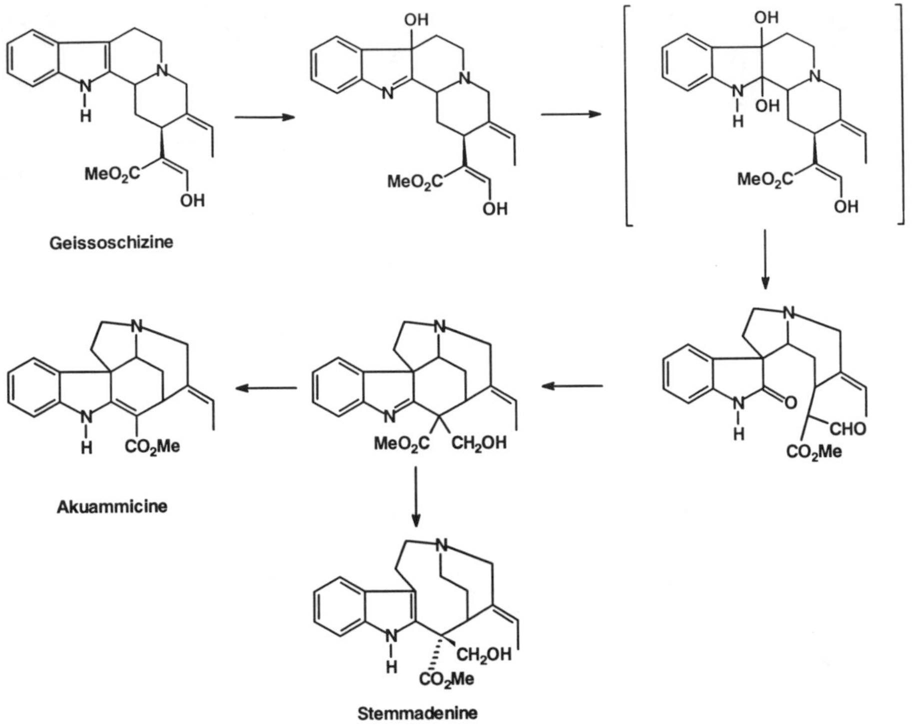

Scheme 10.8

Enantiofacially selective deprotonation of the bicyclo[3.3.0]octane-3,7-dione mono-ketal followed by alkylation gives the enantiomerically enhanced, methylated product (Scheme 10.10).[14,15] Bayer–Villiger oxidation converts the cyclopentanone ring to the E-ring of the heteroyohimbanes.

DIBAL reduction of the lactone to the lactol followed by acid catalysed dehydration sets the unsaturation in the E-ring precursor and also liberates the ketone. Cleavage of the five membered ring was finally achieved through conversion of the ketone to the enol ether followed by oxidative scission of the enol double bond (Scheme 10.11).

Tryptophan was used as the chiral originator in a total synthesis of koumine, the principal medicinal constituent of the Chinese plant *Gelsemium elegans*.[16,17] Pictet–Spengler cyclization of the protected tryptophan with ketoglutaric acid gives the *trans* product as the major isomer, as expected (Scheme 10.12). Dieckmann cyclization preceded by epimerization at the carboxylate bearing carbon, gives the tetracyclic intermediate after appropriate adjustment of functionalities. Installation of the propargylic acid side-chain completes the assembly of the skeleton for the final ring closures.

Ring closure of the fifth ring is achieved by a Michael type process triggered by pyrrolidine and trifluoroacetic acid (TFA) in refluxing benzene (Scheme 10.13). Homologation of the ketone and adjustment of the oxidation stages sets up the precursor for a ring rearrangement–cyclization of the diol to the koumine skeleton. This cyclization closely resembles the route by which Nature also assembles the koumine skeleton.

Our final example utilizes a chiral auxiliary. Meyers has developed the chemistry of chiral amidines, especially in asymmetric alkylation reactions.[18,19] Metallation of the

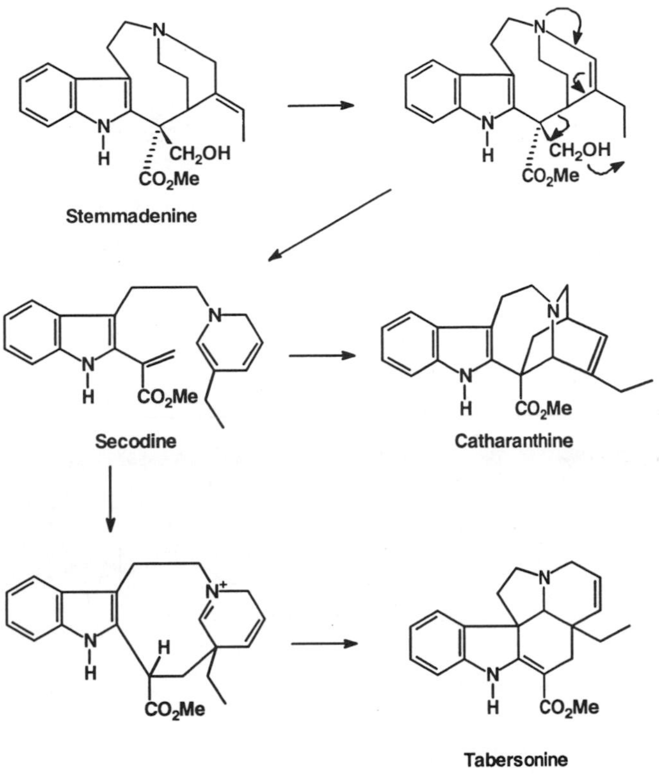

Stemmadenine

Secodine

Catharanthine

Tabersonine

Scheme 10.9

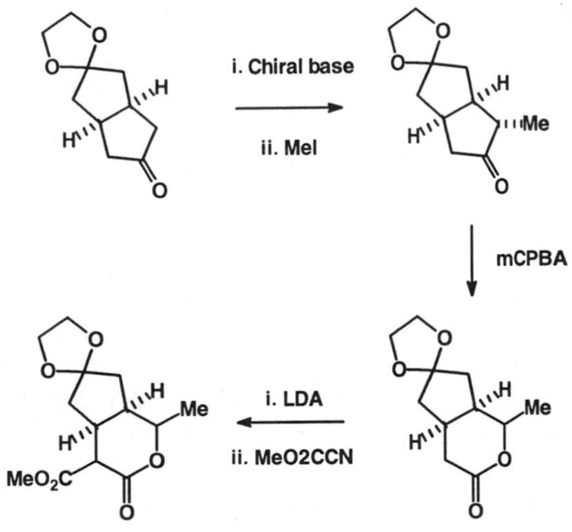

i. Chiral base

ii. MeI

mCPBA

i. LDA

ii. MeO2CCN

Scheme 10.10

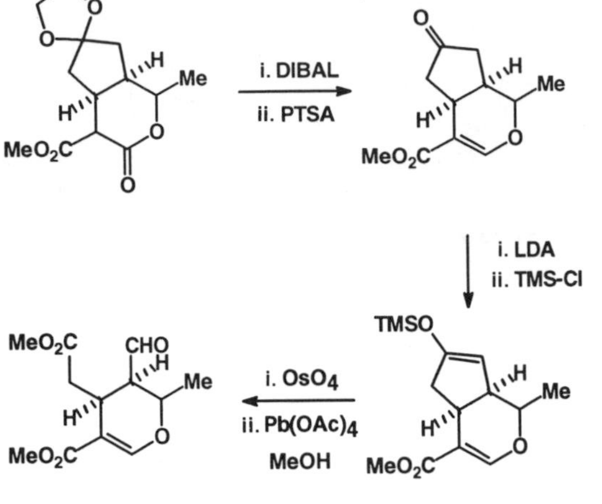

Scheme 10.11

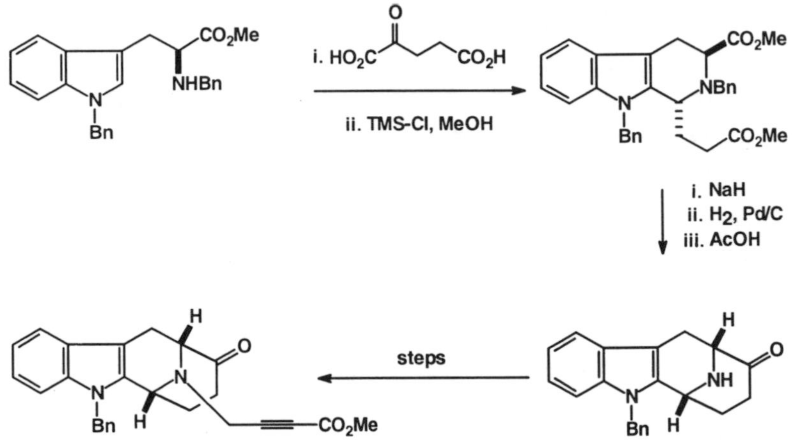

Scheme 10.12

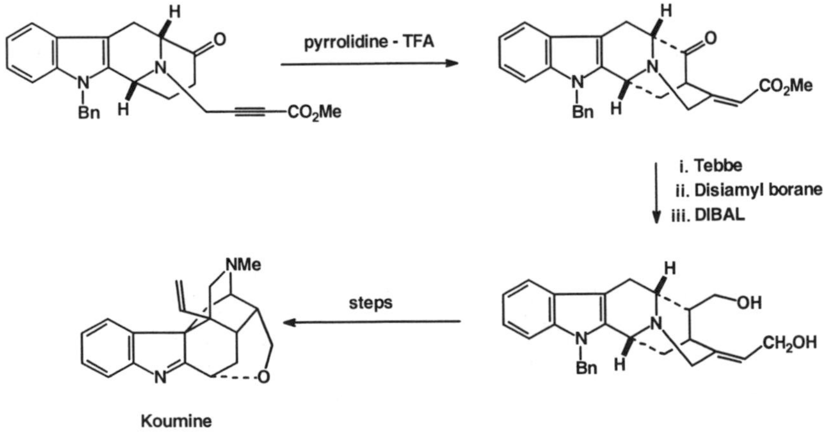

Koumine

Scheme 10.13

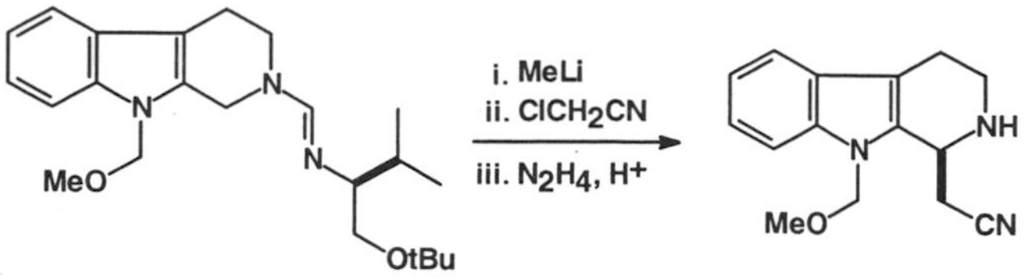

Scheme 10.14

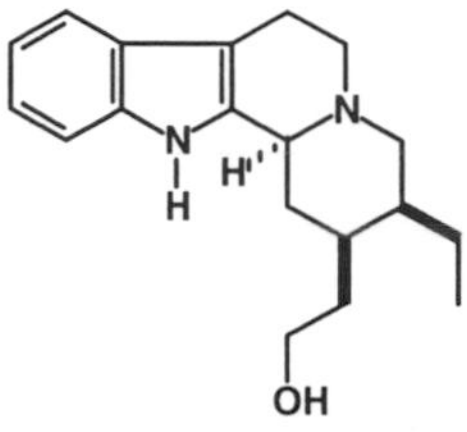

Corynantheidol

Figure 10.7

β-carboline equipped with the leucine derived formamidine auxiliary followed by trapping of the anion with chloroacetonitrile gives rise to the C-3 chiral centre (Scheme 10.14). This intermediate was utilized in the asymmetric synthesis of corynantheidol (Figure 10.7).[20]

10.1.2 Pyrrolidine and Tropane Alkaloids

Pyrrolidine alkaloids are biogenetically formed from a diamino acid, ornithine (Scheme 10.15). Pyridoxal phosphate functions as an activator in these reactions, first leading to decarboxylation and simultaneous formation of the electrophilic imine (or iminium ion after protonation). This reacts with an activated acyl coenzyme A derivative, releasing pyridoxamine. Decarboxylation finally gives hygrine.

Participation of the doubly activated acyl coenzyme A leads to a β-keto acid derivative, which can further cyclize to give the tropane skeleton (Scheme 10.16). The cyclization is presumably preceded by an oxidative transformation of the amino group into an iminium ion. Tropinone itself is reduced to tropine, which is the precursor for many more tropane alkaloids.

The tropane alkaloids occur in two plant families, *Erythroxylonaceae* and *Solanaceae*. Cocaine (Scheme 10.16) is isolated from the leaves of the South American bush *Erythroxylon coca*, and the natives of Peru and Bolivia have used for centuries these leaves to improve endurance and promote a sense of well-being. Cocaine has local anaesthetic effects, and it has been used as a lead structure for the development of more potent local anaesthetics with fewer side effects. Such molecules are in wide use today, including lidocaine and procaine (Figure 10.8).

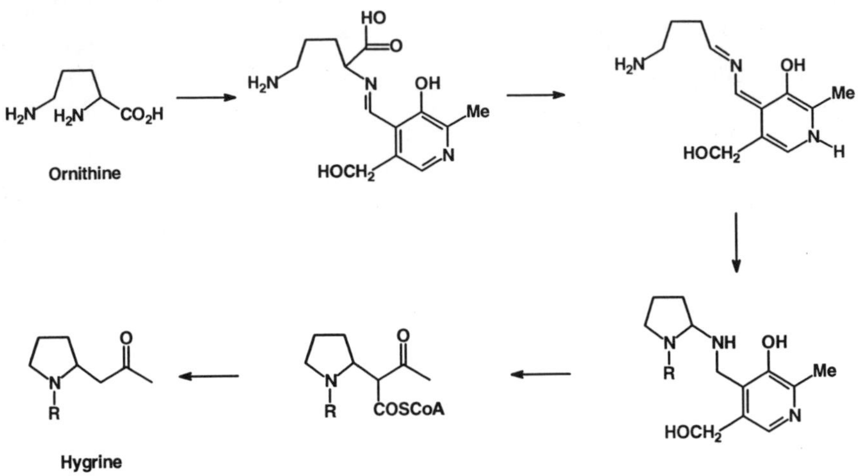

Ornithine

Hygrine

Scheme 10.15

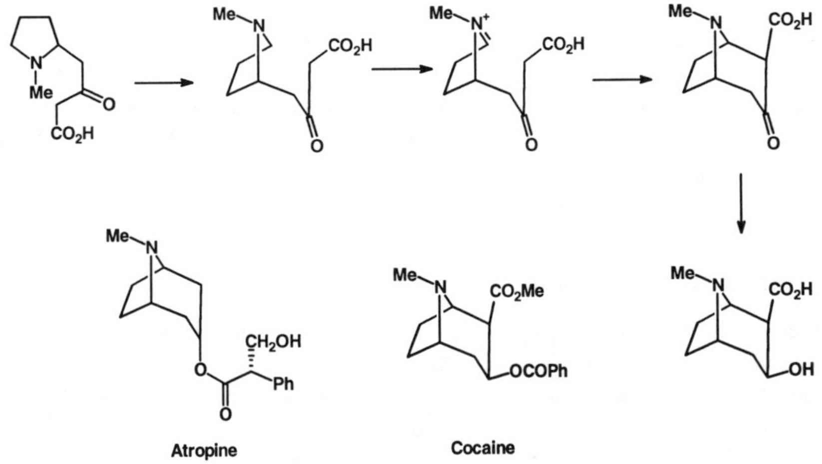

Atropine

Cocaine

Scheme 10.16

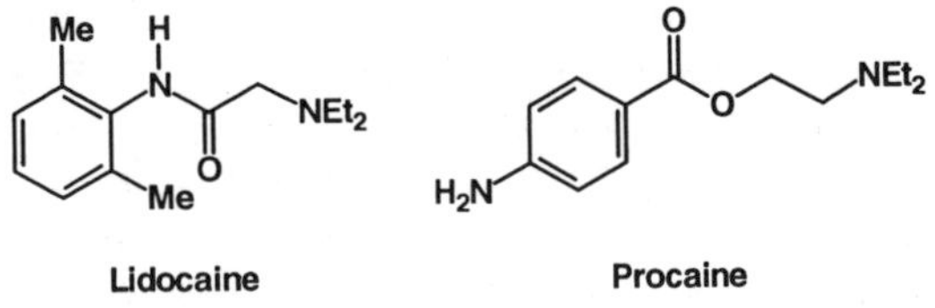

Lidocaine

Procaine

Figure 10.8

10.1.3 Quinoline and Isoquinoline Alkaloids

Approximately 1000 members of this broad class of alkaloids are currently known. Their structures vary quite widely, but they are mostly derived from phenylalanine or tryptophan. The proposed biogenetic pathway leading to quinine exemplifies these sometimes quite remarkable transformations (Scheme 10.17).[21] Oxidation of the pregeissoschizine derivative yields an iminium ion which is cleaved and recyclized to the quinuclidine. Further oxidative cleavage of the pyrrole ring of the indole unit gives an amino aldehyde which ring closes to quinine.

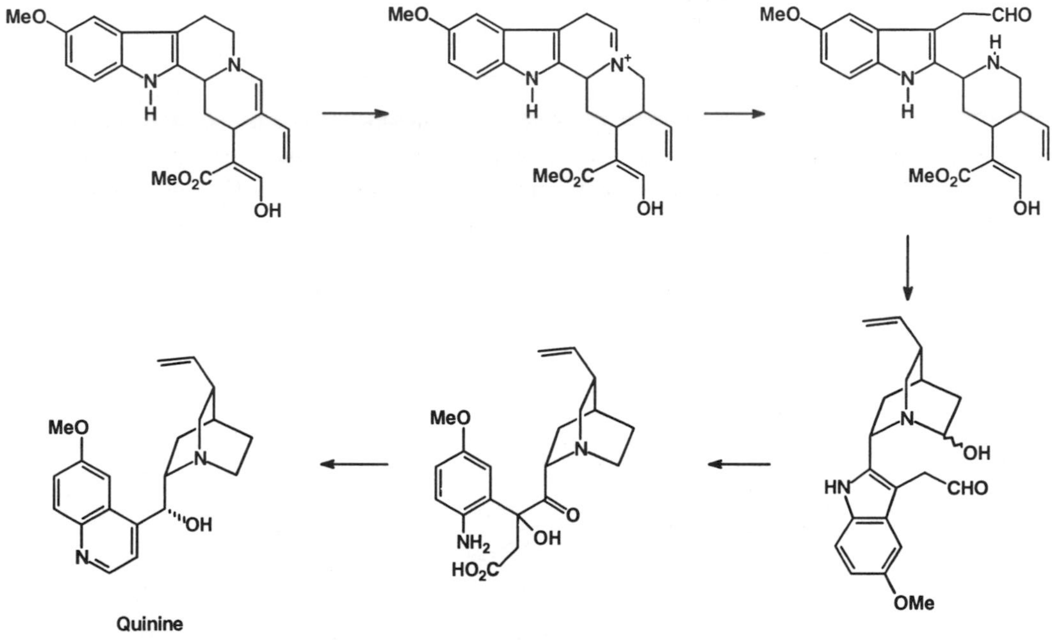

Scheme 10.17

Quinine was originally introduced to European and Western medicine by Spanish seafarers and Jesuit monks. A popular story tells that the bark of the *cinchona* tree was used to treat fevers and tertians, and that the bark was used to treat Countess Anna del Chinchon, the wife of the viceroy to Peru, in 1638. The bark extract was the sole source of the cure for malaria for nearly two centuries, until Pelletier and Caventou, in 1820, were able to isolate the pure compounds quinine and cinchonine from *cinchona*.

The *Ipecac* root extract is used as an emetic to treat, for example, drug poisonings. This extract contains several isoquinoline alkaloids including emetine (Figure 10.9), which is used as an effective amoebicide against both intestinal and extraintestinal amoebiasis. The source of *Ipecac* (Brazil root) is the dried root or rhizome of *Cephaelis ipecacuanha* or *acuminata*, plants native to Brazil and Central America but cultivated also in Malaysia and India. The obvious biogenetic relationship to the heteroyohimbanes is evident from the structure of ipecoside (Figure 10.9).

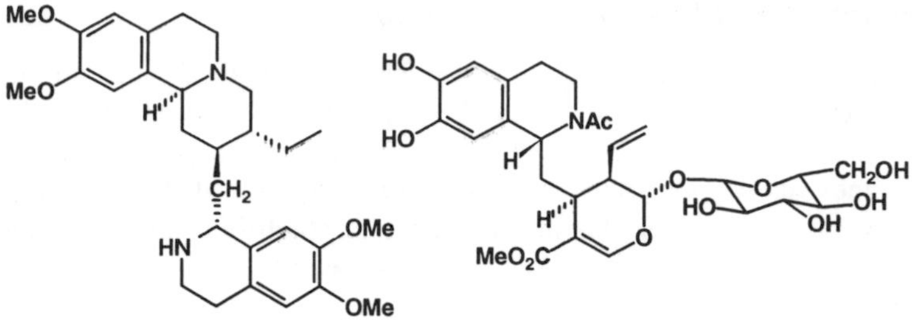

Emetine   Ipecoside

Figure 10.9

Morphine alkaloids and the structurally rather similar hasubanan and homomorphine alkaloids (e.g. androcymbine) also belong to the isoquinoline alkaloids (see Figure 10.10). Morphine alkaloids are also known as opium alkaloids owing to their natural origin; they are isolated from the seeds of the opium poppy, *Papaver somniferum*. The dried latex and the seed capsules contain some two dozen alkaloids, of which morphine covers nearly 10 %. Although pure morphine has been available since 1803 (isolated by Serturner), its structure was not elucidated until 1925 by Sir Robert Robinson.

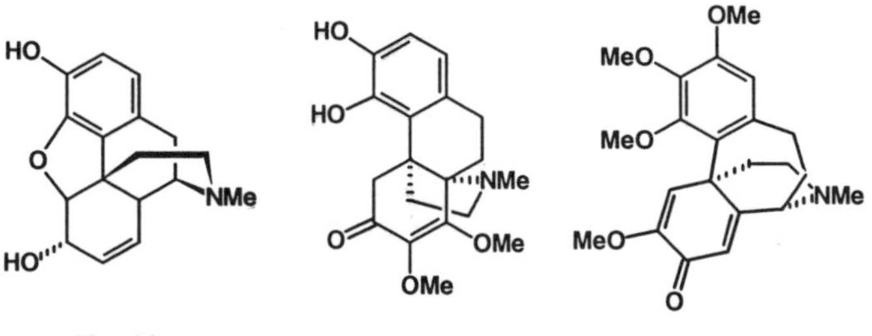

Morphine   Hasubananonine   Androcymbine

Figure 10.10

Morphine itself is a powerful pain killer, and its action is based on its ability to bind to the opiate receptors. It has been suggested[22] that the tyramine ring of morphine and the tyrosine ring of Leu-enkephalin coincide, giving the alkaloid a good affinity towards the receptor. The principal action is through the inhibition of adenyl cyclase which produces cyclic AMP (cAMP), a common second messenger. Reduced levels of cAMP will be compensated for by the cell by increased production of acetylcholine. This will still cause the production of sufficient levels of cAMP to maintain the normal functioning of the cell. However, if the opiate treatment has been used for a prolonged period and then suddenly interrupted, the acetylcholine induced cAMP production cannot accommodate rapidly enough, and the suddenly increased cAMP levels can trigger a multitude of withdrawal symptoms which may prove fatal. The dangerous addictiveness of morphine in continued use is the reason why a large number of analogues have been synthesized and tested for their analgesic abilities with the hope of fewer side effects. Levorphanol

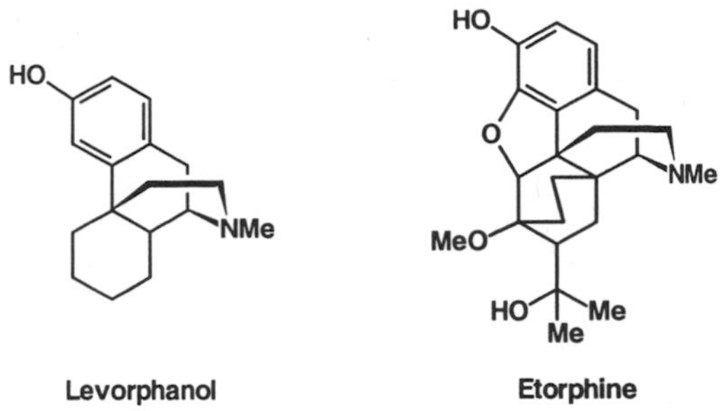

Levorphanol

Etorphine

Figure 10.11

(Figure 10.11), a morphinan with the furan ring cleaved, is a typical example of such a narcotic analgesic, and etorphine (Figure 10.11), an oripavine type compound, is among the most potent opiate agonists. Owing to its high potency, it is used primarily for the immobilization of big animals, such as elephants.

10.1.4 Izidine Alkaloids

These alkaloids include some 500 compounds containing a pyrrolizidine, indolizidine or quinolizidine skeleton (Figure 10.12). Several of them have interesting physiological and pharmacological activities. Some of these compounds also bear a close resemblance to the amino sugars discussed in Section 4.1.3.

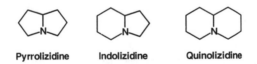

Pyrrolizidine   Indolizidine   Quinolizidine

Figure 10.12

The pyrrolizidines (e.g. Figure 10.13) typically contain a necine base (the hydroxylated pyrrolizidine) esterified with a carboxylic acid either as a monoester, a diester or a cyclic diester with a dicarboxylic acid. The fourth class of these alkaloids is made up of the N-oxides of the above. Most of these alkaloids are toxic and affect the liver.

The indolizidine alkaloids are a broad and varied class of compounds, including such alkaloids as slaframine, elaeocanine, securitinine, tylophorine and the polyhydroxylated indolizidines related to castanospermine, a potent glycosidase inhibitor and thereby a potential AIDS drug, and a number of poisonous-frog alkaloids exemplified by pumiliotoxin B (Figure 10.14).

The syntheses of pyrrolizidines have been reviewed.[23] A common synthetic approach for the izidine alkaloids relies on electrophilic cyclization. One early example is the phenylsulfenyl chloride initiated cyclization of an amino olefin (Scheme 10.18).[24,25] The episulfonium ion is formed from the less hindered face of the double bond, thus directing the formation of the bicyclic array.

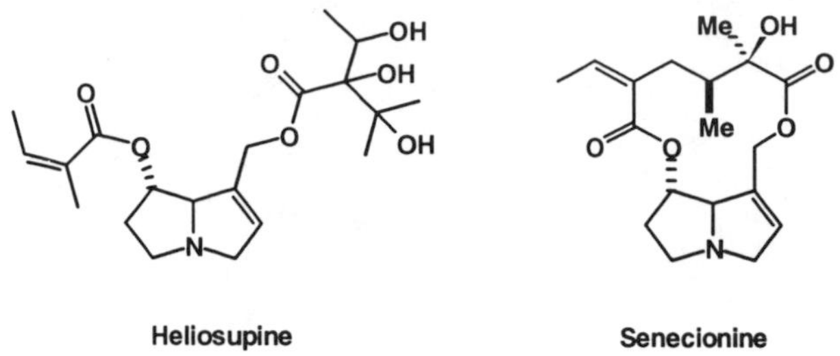
OH
OH
OH
Me OH
Me
Heliosupine
Senecionine

Figure 10.13

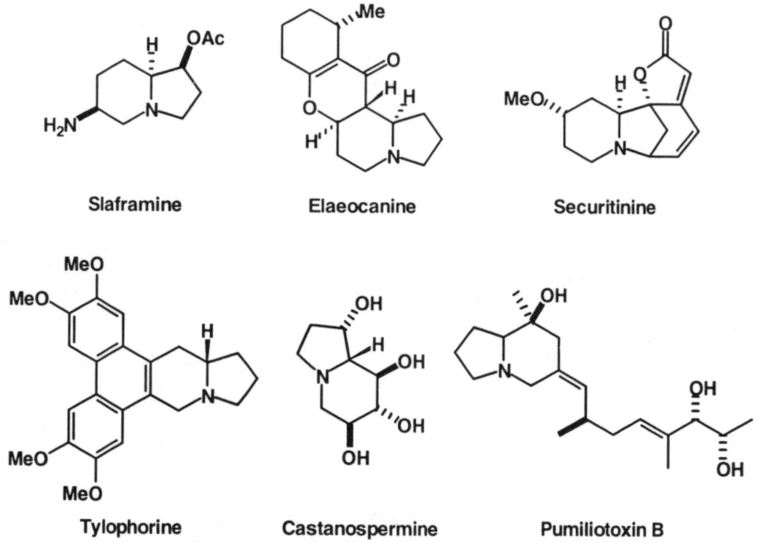
OAc
H_2N
Me
MeO
Slaframine
Elaeocanine
Securitinine
MeO
MeO
MeO
MeO
OH
OH
OH
OH
OH
OH
Tylophorine
Castanospermine
Pumiliotoxin B

Figure 10.14

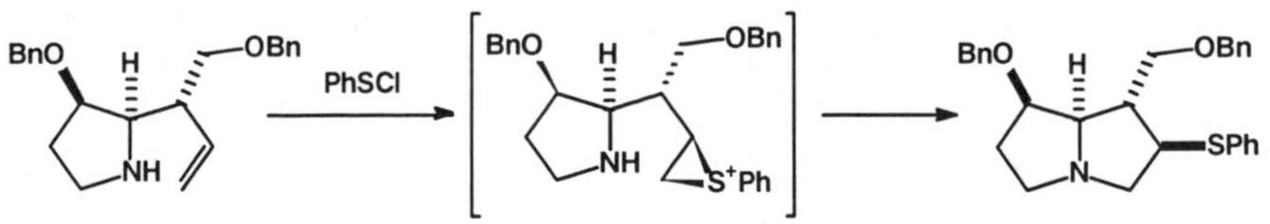
BnO
OBn
PhSCl
NH
S+Ph
SPh

Scheme 10.18

Overman has been very active in developing a methodology based on electrophilic Mannich cyclizations of the appropriate formaldiminium ions. This is exemplified by the synthesis of pumiliotoxin 251D (Scheme 10.19).[26] The epoxycarbamate derived from proline was reacted with the vinylalanate derived from the silylalkyne to give the cyclic carbamate. This was in turn hydrolysed and treated with formalin to yield

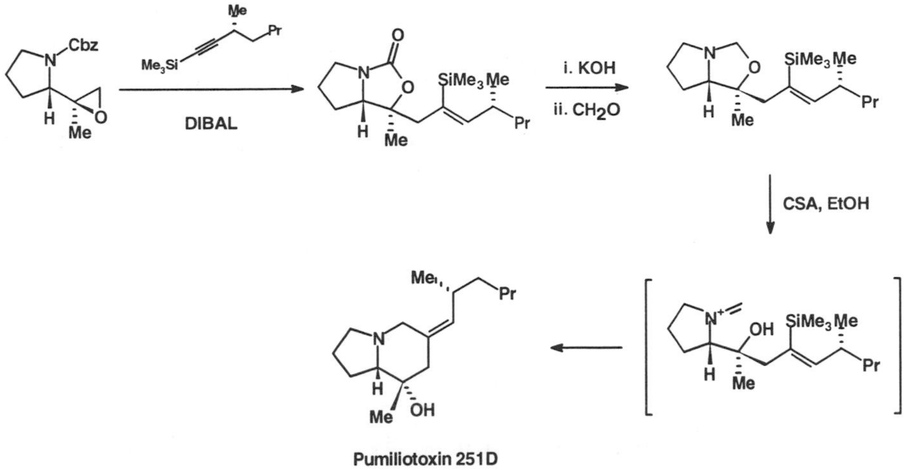

Scheme 10.19

a cyclopentaoxazoline. Heating of this intermediate in ethanol with camphorsulfonic acid (CSA) cleaved the aminal and the resulting iminium ion cyclized to pumiliotoxin 251D in high yield. The vinylsilanes undergo electrophilic cyclization with retention of configuration,[27] thus securing the double bond geometry in the product.

In the above cases the asymmetric information already resided in the starting materials. Methods employing kinetic resolution have also been used in the formation of the pyrrolidine ring, as exemplified by the synthesis of the 1-hydroxyindolizidines (Scheme 10.20), the biosynthetic precursors of slaframine and swainsonine.[28,29] The kinetic resolution was based on the Sharpless epoxidation.

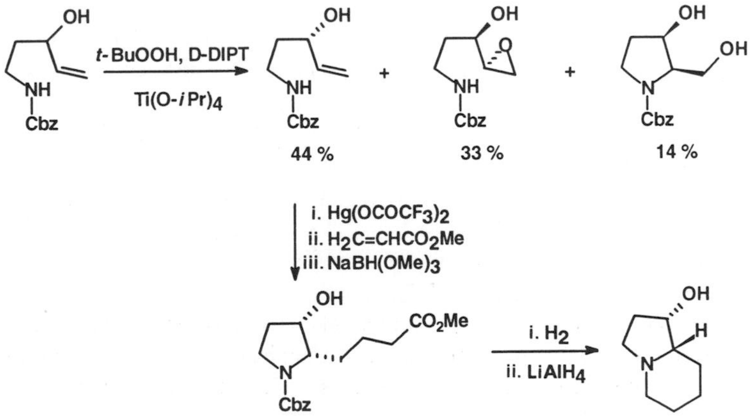

Scheme 10.20

The synthesis of croomine (Scheme 10.21), an alkaloid isolated from Chinese herbal tea and used to treat tuberculosis, bronchitis, pertussis and other ailments, provides another example of the utilization of electrophilic cyclization. In this case iodine is used as the electrophile, giving rise to an iodoamine. The *anti* stereochemistry of the iodine and the amino group are secured by the involvement of an iodonium intermediate. Neighbouring group participation of the amino group through the formation of an aziridine intermediate is the reason for the subsequent transformation giving the *anti* array between the amino and acyloxy functions.

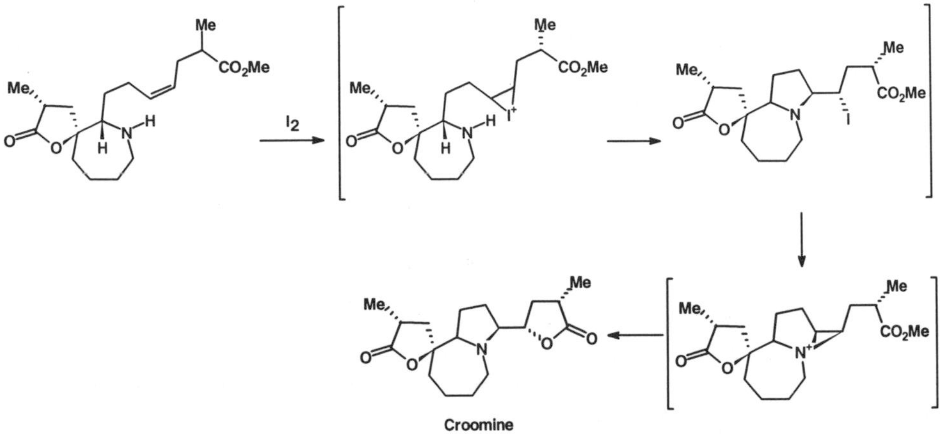

Croomine

Scheme 10.21

10.2 Alkaloids with an Exocyclic Nitrogen

This rather diverse group of natural products contains compounds which biogenetically are derived from several of the routes (mainly polyketide and isoprenoid pathways) discussed in previous chapters. The diterpene derivative taxol belongs to this class of alkaloids.

Taxane diterpenes (e.g. Figure 10.15), isolated from various yew (*Taxus*) species, have recently gained widespread interest mainly owing to the singular antitumour activity of taxol (from the Pacific yew tree *Taxus brevifolia*). The history of this compound is dreadful. Pliny, in ancient Rome, described the Mediterranean cousin of the plant producing taxol to be so toxic that 'already its shadow kills'. Taxol is produced by the Pacific yew tree, *Taxus brevifolia*, and the Mediterranean cousin, *Taxus baccata*, produces baccatine, structurally closely related to taxol. Taxol interacts with tubulin during the mitotic phase of the cell cycle, and thus prevents the disassembly of the microtubules and thereby interrupts the cell division. This phenomenon is being widely applied in the development of new anticancer agents based on the taxol structure, and already promising results have been obtained for the treatment of ovarian cancer. The plant source, however, is insufficient to satisfy even the need for clinical trials, and the efforts to achieve the total synthesis of this challenging compound will play a dominant role in the synthetic efforts during the 1990s.

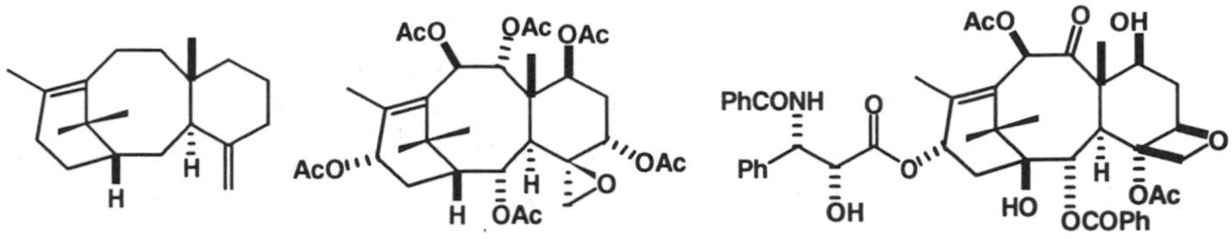

Taxane   Baccatin I   Taxol

Figure 10.15

Taxol, a diterpene, has been proposed to be derived from geranylgeranyl pyrophosphate.[30] Cyclization leads to the bicyclic intermediate of the basic verticillene skeleton, and further cyclization of the C-ring, through a formal 1,5-hydrogen shift, concludes the assembly of the tricyclic core of the taxanes (Scheme 10.22). Taxol itself is suggested to be formed through a number of oxidative transformations from the verticillene structure.

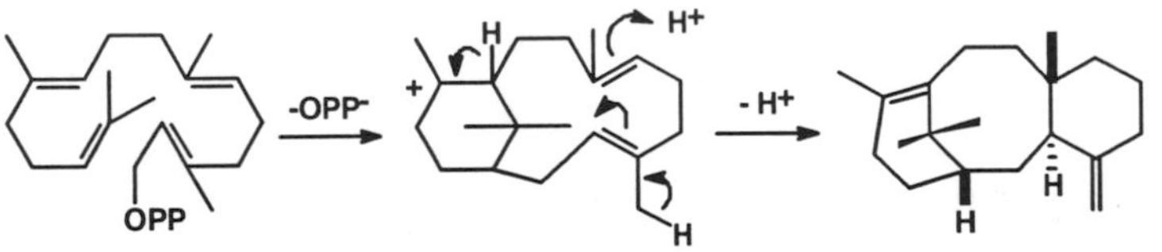

Verticillene   Taxane

Scheme 10.22

The phenylethylamine derivatives also belong to this class of alkaloids (e.g. Figure 10.16). These include the hallucinogens mescaline (from the peyote cactus *Lophophora williamsii*) and psilocybine (from *Psilocybe* mushrooms) and the highly toxic colchicine (from the autumn crocus, *Colchicum autumnale*), which is used in the treatment of gout.

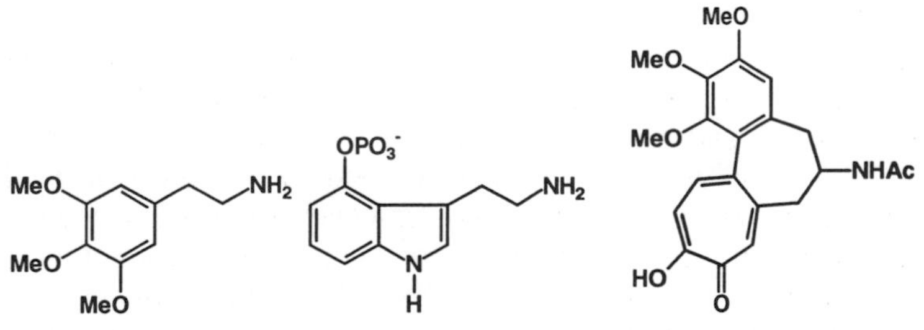

Mescaline   Psilocybine   Colchicine

Figure 10.16

10.3 Polyamine Alkaloids

Putrescine, spermidine and spermine are diamines and members of the so-called biogenic amines. They occur as such, and can also be incorporated into more complex structures, such as chaenorrhine (Figure 10.17).

$H_2N(CH_2)_3NH(CH_2)_4NH(CH_2)_3NH_2$ Spermine

$H_2N(CH_2)_4NH(CH_2)_3NH_2$ Spermidine

$H_2N(CH_2)_4NH_2$ Putrescine

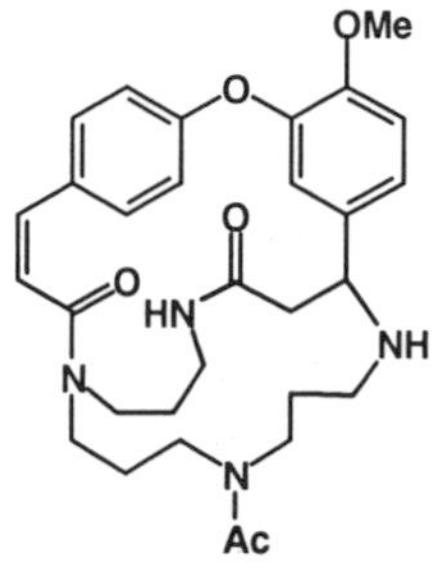

Chaenorrhine

Figure 10.17

10.4 Peptide Alkaloids

The ergot alkaloids are produced by the fungus *Claviceps purpurea* which grows upon rye and other grains. Ergotamine (Figure 10.18) is a typical example of a peptide alkaloid, although most of the known analogues of ergot alkaloids are their non-peptide derivatives. The ergot alkaloids are highly toxic, and as early as 600 BC an Assyrian tablet described a 'noxious pustule in the ear of rye'. In one of the sacred books of Parsees, ergot is also alluded to: 'Among the evil things created by Angro Maynes are noxious grasses that cause pregnant women to drop the womb and die in childbed'. The Greeks and Romans rejected rye, and it was not until the Middle Ages that rye was introduced into southwest Europe. Strange epidemics were described, involving gangrene of the limbs. In severe cases the tissue became dry and black and the mummified limbs separated off without bleeding. The limbs were said to be consumed by the Holy Fire, and the disease was called Holy Fire or St. Anthony's Fire. Ergot was also used as an obstetrical herb, producing pains in the womb and thereby speeding childbirth. The interest towards the ergot alkaloids started to widen in the 1920s, and this led to the development of lysergic acid diethylamide, LSD (Figure 10.18), for the treatment of schizophrenia. LSD and many other agents that act similarly are hallucinogenic, and in small doses they cause psychedelic effects. Although LSD has been defended because it does not cause physiological changes, it is evident that its effects on the psyche can be irreparable. It is known that a person who has used LSD only once can take a 'short trip' decades later, without having taken the compound ever again!

Some natural, cyclic tetrapeptides and higher peptides are also considered to belong to the peptide alkaloids. The peptide analogue mucronine (Figure 10.18) is an example.

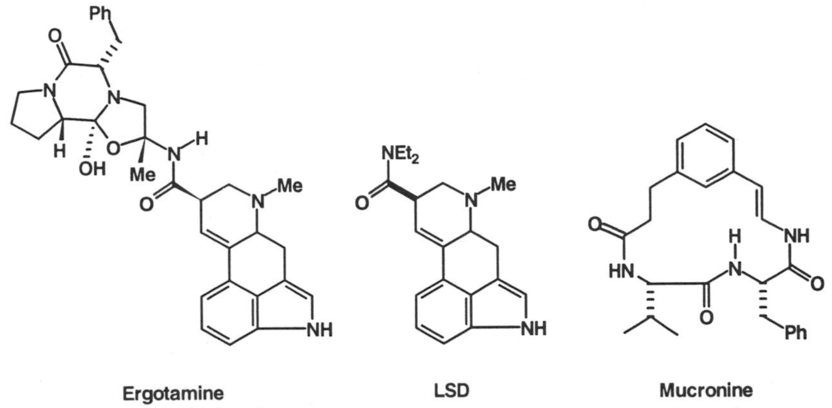

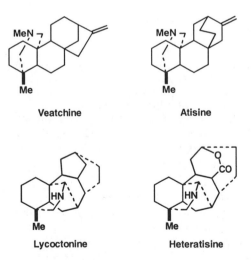

Ergotamine **LSD** **Mucronine**

Figure 10.18

10.5 Terpene Alkaloids

The pyrrolidine, piperidine and the *Nuphar* alkaloids represent the mono- and sesquiterpene alkaloids. Diterpene alkaloids are formed from four main groups: the veatchine, atisine, lycoctonine and heterazine type alkaloids (Figure 10.19). Altogether only some 150 terpene alkaloids (excluding steroidal alkaloids) are known.

Veatchine **Atisine**

Lycoctonine **Heteratisine**

Figure 10.19

In steroidal alkaloids the steroid skeleton is further transformed into an alkaloid structure by adding a unit containing the nitrogen atom. Holarrhimine (Figure 10.20) is the most obvious example of these alkaloids. Conessine and buxenine (Figure 10.20) are examples where more profound changes in the steroid skeleton have taken place.

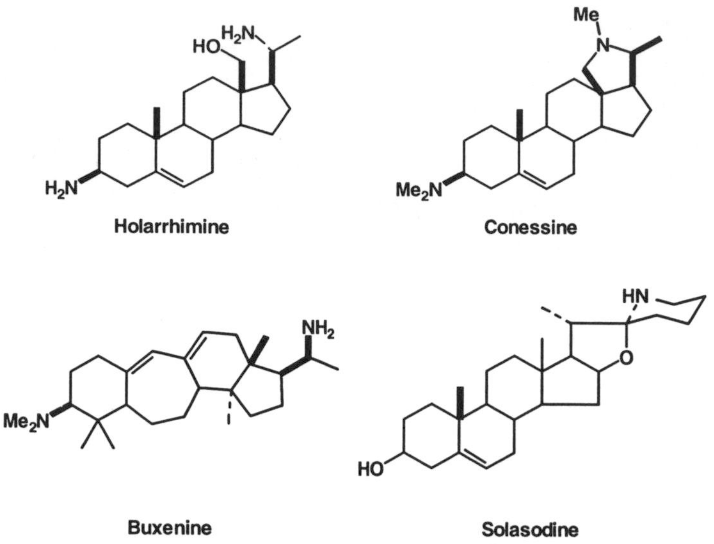

Holarrhimine

Conessine

Buxenine

Solasodine

Figure 10.20

The steroid drugs are usually produced from pregnenolone. The steroid alkaloid solasodine (Figure 10.20) can be chemically degraded into pregnenolone, and this method for the production of synthetic steroids is widely used, especially in the Eastern European countries.

References

1. Hesse, M. *Alkaloid Chemistry* John Wiley & Sons: New York, 1981.
2. Jackson, A.H. and Smith, P. *J. Chem. Soc., Chem. Commun.* 264–266 (1967).
3. Ungemach, F. and Cook, J.M. *Heterocycles* **9**, 1089–1119 (1978).
4. Bailey, P.D. *Tetrahedron Lett.* **28**, 5181–5184 (1987).
5. Bailey, P.D. and Hollinshead, S.P. *J. Chem. Soc. Perkin Trans. 1* 739–745 (1988).
6. Perkin, W.H. and Jr., Robinson, R. *J. Chem. Soc.* **115**, 933–967 (1919).
7. Barger, G. and Scholz, C. *Helv. Chim. Acta*, **16**, 1343–1354 (1933).
8. Hahn, G. and Werner, H. *Justus Liebigs Ann. Chem.* **520**, 123–133 (1933).
9. Woodward, R.B. *Nature* **162**, 155–156 (1948); *Angew. Chem.* **68**, 13–20 (1956).
10. Thomas, R. *Tetrahedron Lett.* 544–553 (1961).
11. Wenkert, E. *J. Am. Chem. Soc.* **84**, 98–102 (1962).
12. Stöckigt, J. In *Indole and Biogenetically Related Alkaloids* (Phillipson, J.D. and Zenk, M.H., Eds.) Academic Press: New York, 6 (1980).
13. LeMen, J. and Taylor, W.I. *Experientia*, **21**, 508–510 (1965).
14. Leonard, J., Ouali, D., and Rahman, S.K. *Tetrahedron Lett.* **31**, 739–742 (1990).
15. Izawa, H., Shirai, R., Kawasaki, H., Kim, H., and Koga, K. *Tetrahedron Lett.* **30**, 7221–7223 (1989).
16. Magnus, P., Mugrage, B., DeLuca, M., and Cain, G.A. *J. Am. Chem. Soc.* **111**, 786–789 (1989).
17. Magnus, P., Mugrage, B., DeLuca, M., and Cain, G.A. *J. Am. Chem. Soc.* **112**, 5220–5230 (1990).
18. Meyers, A.I. *Aldrichimica Acta* **18**, 59–68 (1985).
19. Meyers, A.I. and Guiles, J. *Heterocycles* **28**, 295–301 (1989).
20. Beard, R.L. and Meyers, A.I. *J. Org. Chem.* **56**, 2091–2096 (1991).

21. Leete, E. *Acc. Chem. Res.* **2**, 59–64 (1961).
22. Aubry, A., Birlirakis, N., Sakarellos-Daitsiotis, M., Sakarellos, C., and Marraud, M. *J. Chem. Soc., Chem. Commun.* 963–964 (1988).
23. Ikeda, M., Sato, T., and Ishibashi, H. *Heterocycles* **27**, 1465–1487 (1988).
24. Ohsawa, T., Ihara, M., Fukumoto, K., and Kametani, T. *J. Org. Chem.* **48**, 3644–3648 (1983).
25. Toshimitsu, A., Terao, K., and Uemura, S. *J. Org. Chem.* **51**, 1724–1729 (1986).
26. Overman, L.E., Bell, K.L., and Ito, F. *J. Am. Chem. Soc.* **106**, 4192–4201 (1984).
27. Chan, T.H. and Fleming, I. *Synthesis*, 761–786 (1979).
28. Takahata, H., Banba, Y., and Momose, T. *Tetrahedron: Asymmetry* **1**, 763–764 (1990).
29. Takahata, H., Yamazaki, K., Takamatsu, T., Yamazaki, T., and Momose, T. *J. Org. Chem.* **55**, 3947–3950 (1990).
30. Harrison, J.W., Scrowston, R.M., and Lythgoe, B. *J. Chem. Soc. C* 1933–1945 (1966).

Subject Index

Index compiled by P. Nash

Index of Compounds

Index compiled by P. Nash